Advances in Anthocyanin Research 2018

Advances in Anthocyanin Research 2018

Special Issue Editors

M. Monica Giusti
Gregory T. Sigurdson

MDPI • Basel • Beijing • Wuhan • Barcelona • Belgrade

MDPI

Special Issue Editors
M. Monica Giusti
The Ohio State University
USA

Gregory T. Sigurdson
The Ohio State University
USA

Editorial Office
MDPI
St. Alban-Anlage 66
4052 Basel, Switzerland

This is a reprint of articles from the Special Issue published online in the open access journal *Molecules* (ISSN 1420-3049) in 2018 (available at: https://www.mdpi.com/journal/molecules/special_issues/ anthocyanin)

For citation purposes, cite each article independently as indicated on the article page online and as indicated below:

LastName, A.A.; LastName, B.B.; LastName, C.C. Article Title. *Journal Name* **Year**, *Article Number*, Page Range.

ISBN 978-3-03897-523-6 (Pbk)
ISBN 978-3-03897-524-3 (PDF)

Contents

About the Special Issue Editors

M. Monica Giusti is a Professor and the Graduate Studies Chair at the Food Science and Technology Department, The Ohio State University. She is also a member of the graduate faculty of the Facultad de Industrias Alimentarias, Universidad Nacional Agraria, La Molina, Perú. Her research is focused on the chemistry and functionality of flavonoids, with an emphasis on anthocyanins. Together with her collaborators, she investigates polyphenols including their incidence and concentration in plants, stability and interactions with food matrices, novel analytical procedures, and the bioavailability, bio-transformations and potential bioactivity of these wonderful plant pigments. To date, Dr. Giusti's research has generated 100 peer-reviewed publications and 20 book chapters. She is also the co-editor of three books in the field of anthocyanins, and the co-inventor of six USA and international patents. For her innovative work, she was named the 2010 Ohio Agricultural Research and Development Center Director's Innovator of the Year, the 2011 TechColumbus Outstanding Woman in Technology, and the 2013 OSU Early Career Innovator of the Year. Dr. Giusti is a member of the American Chemical Society and the Institute of Food Technologists (IFT). Before joining The Ohio State University, Dr. Giusti was a faculty member at the Department of Nutrition and Food Science at the University of Maryland. Dr. Giusti, born in Lima, Peru, received a Food Engineer degree from the Universidad Nacional Agraria, La Molina, Peru as well as Master's and Doctorate degrees in Food Science from Oregon State University, Corvallis, Oregon.

Gregory T. Sigurdson, Ph.D., is a Researcher Scientist and Laboratory Manager of the Phytochemicals Lab at the Department of Food Science and Technology, The Ohio State University, where he obtained his doctorate degree in Food Science and Technology after working as professional chef. His research work has focused on the chemistry and application of flavonoid compounds, with a focus on anthocyanins. His research has included aspects on the evaluation of raw materials for their pigment profiles and concentrations, the isolation of individual pigments from mixed matrices, understanding the role of chemical substitutions in color and reactivity, methods of modifying and stabilizing these hues, and the investigation of plant and human metabolites as well as their relations with health. He has a special interest in the development of naturally derived alternatives for synthetic blue food colorants; his work in the production and stabilization of blue colors produced by anthocyanins has led to applications for two patents. In his young academic career, he has generated 17 peer-reviewed publications and one book chapter and he has presented at international conferences on anthocyanin and natural colorants chemistry. At The Ohio State University, he also serves as coordinator in an initiative that aims to engage undergraduate students in research, called FoodSURE (Food Science Undergraduate Research Experience).

Preface to "Advances in Anthocyanin Research 2018"

Interest in and research on anthocyanin-based pigments have been increasing considerably in recent years. PubMed shows an almost exponential growth curve in the number of anthocyanins publications, having increased from 55 (in 1996) to 264 (in 2006) and again to 910 (in 2017). Anthocyanins have long been identified as important pigments responsible for many flower, fruit, and vegetable colorations, producing a complex variety of hues ranging from yellow to red to purple to blue. More recently, colorants are being studied, not only for their biochemical roles in plants, but also for their applications in human products and contributions to health. Works specifically on anthocyanins, as well as epidemiological findings, further indicate these plant-produced pigments to be beneficial in the reduction of chronic inflammatory diseases, such as type 2 diabetes and cardiovascular disease. Considerable advances in the identification, analysis, application, and biological activities of anthocyanins have been made in recent years. However, the pigments exhibit diverse natural chemistries. Currently, more than 700 unique anthocyanin structures have been identified in nature, and many more in processed foods, each generally having distinctive reactivity and colorimetric properties. Recent advances related to anthocyanin chemistry, such as composition, degradative reactions, and biosynthesis; applications in agricultural, cosmetic, and food chemistry industries; use as natural colorants; and aspects or mechanisms of nutrition or reducing the risks of chronic diseases are discussed in this Special Issue of the journal *Molecules*: "Advances in Anthocyanin Research 2018".

M. Monica Giusti, Gregory T. Sigurdson
Special Issue Editors

molecules

MDPI

Review

The Chemical Reactivity of Anthocyanins and Its Consequences in Food Science and Nutrition

Olivier Dangles * and Julie-Anne Fenger

University of Avignon, INRA, UMR408, 84000 Avignon, France; julie-anne.fenger@univ-avignon.fr
* Correspondence: olivier.dangles@univ-avignon.fr; Tel.: +33-490-144-446

Academic Editors: M. Monica Giusti and Gregory T. Sigurdson
Received: 6 July 2018; Accepted: 31 July 2018; Published: 7 August 2018

Abstract: Owing to their specific pyrylium nucleus (C-ring), anthocyanins express a much richer chemical reactivity than the other flavonoid classes. For instance, anthocyanins are weak diacids, hard and soft electrophiles, nucleophiles, prone to developing π-stacking interactions, and bind hard metal ions. They also display the usual chemical properties of polyphenols, such as electron donation and affinity for proteins. In this review, these properties are revisited through a variety of examples and discussed in relation to their consequences in food and in nutrition with an emphasis on the transformations occurring upon storage or thermal treatment and on the catabolism of anthocyanins in humans, which is of critical importance for interpreting their effects on health.

Keywords: anthocyanin; flavylium; chemistry; interactions

1. Introduction

Anthocyanins are usually represented by their flavylium cation, which is actually the sole chemical species in fairly acidic aqueous solution (pH < 2). Under the pH conditions prevailing in plants, food and in the digestive tract (from pH = 2 to pH = 8), anthocyanins change to a mixture of colored and colorless forms in equilibrium through acid–base, water addition–elimination, and isomerization reactions [1,2]. Each chemical species displays specific characteristics (charge, electronic distribution, planarity, and shape) modulating its reactivity and interactions with plant or food components, such as the other phenolic compounds. This sophisticated chemistry must be understood to interpret the variety of colors expressed by anthocyanins and the color changes observed in time and to minimize the irreversible color loss signaling the chemical degradation of chromophores. The chemical reactivity of anthocyanins is also important to interpret their fate after ingestion and their effects on health, as anthocyanins may be consumed as a complex mixture of native forms, derivatives, and degradation products, which themselves can evolve in the digestive tract [3].

2. The Basis of Anthocyanin Chemistry

2.1. Anthocyanins Are Weak Diacids

Due to conjugation with the electron-withdrawing pyrylium ring, the phenolic OH groups of the flavylium ion at C4′, C5, and C7 are fairly acidic [1,2]. In terms of structure–acidity relationships, it is clear that C7-OH is the most acidic group with a pK_{a1} of ca. 4, i.e., 6 pK_a units below the phenol itself. The corresponding neutral quinonoid base (Figure 1) can thus be considered to be the prevailing tautomer. At higher pH levels, a second proton loss from C4′-OH ($pK_{a2} \approx 7$ for common anthocyanins) yields the anionic base with maximized electron delocalization over the three rings. Along this deprotonation sequence, the wavelength of maximal visible absorption typically shifts by 20–30 nm ($AH^+ \rightarrow A$), then by 50–60 nm ($A \rightarrow A^-$) (Figure 2), and the corresponding color turns from red to purple-blue [4].

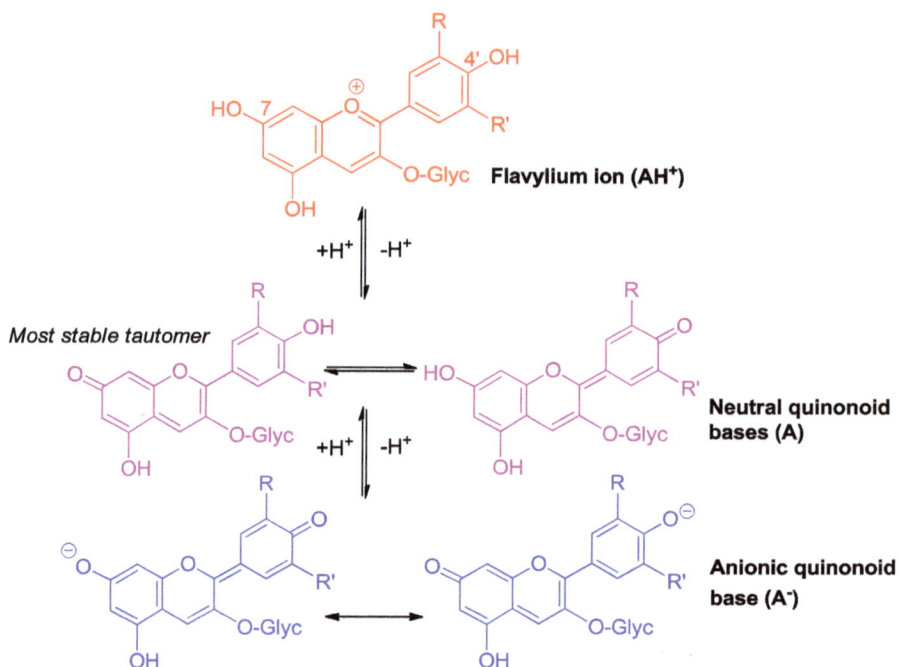

Figure 1. Flavylium ions are weak diacids.

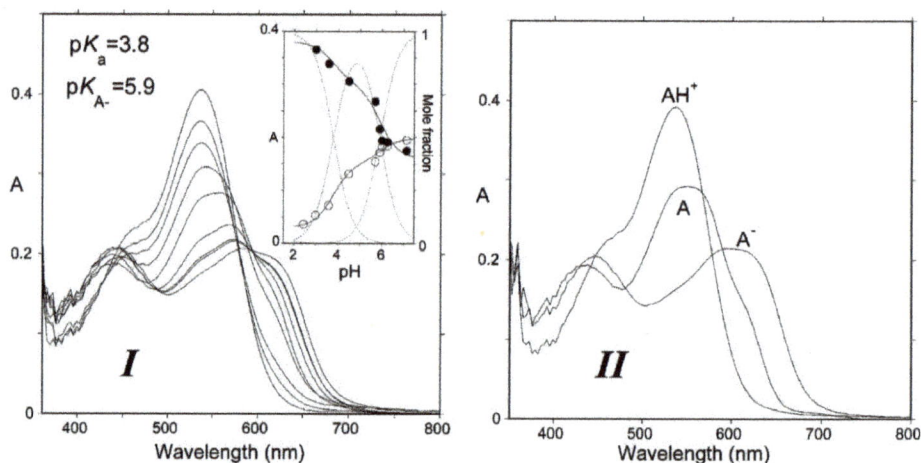

Figure 2. (**I**) Absorption spectra of Cat-Mv3Glc: pH jump from pH = 1.0 (100% flavylium) to pH 3.00, 3.59, 4.50, 5.70, 5.96, 6.25, and 7.15, respectively. Spectra recorded 10 ms after mixing (negligible water addition). (**II**) Spectra of the components obtained by mathematical decomposition. From [4] with permission of the *American Chemical Society*.

2.2. Anthocyanins Are Hard and Soft Electrophiles

By analogy with enones, the C2 and C4 atoms of the pyrylium ring can be regarded as hard and soft electrophilic centers, respectively. Hence, they respectively react with hard (O-centered) and soft (S- and C-centered) nucleophiles, the first mechanism being driven by local charges and the second one by interactions between the frontier molecular orbitals (HOMO of nucleophiles and LUMO of electrophiles).

2.2.1. Nucleophilic Addition at C2

Water addition is the ubiquitous process taking place within aqueous anthocyanin solutions [1,2]. It leads to the colorless hemiketal (Figure 3) and can be characterized by the thermodynamic hydration constant K_h, or as an acceptable approximation (chalcones making only a minor contribution, typically less than 20%, of the total pool of colorless forms), by the apparent constant K'_h connecting the flavylium ion and the colorless forms taken collectively. With common anthocyanins, pK'_h lies in the range of 2–3, which means that hydration is thermodynamically more favorable than proton transfer ($pK'_h < pK_{a1}$). Fortunately, it is also much slower, and its pH-dependent kinetics can be quantified by the apparent rate constant of hydration (k_{obs}) (Equation (1), $h = [H^+]$, χ_{AH} = mole fraction of AH^+ within the mixture of colored forms [2,5]:

$$k_{obs} = k_h \chi_{AH} + k'_{-h} h = \frac{k_h}{1 + K_{a1}/h + K_{a1}K_{a2}/h^2} + k'_{-h} h. \tag{1}$$

k_h is the absolute rate constant of water addition, k'_{-h} is the apparent rate constant of water elimination (from the mixture of hemiketal and *cis*-chalcone in fast equilibrium), and $K'_h \approx k_h/k'_{-h}$ (*trans*-chalcone neglected). Equation (1) can be easily understood by keeping in mind that the flavylium ion is the sole colored form that is electrophilic enough to directly react with water.

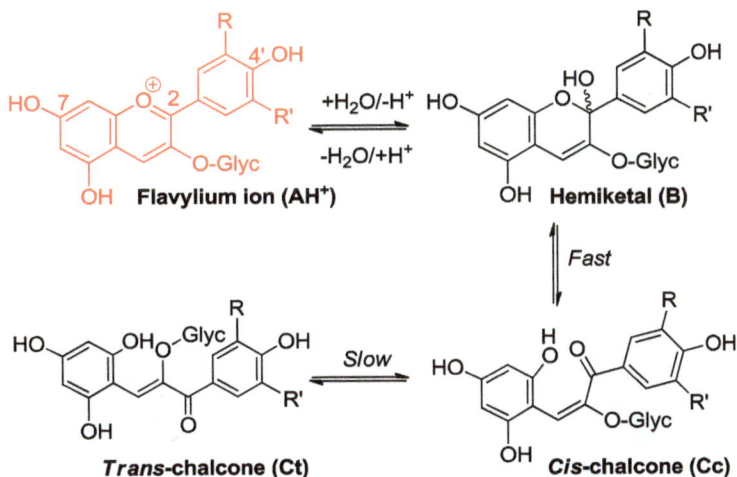

Figure 3. Flavylium ions are hard electrophiles reacting at C2 with O-centered nucleophiles, such as water (water addition followed by formation of minor concentrations of chalcones).

At a given pH, the initial visible absorbance (A_0) (no colorless forms) and the final visible absorbance (A_f) (hydration equilibrium established) can be easily related through Equation (2):

$$\frac{A_f}{A_0} = \frac{1 + K_{a1}/h + K_{a1}K_{a2}/h^2}{1 + (K_{a1} + K'_h)/h + K_{a1}K_{a2}/h^2}. \tag{2}$$

Thus, the magnitude of color loss can be expressed as (Equation (3)):

$$\frac{A_0 - A_f}{A_0} = \frac{K'_h/h}{1 + (K_{a1} + K'_h)/h + K_{a1}K_{a2}/h^2} \tag{3}$$

From typical values for the rate and thermodynamic constants of common anthocyanins, simulations of the pH dependence of the apparent rate constant and percentage of color loss can be constructed (Figure 4). The plots clearly show that the reversible color loss due to water addition to the flavylium ion becomes slower at higher pH (less flavylium in solution), whereas its magnitude becomes larger because of the higher stability of the colorless forms. The typical time-dependence of the visible spectrum during water addition is shown in Figure 5 [4].

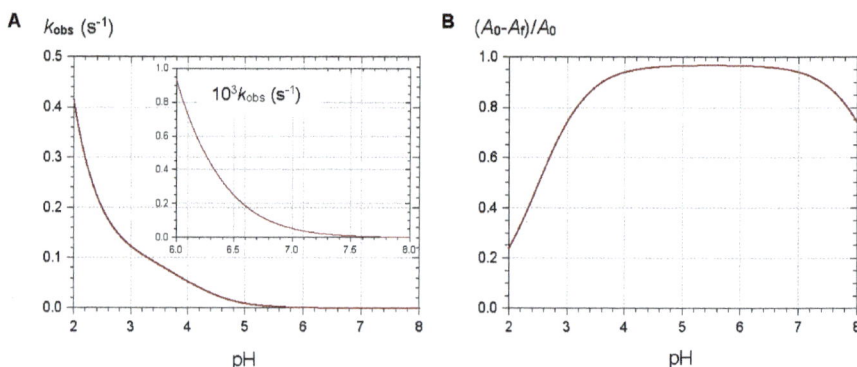

Figure 4. Simulations of the pH dependence of the apparent rate constant (**A**) and relative magnitude (**B**) of color loss. Selected values for parameters: pK_{a1} = 4, pK_{a1} = 7, pK'_h = 2.5, k_h = 0.1 s^{-1}, $k'_{-h} \approx k_h/K'_h$.

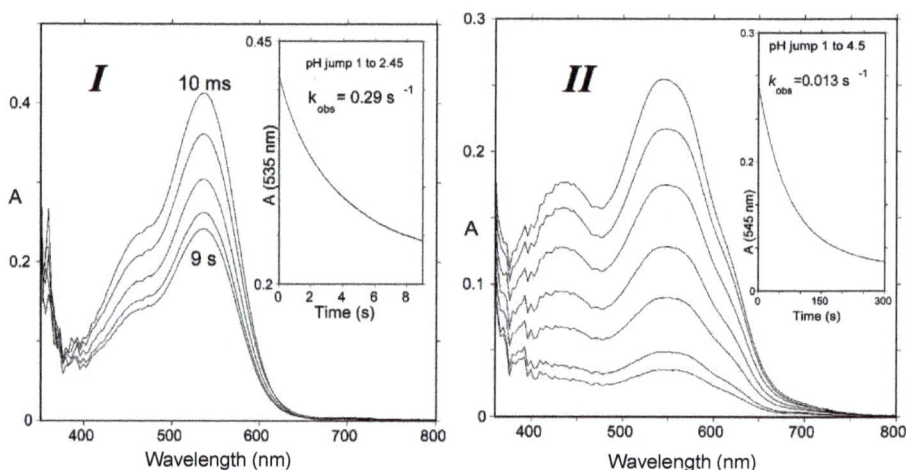

Figure 5. (**I**) Spectral changes of Cat-Mv3Glc between 10 ms and 9 s following a pH jump from pH = 1 to pH = 2.45; half-life of flavylium = 2.4 s. (**II**) pH jump from pH = 1 to pH = 4.5; half-life of quinonoid bases = 53.3 s. At pH = 6, the half-life of quinonoid bases ≈ 30 min. From reference [4] with permission of the *American Chemical Society*.

Near neutrality water addition is so slow (fraction of flavylium ion < 0.1%) that the colored forms (mixtures of neutral and anionic bases) can, in principle, persist for hours. However, such a reasoning ignores the irreversible mechanisms of color loss taking place near neutrality as the anionic base is obviously much more sensitive to autoxidation (non-enzymatic oxidation by O_2 triggered by transition metal traces) than the flavylium ion. These mechanisms will be addressed in Section 2.4.1.

2.2.2. Nucleophilic Addition at C4

Bisulfite is an antimicrobial and anti-browning agent that is frequently used in the food industry. As a S-centered nucleophile, it reversibly reacts with the flavylium ion at C4, thus yielding colorless adducts (Figure 6) [6]. No such adducts have been identified so far by simply reacting anthocyanins with natural thiols such as cysteine and glutathione (GSH). Unlike bisulfite, which is actually the conjugated base of SO_2 ($pK_a \approx 1.8$) and can coexist with the flavylium ion under acidic conditions, thiolate anions ($pK_a = 8$–9) are usually formed at much higher pH levels where the flavylium ion is only present as traces.

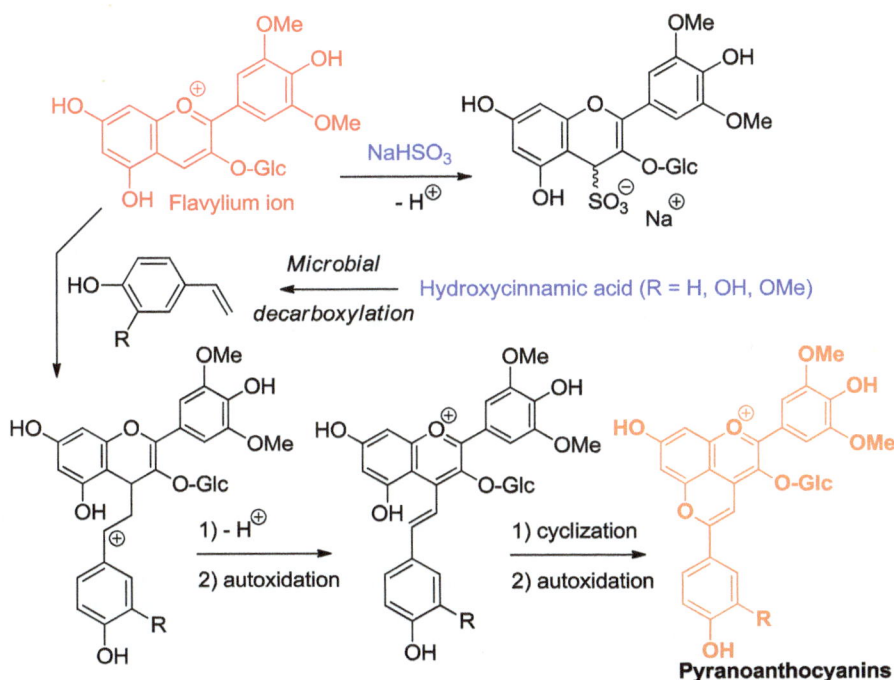

Figure 6. Flavylium ions are soft electrophiles that react at C4 with S- and C-centered nucleophiles, such as bisulfite and 4-vinylphenols.

A variety of C-centered nucleophiles are also known to add to the flavylium ion, and this chemistry underlies the color changes observed in red wine upon aging [7]. In this context, the most important C-centered nucleophiles are electron-rich C–C double bonds, such as 4-vinylphenols (4-hydroxystyrenes), formed upon microbial decarboxylation of 4-hydroxycinnamic acids (Figure 6) and the enol forms of various aldehydes and ketones such as pyruvic acid and ethanal (acetaldehyde) [8,9]. In the process, new pigments, called pyranoanthocyanins, are formed, which are resistant to nucleophilic addition at C2 and C4 [10–12]. Their color (shifted to orange-red, compared to the corresponding flavylium ion) is thus more stable. Through their nucleophilic C6- and C8-atoms, flavanols and proanthocyanidins can also add to the electrophilic C4 center of anthocyanins [13]. However, the flavene intermediate thus formed is not accumulated and evolves through two possible routes: (a) under strongly acidic conditions (pH = 2), protonation at C3 allows a second nucleophilic attack by a nearby phenolic OH group of the tannin to yield a colorless product (see Section 2.3 for a similar mechanism); or (b) under moderately acidic conditions (pH = 3–6), dehydration with concomitant formation of an additional pyrane ring is favored and a new pigment bearing a xanthylium chromophore is formed.

With its enediol structure, ascorbate (vitamin C) can also react with flavylium ions at C4 but the corresponding adducts have not been reported so far.

2.3. Anthocyanin Hemiketals Are Nucleophiles

Basic organic chemistry teaches that electron-donating substituents of benzene rings accelerate aromatic electrophilic substitutions ($S_E Ar$) and orient the entering electrophiles to the *ortho* and *para* positions. In that perspective, the phloroglucinol (1,3,5-trihydroxybenzene) ring (A-ring) of anthocyanins must be especially favorable to $S_E Ar$ as the three O-atoms combine their electronic effects to increase the reactivity of C6 and C8. However, the pyrylium ring (C-ring) of the flavylium ion (and, to a lesser degree, the enone moiety of chalcones) is strongly electron-withdrawing, so that only the hemiketal is expected to react by $S_E Ar$.

Here, again, wine chemistry provides interesting examples of $S_E Ar$ between anthocyanins and various carbocations derived from other wine components (Figure 7) [7]. For instance, wine pigments in which anthocyanins and flavanols are linked though an ethylidene bridge between their C6- and/or C8-atoms are formed by double $S_E Ar$ between A-rings and ethanol [14,15]. The likely intermediates in the reaction are the 6- or 8-vinyl-flavanol and the 6- or 8-vinyl-anthocyanin hemiketals, the protonation of which delivers a benzylic cation that is directly involved in the $S_E Ar$ reaction. Of course, in addition to the cross reaction products, anthocyanin–ethylidene–anthocyanin and flavanol–ethylidene–flavanol adducts can also form oligomers and mixed oligomers [16]. Even, pyranoanthocyanins stemming from the nucleophilic attack of vinyl-phenols at C4 of anthocyanins can be produced.

Flavanol carbocations formed by acidic cleavage of the inter-flavan linkage of proanthocyanidins also react with anthocyanin hemiketals by $S_E Ar$ [17]. Interestingly, both direct and ethylidene-bridged flavanol–anthocyanin adducts are more purple than the native anthocyanin, but only the latter expresses a color that is stable, i.e., a flavylium nucleus that is less sensitive to water addition [4,18]. A possible explanation is that ethylidene-bridged flavanol–anthocyanin adducts are prone to non-covalent self-association by π-stacking, which provides a less aqueous environment for the flavylium nuclei.

Water elimination from the anomeric C-atom of the ellagitannin vescalagin (abundant in oak barrels) also delivers a carbocation for direct coupling with wine anthocyanins [19] and subsequent modest protection against water addition [20]. Finally, the anthocyanin hemiketal can react with the flavylium ion itself, and this pathway provides a route for anthocyanin oligomerization, a poorly documented mechanism as the corresponding oligomers are probably difficult to evidence and quantify. However, an oenin trimer has been found in Port wine, and its structure has been fully elucidated by NMR [21]. The two linkages are of the C4–C8 type. As in the direct flavanol–anthocyanin coupling (see Section 2.2.2), flavene intermediates evolve by C–O coupling and only the lower unit remains colored. Similar oligomers also occur with 3-deoxyanthocyanidins, e.g., in red sorghum, but the detailed structures remain unknown [22].

Anthocyanin hemiketals can also react by Michael addition with o-quinones formed by two-electron oxidation of catechols, such as epicatechin [13] and caffeoyltartaric acid [23].

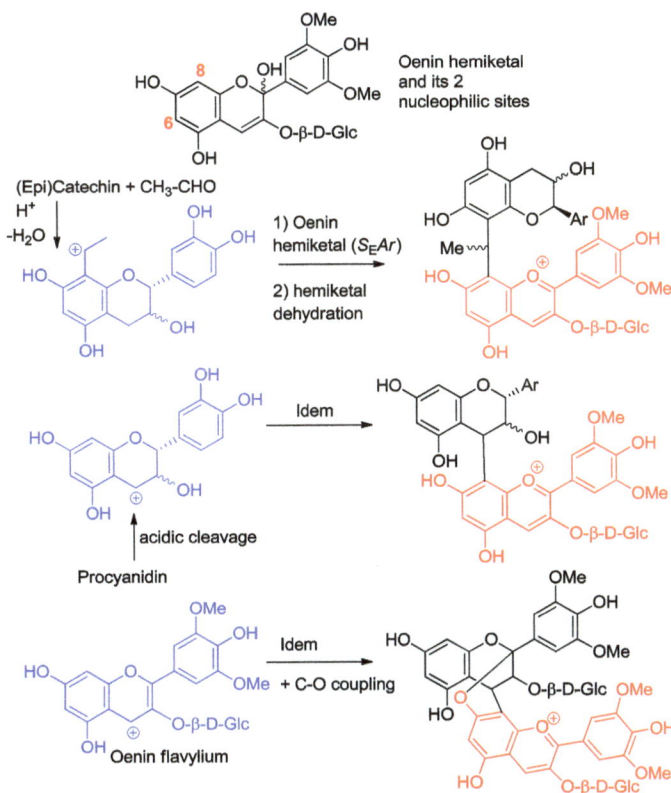

Figure 7. Anthocyanin hemiketals are nucleophiles reacting with carbocations (Ar = catechol ring).

2.4. Anthocyanins Are Electron-Donors

Many polyphenols, especially those containing electron-rich catechol (1,2-dihydroxybenzene) or pyrogallol (1,2,3-trihydroxybenzene) nuclei are good electron- or H-donors. Electron transfer is typically faster when the pH is raised, i.e., when the fraction of phenolate anion (a much better electron-donor than the parent phenol) increases. Electron transfer from phenols is involved in their oxidation mechanisms and also underlies the most common mechanism of antioxidant activity, i.e., the reduction of reactive oxygen species (ROS) involved in oxidative stress from plants to humans. Anthocyanins are known to be thermally unstable, especially under neutral conditions, and various degradation products have been identified. Their antioxidant activity has been also established in various chemical models. However, detailed knowledge on the mechanisms involved and on the relative contributions of the different colored and colorless forms is still missing.

2.4.1. Oxidation

Anthocyanins are among the least thermally stable flavonoids. Anthocyanidins, the corresponding aglycones, are actually only stable under highly acidic conditions and are extensively degraded in less than one hour under physiological conditions (pH = 7.4, 37 °C) [24,25]. From the structure of the degradation products, it is clear that a combination of hydrolytic and autoxidative pathways operate, leading to cleavage of the C2–C1′, C2–C3 and C3–C4 bonds (Figure 8) [13,26,27]. A mechanism involving pre-formed hydrogen peroxide actually accounts for the formation of some cleavage products (Figure 9). The critical step is the addition of H_2O_2 (a hard nucleophile) at C2 of the flavylium ion, followed by Baeyer–Villiger rearrangement, which opens routes for cleavage of the C2–C1′ and C2–C3 bonds [13,26]. However, the preliminary formation of H_2O_2 remains unclear and must involve the direct autoxidation of anthocyanins. Thus, an alternative mechanism beginning by electron or H-atom transfer (mediated by unidentified transition metal traces) from the anionic or neutral base to O_2 would deliver a highly delocalized radical that is susceptible to O_2 addition at different centers (Figure 10). The cleavage of hydroperoxide intermediates thus formed could also yield the degradation products detected.

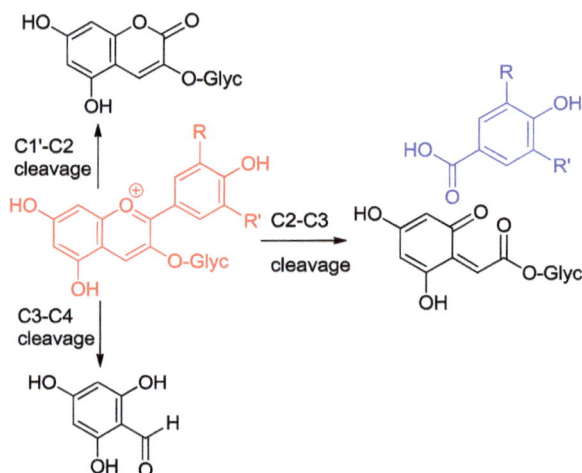

Figure 8. Pathways of anthocyanin degradation.

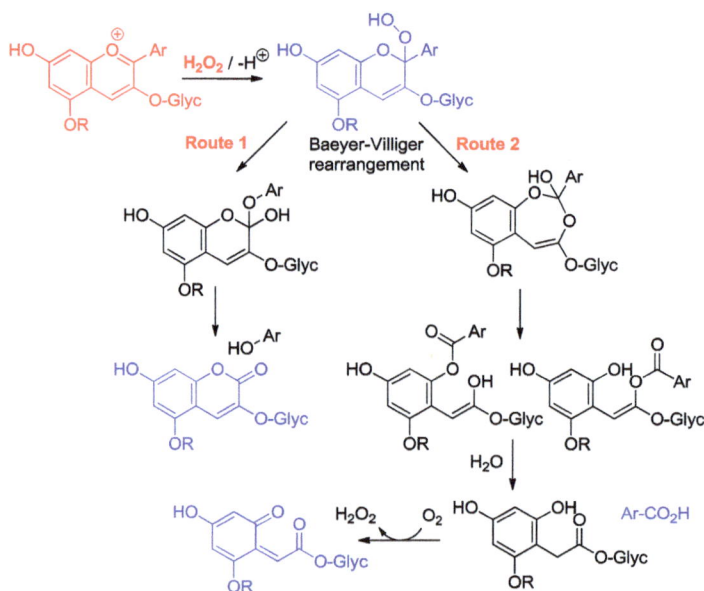

Figure 9. Possible mechanisms of anthocyanin degradation with pre-formed hydrogen peroxide.

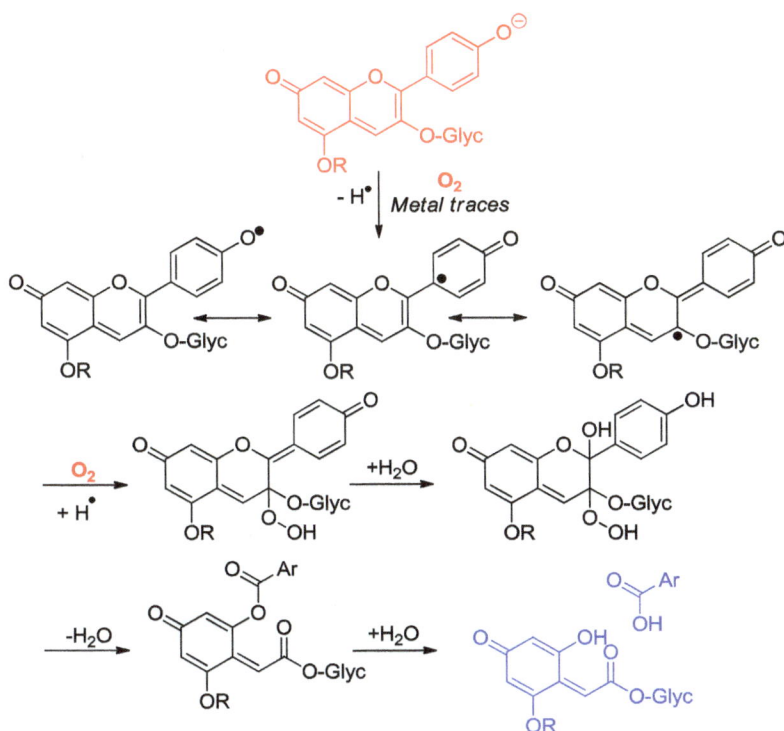

Figure 10. Possible mechanisms of anthocyanin degradation without pre-formed hydrogen peroxide.

2.4.2. Antioxidant Activity

Anthocyanins under their native forms can transfer electrons to ROS and could, therefore, provide protection to important oxidizable biomolecules, such as polyunsaturated fatty acids (PUFAs), proteins, and DNA. The relevance of such phenomena is probably much higher in food preservation than in nutrition and health, given the current knowledge on anthocyanin bioavailability (see Section 3). In this section, we simply mention that anthocyanins can indeed effectively reduce one-electron oxidants such as the stable radical DPPH (2,2-diphenyl-1-picrylhydrazyl). Structure–activity relationships show that hydroxylation at C3′ and C5′ increases the H-donating capacity, thus suggesting that the B-ring is primarily involved in electron donation [28]. Comparing oenin and the flavanol catechin shows that the transfer of the first (most labile) H-atom to DPPH is roughly as fast for both flavonoids but that oenin reduces at least twice as many radicals than catechin (Table 1) [29]. This advantage must be rooted in the extensive oxidative degradation undergone by oenin during the DPPH-scavenging process with the transient formation of intermediates (possibly, syringic acid) retaining a substantial electron-donating activity. It is also remarkable that the wine pigments combining the oenin and catechin units retain a high but contrasting DPPH-scavenging activity [29]: the direct coupling between the two flavonoid units results in a faster first H-atom transfer (higher k_1) but markedly lowers the total number of radicals reduced (n_{tot}), whereas the coupling through an ethylidene bridge apparently leaves each unit free to independently react with DPPH (k_1 almost unchanged, approximate additivity in the n_{tot} value), as observed with the equimolar oenin–catechin mixture (Table 1).

Table 1. Antioxidant activity of malvidin 3-*O*-β-D-glucoside (oenin) and related pigments: reduction of the DPPH (2,2-diphenyl-1-picrylhydrazyl) radical (MeOH, 25 °C, [1] and [2]) and inhibition of heme-induced peroxidation of linoleic acid (0.1 mM linoleic acid in acetate buffer + 2 mM Tween-20, 0.1 μM metmyoglobin, pH = 4, 37 °C, [3]). From reference [29].

Antioxidant	n_{tot} [1]	k_1/s^{-1} [2]	$IC_{50}/\mu M$ [3]
Oenin	11.26 (±0.08)	910 (±70)	0.68
Catechin	4.86 (±0.03)	1200 (±110)	0.27
Oenin + Catechin (1:1)	14.04 (±0.10)	1160 (±330)	nd
(*R*)-Catechin-8-CHMe-8-Oenin	14.56 (±0.03)	1000 (±320)	0.15
(*S*)-Catechin-8-CHMe-8-Oenin	14.61 (±0.18)	600 (±120)	0.41
Catechin-4,8-Oenin	7.16 (±0.08)	5120 (±1050)	0.60

[1] Antioxidant stoichiometry (number of DPPH radicals reduced per antioxidant molecule). [2] Rate constant for the transfer of the first H-atom from antioxidant to DPPH. [3] Antioxidant concentration for a doubling of the period of time required for the accumulation of a fixed concentration of polyunsaturated fatty acid (PUFA) hydroperoxides (conjugated dienes).

Oenin, catechin, and wine pigments were also compared for their ability to inhibit the peroxidation of linoleic acid induced by dietary heme iron in acidic micelle solutions, a chemical model of postprandial oxidative stress in the stomach [29]. As hydrophilic antioxidants, polyphenols are known to act at the initiation stage by reducing the hypervalent iron species (Fe^{IV}) involved in the generation of propagating lipid peroxyl radicals (Figure 11) [30] which, on the other hand, are directly reduced by the typical chain-breaking amphiphilic antioxidant α-tocopherol (vitamin E). The highly hydrophilic oenin was found to be less potent than catechin in the inhibition, but coupling both flavonoids via an ethylidene bridge improves their efficiency (Table 1).

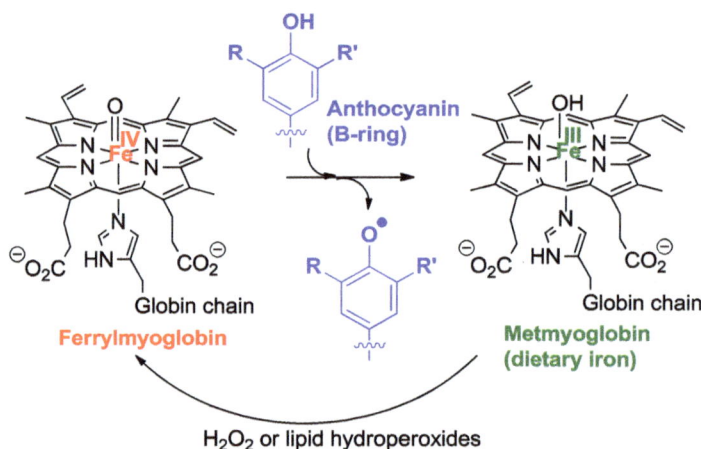

Figure 11. Possible mechanisms for the antioxidant activity of anthocyanins in food and in the gastro-intestinal tract.

Acylation by electron-rich hydroxycinnamic acids, such as sinapic and ferulic acids, potentiates the capacity of anthocyanins to inhibit the diazo-initiated autoxidation of styrene in acetonitrile. In particular, a higher rate constant and stoichiometric factor of radical scavenging were obtained for acylated (*vs.* non-acylated) anthocyanins [31]. Curiously, this trend could not be confirmed for the peroxidation of linoleic acid in micelles, as if the intrinsic differences in electron-donating activity were counterbalanced by differences in the partition of anthocyanins between micelles and the aqueous phase.

2.5. Anthocyanin Complexes

Phenolic nuclei have an intrinsic ability to develop molecular (non-covalent) interactions as they combine flat polarizable apolar surfaces (the aromatic nuclei) for strong dispersion interactions and polar OH groups that are susceptible to acting as H-bond donors and acceptors.

2.5.1. Self-Association and Co-Pigmentation

One of the most remarkable properties of the anthocyanin chromophores is their ability to develop π-stacking interactions [32–34], mostly driven by dispersion interactions and the concomitant favorable release of water molecules from the solvation shells of the interacting nuclei, known as the hydrophobic effect. Owing to their planar structures and extended electron delocalization over the three rings, the colored forms are much more prone to π-stacking interactions than the colorless forms, for which such interactions, although not necessarily absent, are typically neglected. Examples of π-stacking interactions with anthocyanins are self-association and binding between anthocyanins and other phenols, a phenomenon called co-pigmentation. The affinity of co-pigments for a given anthocyanin (as measured by the corresponding thermodynamic binding constant) decays along the series: planar flavonoids (flavones, flavonols) > non-planar flavonoids (catechins), hydroxycinnamic acids > hydroxybenzoic acids [32]. As for self-association, it is stronger for the neutral base than for the flavylium ion and the anionic base, as the latter stacks are destabilized by charge repulsion.

The spectral consequences of co-pigmentation are summarized in Figure 12 with malvin (malvidin 3,5-di-*O*-β-glucoside) and a highly water-soluble rutin (quercetin 3-*O*-β-rutinoside) derivative [35]. In strongly acidic solutions (negligible water-to-flavylium addition), π-stacking interactions between the two partners promote bathochromism as a consequence of co-pigment-to-pigment charge transfer. Changes in color intensity simply reflect differences between the molar absorption coefficients of free and bound pigments. Under the mildly acidic conditions typically encountered in natural media, pigment–co-pigment interactions also promote hyperchromism, which can be understood as a shift in the now established flavylium–hemiketal equilibrium toward the colored form, which is selectively stabilized by its association with the co-pigment. This combination of bathochromic and hyperchromic shifts makes co-pigmentation one of the most important mechanisms for color variation and stabilization in plants. It can also be noted that heating usually attenuates the hyperchromic shift (Figure 12) as a consequence of the exothermic character of co-pigmentation ($\Delta H^0 < 0$).

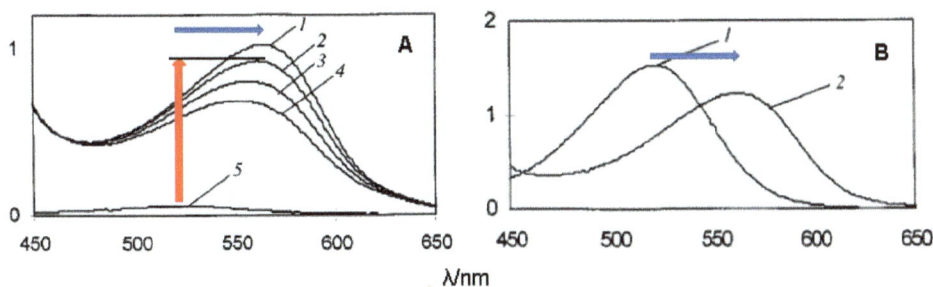

Figure 12. Co-pigmentation of malvin (malvidin 3,5-diglucoside, 50 μM) by rutin bis(hydrogensuccinate) (mixture of 3 regioisomers, 200 equiv.). (**A**) pH = 3.5, malvin + co-pigment at T = 15.5 (1), 25.0 (2), 35.0 (3), 44.2 (4) °C, malvin alone at T = 25.3 °C (5). (**B**) pH = 0.9, T = 25.0 °C, malvin alone (1), malvin + co-pigment (2). Adapted from reference [35].

The possibility of developing π-stacking interactions increases with the acylation of anthocyanins on their glycosyl moieties by hydroxycinnamic acid (HCA) residues. Indeed, depending on the location and number of HCA residues, different spatial arrangements can be observed (Figure 13) [34]:

- Intramolecular co-pigmentation: π-stacking interactions promote a conformational folding of the pigment bringing one or more HCA residue(s) into contact with the chromophore;
- Enhanced self-association: the HCA residues can stabilize the chiral stacking of chromophores evidenced by circular dichroism.

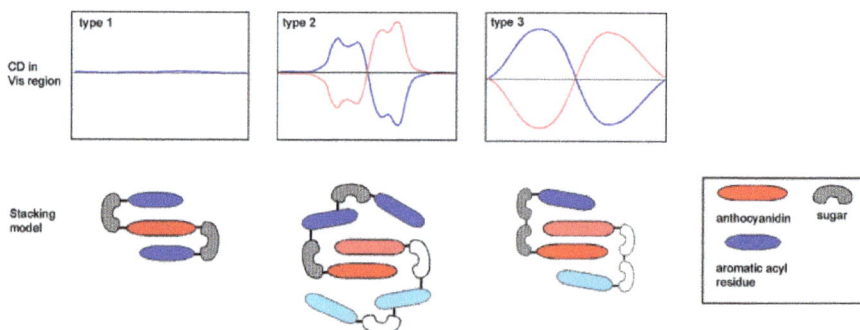

Figure 13. Acylated anthocyanins: discrimination of intramolecular co-pigmentation (type 1) and self-association (types 2 and 3) by circular dichroism (pink or blue CD spectra depending on the chirality of the stacks). From [34] with permission of the *Royal Society of Chemistry*.

In such assemblies, the flavylium nucleus has restricted access to the water solvent. Consequently, the thermodynamics of water addition are less favorable (increased pK'_h), and the percentage of colored forms at equilibrium increases [5,36–38]. For instance, at pH = 3, ca. 90% of the triacylated *Morning glory* pigment is still in colored form (mostly flavylium) vs. 15% for its non-acylated counterpart (Figure 14). Its vulnerability to water addition prevents the non-acylated pigment from accumulating the neutral quinonoid base at higher pH levels, and the corresponding solutions are almost colorless. In contrast, 30% of the triacylated pigment is present as the colored neutral base at pH = 5. Moreover, the π-stacking interactions developed by the triacylated flavylium ion induce a 20 nm bathochromic shift of its λ_{max} compared to its non-acylated counterpart.

Anthocyanins with an *o*-dihydroxy substitution on their B-ring (cyanidin, delphinidin, and petunidin derivatives) also bind hard metal ions, such as Al^{3+} and Fe^{3+}, in mildly acidic to neutral solution. As the anthocyanin binds as the quinonoid base with additional proton loss from C3'-OH, bathochromism is observed with additional ligand-to-metal charge transfer with Fe^{3+} (Figure 15).

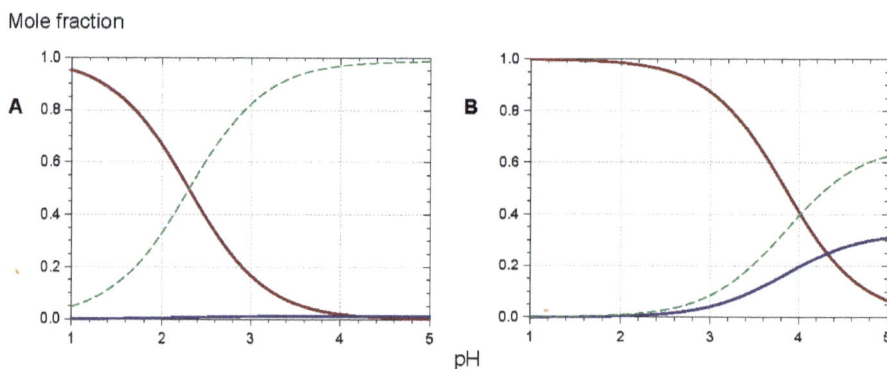

Figure 14. *Cont.*

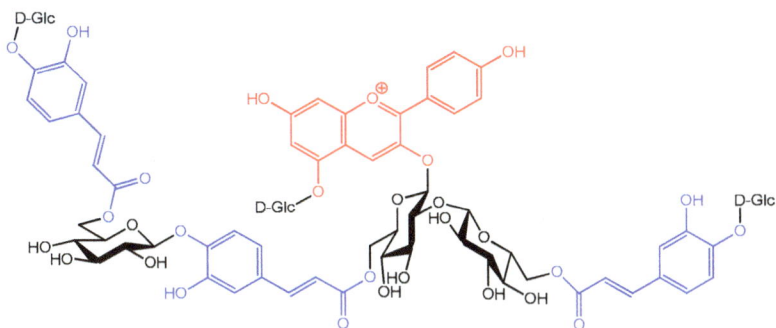

Figure 14. Triacylated (**B**) vs. non-acylated (**A**) *Morning glory* (*Pharbitis nil*) anthocyanins: equilibrium distribution of anthocyanin species in aqueous solution. Red solid line: flavylium ion, blue solid line: neutral base, dotted green line: total colorless forms. Parameters for plots are $pK'_h = 2.30$, $pK_{a1} = 4.21$ (**A**); $pK'_h = 4.01$, $pK_{a1} = 4.32$ (**B**). From [36,37].

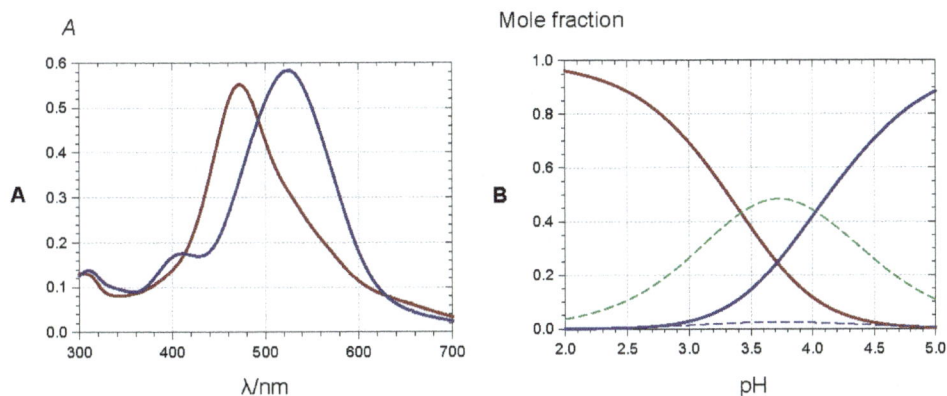

Figure 15. (**A**) 3′,4′-Dihydroxy-7-O-β-D-glucopyranosyloxyflavylium (50 μM) in a pH 4 buffer (0.1 M acetate), red spectrum: before hydration, blue spectrum: 10 min after addition of Al^{3+} (4 equiv.); (**B**) equilibrium distribution of species in aqueous solution. Red solid line: flavylium ion, blue dotted line: neutral base, dotted green line: total colorless forms, blue solid line: Al^{3+} complex. Parameters for plots are $pK'_h = 3.42$, $pK_{a1} = 4.72$, $K_M = 2 \times 10^{-4}$. From [39].

At least in mildly acidic solution, metal binding is restricted to the colored forms and thus efficiently competes with the hydration equilibrium, thereby preventing the formation of the colorless forms. In the most sophisticated assemblies, metal binding and π-stacking interactions combine, thus providing the most common mechanism towards the formation of stable blue colors [34,40,41]. In the so-called metalloanthocyanins, a fixed metal–pigment–co-pigment stoichiometry of 2:6:6 is observed: three anthocyanins bind to each metal ion and two equivalent complexes assemble by left-handed π-stacking interactions between the chromophores. Then, three pairs of flavone or flavonol co-pigments in left-handed π-stacking intercalate between the pairs of stacked anthocyanins. In this intercalation, right-handed pigment–copigment π-stacking occurs. Large-scale aggregation of acylated anthocyanins can also result in the formation of highly colored assemblies within the vacuole (the so-called anthocyanin vacuolar inclusions) [42], the organelle where anthocyanins are stored in plant cells.

2.5.2. Binding to Biopolymers

Despite the potential significance of such associations in food chemistry and nutrition, the ability of anthocyanins to bind proteins and polysaccharides is still poorly documented at the molecular level. This paragraph focuses on anthocyanins (glycosides), although anthocyanidins are also commonly investigated. Indeed, aglycones are chemically unstable in mildly acidic and neutral conditions and may be substantially degraded over the duration of analysis.

Saturation transfer difference (STD)-NMR was used to probe the binding of cyanidin and delphinidin 3-glucosides to pectin from citrus fruits (MM = 111 kDa) [43]. Indeed, magnetization transfer (requiring proton pairs distant by less than 0.5 nm) from irradiated pectin protons to anthocyanin protons provided direct evidence that the two partners are in close contact. STD titrations at pH = 4.0 and pH = 1.5 suggest that the flavylium ion has a higher affinity for pectin than the hemiketal. Assuming the Scatchard model (*n* identical binding sites having the same binding constant, K_b), pectin was found to bind 180–600 anthocyanin units depending on the selected anthocyanin and pH. The corresponding K_b values are very weak ($<10^3$ M^{-1}). Thus, the picture emerging from this study is that anthocyanins (as individual species or non-covalent oligomers) provide a coating of the pectin's surface through the development of very weak interactions (van der Waals contacts, H-bonds).

The quenching of intrinsic protein fluorescence by increasing ligand concentrations is a classical method to probe ligand–protein binding and to extract binding parameters. As anthocyanins typically absorb light at the protein's excitation and emission wavelengths, corrections for these inner-filter effects should be applied [44], which are not systematic [45] and thus lead to discrepancies in K_b values as well as in enthalpy and entropy changes. With human serum albumin (HSA), a globular protein, 1:1 binding is observed with a K_b in the order of 10^5 M^{-1} [44,45], meaning a moderate affinity. The influence of the pH (from pH = 4 to pH = 7.4) on the binding strength is very modest [44]. Competition with probes of a known binding site (ibuprofen, warfarin) enables location of the anthocyanin binding site, a hydrophobic pocket lined by positively charged amino-acid residues (Arg, Lys) for possible accommodation of the anionic base [45]. As for the weakly structured salivary proteins, interaction with malvidin-3-glucoside (probed by STD-NMR) was found to be much weaker ($K_b \approx 500$ M^{-1}) and largely pH-independent (same affinity at pH = 1.0 and pH = 3.4), which suggests that the hemiketal and flavylium ions bind with close affinities [46]. Electrospray ionization MS revealed the formation of soluble aggregates involving 2–6 anthocyanin units and 1–4 peptides (proline-rich proteins or histatin). STD-NMR was also used to investigate the binding of keracyanin (cyanidin 3-rutinoside) to wheat flour gliadins at pH = 2.5 [47]. Protons C2'-H, C5'-H, C6-H and C8-H appear to be primarily involved in the binding. At this low pH, the corresponding aglycone (cyanidin) is stable and can be also investigated. Its affinity for gliadins appears higher based on the strong shielding of its proton signals when gliadins are added (confirmed by the large retention of cyanidin in the centrifugation pellet: up to 80% vs. only 8% for keracyanin). However, STP-NMR did not point protons specifically involved in the interaction. Cyanidin 3-glucoside expresses a rather high affinity for sodium caseinate (NaCas) [48]. Two binding sites were identified at pH = 2 and pH = 7, one of high affinity ($K_b \approx 1$–7×10^6 M^{-1} depending on pH and T) and a second of lower affinity ($K_b \approx 2$–7×10^5 M^{-1}). For both sites, the binding was found to be exothermic at pH = 7 but endothermic at pH = 2 and thus is driven by a favorable entropy, which could point to a large contribution of the hydrophobic effect. Interestingly, NaCl addition gradually cancels cyanidin 3-glucoside–NaCas binding at pH = 7 but has no effect at pH = 2. In contrast to the high affinity of cyanidin 3-glucoside for NaCas, malvidin 3-glucoside only weakly binds to α- and β-caseins [49] and to β-lactoglobulin [50] (1:1 binding with $K_b < 10^3$ M^{-1}).

Unlike co-pigmentation, the binding of anthocyanins to biopolymers does not trigger spectacular spectral changes. For instance, in the presence of various polysaccharides [51], no change in the wavelength of maximal visible absorption (λ_{max}) was observed. Interactions of anthocyanins with cellulose, oat bran, and lignin is associated with a weak hypochromic effect, whereas an opposite effect (weak hyperchromism) is observed with highly methylated apple pectins. Sugar beet pectins have been shown to promote strong bathochromism in solutions of blackcurrant anthocyanins (cyanidin and delphinidin glycosides), but this effect is due to endogenous iron ions (bound to the polysaccharide) forming blue chelates with the pigments [52]. In agreement with the small spectral changes observed, the binding of anthocyanins to pectin does not significantly affect the thermodynamic constants of the acid–base and hydration equilibria [43]. In other words, all anthocyanin forms (colored or colorless) bind pectin with close affinities. This apparent discrepancy with the STD-NMR data (stronger flavylium–pectin binding) might be due to anthocyanin self-association, which probably is significant in the concentrated solutions used in the STD-NMR experiments. In contrast, the flavylium cation of the pyranoanthocyanin portisin is strongly stabilized by interactions with anionic wood lignosulfates as evidenced by its much weaker acidity in the presence of the polysaccharide ($pK_{a1} = 6.6$ vs. 4.6 for portisin alone) [53].

2.6. Anthocyanins in the Excited State

Although their main function is to absorb visible light and express color, anthocyanins are intrinsically poorly fluorescent with quantum yields typically lower than 4×10^{-3} (meaning that less than one photon out of 250 absorbed is actually re-emitted) [54]. Indeed, the fate of anthocyanins after absorption, i.e., once in the excited state, is a difficult question that must be addressed by sophisticated fast techniques, such as time-resolved fluorescence and transient absorption-emission spectroscopies. In the HOMO → LUMO transition accompanying the absorption of visible light by the flavylium ion, electron transfer from the B-ring to the A-/C-rings takes place (Figure 16) [55]. In the excited state, the flavylium ion is a strong acid ($pK_a < 0$) that transfers a proton to the solvent on a picosecond timescale (20 ps for pelargonin at pH = 1) [54,56]. In the next step, the quinonoid base in the excited state is deactivated by a combination of radiative (fluorescence) and non-radiative (heat) processes and then captures a proton in the ground state to form the ground state flavylium ion. In other words, the quinonoid base is responsible for the (weak) fluorescence observed for anthocyanins even in strongly acidic solution. In the presence of a co-pigment, other mechanisms (Figure 17) supersede the fast flavylium deprotonation observed with free anthocyanins [57] in the following ways: (a) within the complex in the excited state, through ultrafast internal conversion (<1 ps) via a low-energy co-pigment-to-pigment charge transfer state, resulting in static fluorescence quenching; and (b) for the fraction of free anthocyanin, diffusion-controlled electron transfer from the co-pigment to the flavylium ion in the excited state, resulting in dynamic fluorescence quenching. The mechanism of energy dissipation by ultrafast internal conversion has been confirmed for the folded conformation of a cyanidin glycoside acylated by *p*-coumaric acid [58]. In addition, fast energy transfer to the chromophore following absorption of UV light by the acyl residue operates (Figure 17), thereby conferring acylated anthocyanins to have an important role in plant photoprotection.

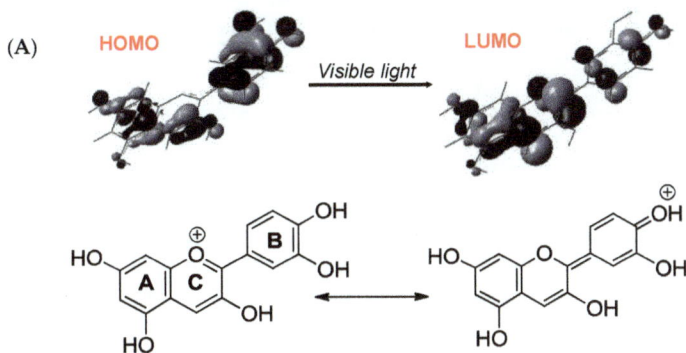

Figure 16. *Cont.*

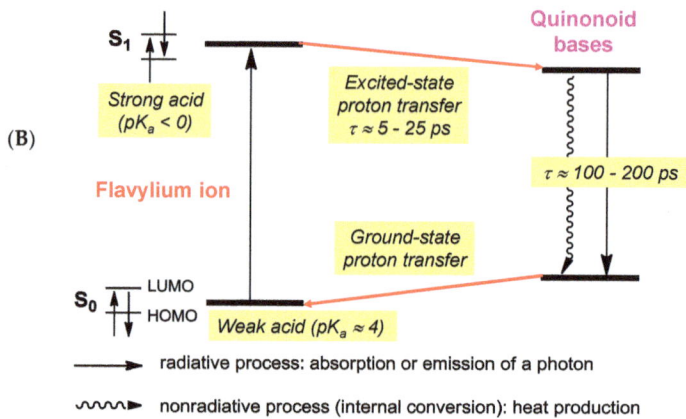

Figure 16. (A) Frontier MOs of the flavylium ion of cyanidin (from reference [55]) and its most representative mesomeric forms in the ground state (left) and first excited state (right). (B) The fate of free anthocyanins in the excited state (from references [54,56]).

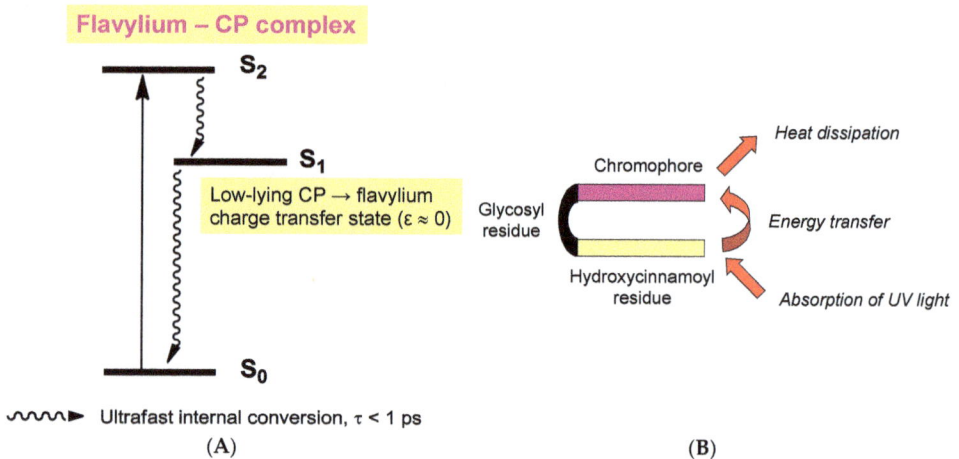

Figure 17. The influence of co-pigmentation on the fate of anthocyanins in the excited state. (A) Intermolecular co-pigmentation (from reference [57]). (B) Intramolecular co-pigmentation (from reference [58]).

3. The Importance of Anthocyanin Chemistry in Food and Nutrition

3.1. Formulation of Anthocyanins for Food Applications

Anthocyanin degradation typically occurs during thermal processing and storage. The knowledge on anthocyanin–biopolymer interactions can be applied to devise formulations for improved chemical stability. Degradation studies aimed at demonstrating the protection afforded by biopolymers may be limited to monitoring the color loss under given conditions of pH, temperature, and light exposure. More information is obtained when samples are also acidified to pH 1–2 for quantification of the residual flavylium ions by HPLC or by UV-visible spectroscopy. With this approach, color loss (directly observed at the monitoring pH), which combines the reversible water addition and irreversible phenomena (hydrolysis, autoxidation), and true anthocyanin loss (irreversible component), can be distinguished.

In the simplest experiments involving modeling beverages, solutions of anthocyanins and soluble biopolymers are heated, and their color or residual anthocyanin concentration is monitored as a function of time. For instance, yeast mannoproteins (0.5% w/w for both anthocyanins and mannoproteins) increase the half-life of color loss by a factor of 5.4 in experiments conducted at pH = 7 and T = 80 °C or 126 °C (modeling pasteurization or sterilization) [59]. Similarly, the color loss in solutions of purple carrot anthocyanins at pH = 3.0 and T = 40 °C (in light) was shown to be inhibited by the addition of gum arabic (0.05–5.0%) with maximal stability observed at 1.5% (50% color retention after 5 days, vs. 20% in control) [60]. Similar observations were made with pectins or whey proteins (1%), the best result being obtained with heat-denatured whey proteins (70% color retention after 7 days at 40 °C, vs. 20% in control) [61]. In these works, fluorescence quenching experiments suggest that color protection involves direct interactions between anthocyanins and proteins (including the glycoprotein of gum arabic). However, the mechanism of protection remains largely unknown. It may be speculated that biopolymers mostly act by providing a more hydrophobic environment to anthocyanins, resulting in slower hydrolysis (despite the weak impact on the hydration equilibrium itself, see Section 2.5.2) and/or by scavenging transition metal traces acting as initiators/catalysts of anthocyanin autoxidation.

A more sophisticated approach consists of preparing solid micro- or nanoparticles as delivery systems for anthocyanins. For instance, nanoparticles of whey proteins and beet pectin can be loaded with anthocyanin extracts with a higher efficiency (55%) when anthocyanins are added prior nanoparticle formation [62]. However, when dispersed in pH 4 solution, these nanoparticles do not show improved color stability. Particles of chitosan and carboxymethylchitosan (CMC) loaded with anthocyanins (size ≈ 200 nm, encapsulation efficiencies ranging from 16 to 44% depending on the CMC/chitosan proportions) can be simply prepared by mixing at pH = 5–6 followed by centrifugation [63]. The thermal stability of encapsulated anthocyanins was shown to greatly improve: 12% degradation after 3 days at 40 °C, vs. 90% in the control (no particles). Similar protection was observed in samples exposed to white light for 10 days (−20% vs. −80%). Sulfonylated polysaccharides, such as dextran sulfate and carrageenans, can also be used to encapsulate bilberry anthocyanins from acidic solutions (pH ≈ 3) with high efficiency and improved stability [64,65]. The binding of isotherms and HPLC analysis showed that the binding is selective of anthocyanins (the other phenols remaining in solution) and is stronger when the sulfonylation degree is higher. These data strongly suggest that the encapsulation is driven by ionic flavylium–sulfate interactions. Interestingly, the nanoparticles are gradually dissociated under near neutral conditions modeling the small intestine, which is desirable for subsequent intestinal absorption. Combining chitosan and cellulose nanocrystals at pH 2–3 also allows the formation of nanoparticles with high affinity for anthocyanins (up to 94% encapsulation) [66]. When cellulose is replaced by sodium tripolyphosphate, a reticulating agent for the polycationic chitosan chain, gel microcapsules (size ≈ 34 μm, encapsulation yield ≈ 33%) are formed. Finally, large hydrogel particles (size ≈ 2–3 mm) combining alginate and pectin can be used for encapsulation of anthocyanin-rich extracts under acidic conditions (pH = 1–3), and they are released upon dissolution at higher pH [67]. When exposed to white light, the half-life values of anthocyanins in hydrogel, hydrogel particles dispersed in pH 3 solution, and in a control solution (pH = 3) were 630 h, 277 h, and 58 h, respectively.

Interestingly, anthocyanin-rich blackcurrant extracts can be incorporated into bread [68]. Replacing wheat flour by a mixture of gluten and starch led to markedly decreased anthocyanin concentrations (especially, for delphinidin glycosides, which are most sensitive to oxidation). This suggests that other flour proteins (e.g., albumins, globulins) and non-starch polysaccharides (e.g., hemicelluloses, β-glucans) may be important to provide chemical stability to anthocyanins in such matrices.

3.2. The Fate of Anthocyanins in Humans, Consequences on the Possible Effects on Health

The bioavailability of phenolic compounds has been largely elucidated over the last decades [69]. This knowledge, which is crucial to the interpretation of the possible effects on health, encompasses the bioaccessibility (the release of phenols from the food matrix during digestion), intestinal absorption, metabolism, transport, distribution to tissues, and excretion of dietary phenols and their metabolites. Anthocyanins have emerged as poorly bioavailable micronutrients as judged from the low concentrations (generally, <0.1 µM) of native forms (mostly, anthocyanidin glucosides) and anthocyanidin conjugates detected in the general blood circulation [70,71]. These derivatives are formed in the small intestine after enzymatic hydrolysis by membrane-bound lactate phlorizin hydrolase or by cytosolic β-glucosidase, and subsequent conjugation by O-glucuronidation, O-methylation, and/or O-sulfonylation. The detection of native forms in the blood circulation is not equivalent to other flavonoid glucosides and could be due to partial absorption from the stomach. This early absorption has been demonstrated in cell and animal models [72–74] and has been proposed to involve the organic anion transporter bilitranslocase in the gastric epithelium [72].

Most importantly, recent investigations, in particular using ^{13}C-labelled compounds [3], have shown that the bulk of the ingested amount of anthocyanins is actually converted into simple phenolic compounds (Table 2), as a consequence of (a) the chemical instability (under near neutral conditions) of anthocyanins and, especially, of anthocyanidins [24] and (b) the extensive catabolism by the colonic microbiota of the fraction reaching the large intestine. These simple metabolites, which themselves can be further conjugated by intestinal and hepatic enzymes, have been found in the blood circulation in much higher concentration than anthocyanidin derivatives [3,75].

Table 2. Serum pharmacokinetic profiles of cyanidin 3-glucoside (C3G) and its metabolites in humans after the consumption of 500 mg ^{13}C-labelled C3G. From reference [3] (in red is the reference compound and its most abundant metabolites).

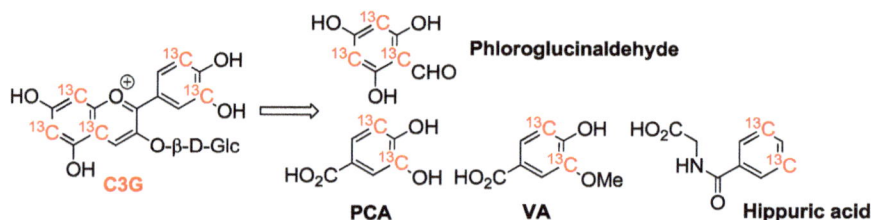

Compound	n	C_{max}/nM	t_{max}/h	$t_{1/2}$/h	AUC_{0-48}/nM h
Cyanidin-3-glucoside (C3G)	5	141 (±70)	1.8 (±0.2)	0.4	279 (±170)
Protocatechuic acid (PCA)	8	146 (±74)	3.3 (±0.7)	9.9 (±3.4)	1377 (±760)
Phloroglucinaldehyde	4	582 (±536)	2.8 (±1.1)	nd	7882 (±7768)
PCA-sulfates	8	157 (±116)	11.4 (±3.8)	31.9 (±19.1)	1180 (±349)
Vanillic acid (VA)	2	1845 (±838)	12.5 (±11.5)	6.4	23319 (±20650)
VA-sulfates	4	430 (±299)	30.1 (±11.4)	nd	10689 (±7751)
Ferulic acid	7	827 (±371)	8.2 (±4.1)	21.4 (±7.8)	17422 (±11054)
Hippuric acid	8	1962 (±1389)	15.7 (±4.1)	95.6 (±77.8)	46568 (±30311)

In agreement with the strong in vivo catabolism of anthocyanins, in vitro digestion models have shown that whereas anthocyanins are readily released into the acidic gastric compartment and relatively stable, they undergo substantial degradation in the near neutral upper intestinal compartment, possibly because of autoxidation [76,77]. However, this chemical instability could be overestimated in in vitro models, as the O_2 content is higher than under real physiological conditions. As a striking example, protocatechuic acid (PCA, 3,4-dihydroxybenzoic acid), recovered in blood and fecal samples, was shown to represent more than 70% of the ingested dose of the cyanidin O-glucosides from blood orange juice [75]. Interestingly, PCA can be formed by chemical oxidative degradation of anthocyanins and anthocyanidins (Figures 8–10). However, it must be noted that anthocyanins bearing an electron-rich B-ring (e.g., cyanidin and delphinidin glycosides) must be much more prone to oxidative degradation than, for instance, pelargonidin derivatives [78], which indeed could be detected in higher concentrations (0.2–0.3 μM) in the blood [79].

In the digestive tract, anthocyanins may also modulate the digestion and uptake of nutrients by interacting with intestinal α-glucosidase [80]. They could, as well, attenuate oxidative stress in the digestive tract, for instance, by inhibiting the peroxidation of dietary lipids induced by heme iron [29,81]. After intestinal absorption, anthocyanin derivatives are probably transported in the blood in moderate association with serum albumin [45] before distribution to tissues, which, again, could involve bilitranslocase, as evidenced in the kidneys of rats [82].

Most importantly, it must be kept in mind that the degradation products of anthocyanins, which are formed in the digestive tract and are generally much more abundant than the residual anthocyanidin derivatives, could mediate most of the potential health effects of anthocyanins [83,84], which remains intriguing given their chemical simplicity [3] (Table 2). However, redox-active compounds, such as PCA, could indeed participate in regulating the expression of genes associated with transcription factors susceptible to redox activation. Such mechanisms could, at least partly, underline the induction of antioxidant defense via the Nrf2 pathway and the reduction of inflammation via NF-κB inhibition observed in cells and in rodents with cyanidin derivatives [85] or PCA itself [86].

Author Contributions: O.D. and J.-A.F. analyzed the literature and wrote the paper.

Funding: This research received no external funding.

Conflicts of Interest: The authors declare no conflict of interest.

References

1. Pina, F.; Melo, M.J.; Laia, C.A.T.; Parola, A.J.; Lima, J.C. Chemistry and applications of flavylium compounds: A handful of colours. *Chem. Soc. Rev.* **2012**, *41*, 869–908. [CrossRef] [PubMed]
2. Pina, F. Chemical applications of anthocyanins and related compounds. A source of bioinspiration. *J. Agric. Food Chem.* **2014**, *62*, 6885–6897. [CrossRef] [PubMed]
3. De Ferrars, R.M.; Czank, C.; Zhang, Q.; Botting, N.P.; Kroon, P.A.; Cassidy, A.; Kay, C.D. The pharmacokinetics of anthocyanins and their metabolites in humans. *Br. J. Pharmacol.* **2014**, *171*, 3268–3282. [CrossRef] [PubMed]
4. Nave, F.; Petrov, V.; Pina, F.; Teixeira, N.; Mateus, N.; de Freitas, V. Thermodynamic and kinetic properties of a red wine pigment: Catechin-(4,8)-malvidin-3-O-glucoside. *J. Phys. Chem. B* **2010**, *114*, 13487–13496. [CrossRef] [PubMed]
5. Moloney, M.; Robbins, R.J.; Collins, T.M.; Kondo, T.; Yoshida, K.; Dangles, O. Red cabbage anthocyanins: The influence of D-glucose acylation by hydroxycinnamic acids on their structural transformations in acidic to mildly alkaline conditions and on the resulting color. *Dyes Pigment.* **2018**, *158*, 342–352. [CrossRef]
6. Berké, B.; Chèze, C.; Vercauteren, J.; Deffieux, G. Bisulfite addition to anthocyanins: Revisited structures of colourless adducts. *Tetrahedron Lett.* **1998**, *39*, 5771–5774. [CrossRef]
7. Fulcrand, H.; Dueñas, M.; Salas, E.; Cheynier, V. Phenolic reactions during winemaking and aging. *Am. J. Enol. Vitic.* **2006**, *57*, 289–297.
8. Oliveira, J.; Mateus, N.; de Freitas, V. Previous and recent advances in pyranoanthocyanins equilibria in aqueous solution. *Dyes Pigment.* **2014**, *100*, 190–200. [CrossRef]
9. Vallverdú-Queralt, A.; Meudec, E.; Ferreira-Lima, N.; Sommerer, N.; Dangles, O.; Cheynier, V.; Guernevé, C.L. A comprehensive investigation of guaiacyl-pyranoanthocyanin synthesis by one-/two-dimensional NMR and UPLC–DAD–ESI–MSn. *Food Chem.* **2016**, *199*, 902–910. [CrossRef] [PubMed]
10. Oliveira, J.; Fernandes, V.; Miranda, C.; Santos-Buelga, C.; Silva, A.; de Freitas, V.; Mateus, N. Color properties of four cyanidin-pyruvic acid adducts. *J. Agric. Food Chem.* **2006**, *54*, 6894–6903. [CrossRef] [PubMed]

11.	Cruz, L.; Petrov, V.; Teixeira, N.; Mateus, N.; Pina, F.; Freitas, V.D. Establishment of the chemical equilibria of different types of pyranoanthocyanins in Aqueous solutions: Evidence for the formation of aggregation in pyranomalvidin-3-*O*-coumaroylglucoside-(+)-catechin. *J. Phys. Chem. B* **2010**, *114*, 13232–13240. [CrossRef] [PubMed]

12.	Vallverdú-Queralt, A.; Biler, M.; Meudec, E.; Guernevé, C.L.; Vernhet, A.; Mazauric, J.-P.; Legras, J.-L.; Loonis, M.; Trouillas, P.; Cheynier, V.; et al. *p*-Hydroxyphenyl-pyranoanthocyanins: An experimental and theoretical investigation of their acid—Base properties and molecular interactions. *Int. J. Mol. Sci.* **2016**, *17*, 1842. [CrossRef] [PubMed]

13.	Dueñas, M.; Fulcrand, H.; Cheynier, V. Formation of anthocyanin–flavanol adducts in model solutions. *Anal. Chim. Acta* **2006**, *563*, 15–25. [CrossRef]

14.	Lee, D.F.; Swinny, E.E.; Jones, G.P. NMR identification of ethyl-linked anthocyanin–flavanol pigments formed in model wine ferments. *Tetrahedron Lett.* **2004**, *45*, 1671–1674. [CrossRef]

15.	Li, L.; Zhang, M.; Zhang, S.; Cui, Y.; Sun, B. Preparation and antioxidant activity of ethyl-linked anthocyanin-flavanol pigments from model wine solutions. *Molecules* **2018**, *23*, 1066. [CrossRef] [PubMed]

16.	Vallverdú-Queralt, A.; Meudec, E.; Eder, M.; Lamuela-Raventos, R.M.; Sommerer, N.; Cheynier, V. The hidden face of wine polyphenol polymerization highlighted by high-resolution mass spectrometry. *ChemistryOpen* **2017**, *6*, 336–339. [CrossRef] [PubMed]

17.	Salas, E.; Fulcrand, H.; Meudec, E.; Cheynier, V. Reactions of anthocyanins and tannins in model solutions. *J. Agric. Food Chem.* **2003**, *51*, 7951–7961. [CrossRef] [PubMed]

18.	Dueñas, M.; Salas, E.; Cheynier, V.; Dangles, O.; Fulcrand, H. UV−visible spectroscopic investigation of the 8,8-methylmethine catechin-malvidin 3-glucoside pigments in aqueous solution: Structural transformations and molecular complexation with chlorogenic acid. *J. Agric. Food Chem.* **2006**, *54*, 189–196. [CrossRef] [PubMed]

19.	Quideau, S.; Jourdes, M.; Lefeuvre, D.; Montaudon, D.; Saucier, C.; Glories, Y.; Pardon, P.; Pourquier, P. The chemistry of wine polyphenolic C-glycosidic ellagitannins targeting human topoisomerase II. *Chem. A Eur. J.* **2005**, *11*, 6503–6513. [CrossRef] [PubMed]

20.	Chassaing, S.; Lefeuvre, D.; Jacquet, R.; Jourdes, M.; Ducasse, L.; Galland, S.; Grelard, A.; Saucier, C.; Teissedre, P.-L.; Dangles, O.; et al. Physicochemical studies of new anthocyano-ellagitannin hybrid pigments: About the origin of the influence of oak C-glycosidic ellagitannins on wine color. *Eur. J. Org. Chem.* **2010**, 55–63. [CrossRef]

21.	Oliveira, J.; da Silva, M.A.; Jorge Parola, A.; Mateus, N.; Brás, N.F.; Ramos, M.J.; de Freitas, V. Structural characterization of a A-type linked trimeric anthocyanin derived pigment occurring in a young Port wine. *Food Chem.* **2013**, *141*, 1987–1996. [CrossRef] [PubMed]

22.	Geera, B.; Ojwang, L.O.; Awika, J.M. New highly stable dimeric 3-deoxyanthocyanidin pigments from sorghum bicolor leaf sheath. *J. Food Sci.* **2012**, *77*, C566–C572. [CrossRef] [PubMed]

23.	Sarni-Manchado, P.; Cheynier, V.; Moutounet, M. Reactions of polyphenoloxidase generated caftaric acid o-quinone with malvidin 3-*O*-glucoside. *Phytochemistry* **1997**, *45*, 1365–1369. [CrossRef]

24.	Fleschhut, J.; Kratzer, F.; Rechkemmer, G.; Kulling, S.E. Stability and biotransformation of various dietary anthocyanins in vitro. *Eur. J. Nutr.* **2006**, *45*, 7–18. [CrossRef] [PubMed]

25.	Cabrita, L.; Petrov, V.; Pina, F. On the thermal degradation of anthocyanidins: Cyanidin. *RSC Adv.* **2014**, *4*, 18939–18944. [CrossRef]

26.	Lopes, P.; Richard, T.; Saucier, C.; Teissedre, P.-L.; Monti, J.-P.; Glories, Y. Anthocyanone a: A quinone methide derivative resulting from malvidin 3-*O*-glucoside degradation. *J. Agric. Food Chem.* **2007**, *55*, 2698–2704. [CrossRef] [PubMed]

27.	Sadilova, E.; Carle, R.; Stintzing, F.C. Thermal degradation of anthocyanins and its impact on color and in vitro antioxidant capacity. *Mol. Nutr. Food Res.* **2007**, *51*, 1461–1471. [CrossRef] [PubMed]

28.	Kähkönen, M.P.; Heinonen, M. Antioxidant activity of anthocyanins and their aglycons. *J. Agric. Food Chem.* **2003**, *51*, 628–633. [CrossRef] [PubMed]

29.	Goupy, P.; Bautista-Ortin, A.-B.; Fulcrand, H.; Dangles, O. Antioxidant activity of wine pigments derived from anthocyanins: Hydrogen transfer reactions to the DPPH radical and inhibition of the heme-induced peroxidation of linoleic acid. *J. Agric. Food Chem.* **2009**, *57*, 5762–5770. [CrossRef] [PubMed]

30. Achat, S.; Rakotomanomana, N.; Madani, K.; Dangles, O. Antioxidant activity of olive phenols and other dietary phenols in model gastric conditions: Scavenging of the free radical DPPH and inhibition of the haem-induced peroxidation of linoleic acid. *Food Chem.* **2016**, *213*, 135–142. [CrossRef] [PubMed]

31. Matera, R.; Gabbanini, S.; Berretti, S.; Amorati, R.; De Nicola, G.R.; Iori, R.; Valgimigli, L. Acylated anthocyanins from sprouts of *Raphanus sativus* cv. Sango: Isolation, structure elucidation and antioxidant activity. *Food Chem.* **2015**, *166*, 397–406. [CrossRef] [PubMed]

32. Trouillas, P.; Sancho-García, J.C.; De Freitas, V.; Gierschner, J.; Otyepka, M.; Dangles, O. Stabilizing and modulating color by copigmentation: Insights from theory and experiment. *Chem. Rev.* **2016**, *116*, 4937–4982. [CrossRef] [PubMed]

33. Mori, M.; Miki, N.; Ito, D.; Kondo, T.; Yoshida, K. Structure of tecophilin, a tri-caffeoylanthocyanin from the blue petals of Tecophilaea cyanocrocus, and the mechanism of blue color development. *Tetrahedron* **2014**, *70*, 8657–8664. [CrossRef]

34. Yoshida, K.; Mori, M.; Kondo, T. Blue flower color development by anthocyanins: From chemical structure to cell physiology. *Nat. Prod. Rep.* **2009**, *26*, 884. [CrossRef] [PubMed]

35. Alluis, B.; Pérol, N.; El hajji, H.; Dangles, O. Water-soluble flavonol (= 3-Hydroxy-2-phenyl-4H-1-benzopyran-4-one) derivatives: Chemical synthesis, colouring, and antioxidant properties. *Helv. Chim. Acta* **2000**, *83*, 428–443. [CrossRef]

36. Dangles, O.; Saito, N.; Brouillard, R. Anthocyanin intramolecular copigment effect. *Phytochemistry* **1993**, *34*, 119–124. [CrossRef]

37. Dangles, O.; Saito, N.; Brouillard, R. Kinetic and thermodynamic control of flavylium hydration in the pelargonidin-cinnamic acid complexation. Origin of the extraordinary flower color diversity of Pharbitis nil. *J. Am. Chem. Soc.* **1993**, *115*, 3125–3132. [CrossRef]

38. Figueiredo, P.; George, F.; Tatsuzawa, F.; Toki, K.; Saito, N.; Brouillard, R. New features of intramolecular copigmentation byacylated anthocyanins. *Phytochemistry* **1999**, *51*, 125–132. [CrossRef]

39. Mora-Soumille, N.; Al Bittar, S.; Rosa, M.; Dangles, O. Analogs of anthocyanins with a 3′,4′-dihydroxy substitution: Synthesis and investigation of their acid–base, hydration, metal binding and hydrogen-donating properties in aqueous solution. *Dyes Pigment.* **2013**, *96*, 7–15. [CrossRef]

40. Kondo, T.; Yoshida, K.; Nakagawa, A.; Kawai, T.; Tamura, H.; Goto, T. Structural basis of blue-colour development in flower petals from *Commelina communis*. *Nature* **1992**, *358*, 515–518. [CrossRef]

41. Shiono, M.; Matsugaki, N.; Takeda, K. Phytochemistry: Structure of the blue cornflower pigment. *Nature* **2005**, *436*, 791. [CrossRef] [PubMed]

42. Kallam, K.; Appelhagen, I.; Luo, J.; Albert, N.; Zhang, H.; Deroles, S.; Hill, L.; Findlay, K.; Andersen, Ø.M.; Davies, K.; et al. Aromatic decoration determines the formation of anthocyanic vacuolar inclusions. *Curr. Biol.* **2017**, *27*, 945–957. [CrossRef] [PubMed]

43. Fernandes, A.; Brás, N.F.; Mateus, N.; de Freitas, V. Understanding the molecular mechanism of anthocyanin binding to pectin. *Langmuir* **2014**, *30*, 8516–8527. [CrossRef] [PubMed]

44. Cahyana, Y.; Gordon, M.H. Interaction of anthocyanins with human serum albumin: Influence of pH and chemical structure on binding. *Food Chem.* **2013**, *141*, 2278–2285. [CrossRef] [PubMed]

45. Tang, L.; Zuo, H.; Shu, L. Comparison of the interaction between three anthocyanins and human serum albumins by spectroscopy. *J. Lumin.* **2014**, *153*, 54–63. [CrossRef]

46. Ferrer-Gallego, R.; Soares, S.; Mateus, N.; Rivas-Gonzalo, J.; Escribano-Bailón, M.T.; Freitas, V.D. New anthocyanin-human salivary protein complexes. *Langmuir* **2015**, *31*, 8392–8401. [CrossRef] [PubMed]

47. Mazzaracchio, P.; Tozzi, S.; Boga, C.; Forlani, L.; Pifferi, P.G.; Barbiroli, G. Interaction between gliadins and anthocyan derivatives. *Food Chem.* **2011**, *129*, 1100–1107. [CrossRef] [PubMed]

48. Casanova, F.; Chapeau, A.-L.; Hamon, P.; de Carvalho, A.F.; Croguennec, T.; Bouhallab, S. pH- and ionic strength-dependent interaction between cyanidin-3-*O*-glucoside and sodium caseinate. *Food Chem.* **2018**, *267*, 52–59. [CrossRef] [PubMed]

49. He, Z.; Xu, M.; Zeng, M.; Qin, F.; Chen, J. Interactions of milk α- and β-casein with malvidin-3-*O*-glucoside and their effects on the stability of grape skin anthocyanin extracts. *Food Chem.* **2016**, *199*, 314–322. [CrossRef] [PubMed]

50. He, Z.; Zhu, H.; Xu, M.; Zeng, M.; Qin, F.; Chen, J. Complexation of bovine β-lactoglobulin with malvidin-3-*O*-glucoside and its effect on the stability of grape skin anthocyanin extracts. *Food Chem.* **2016**, *209*, 234–240. [CrossRef] [PubMed]

51. Mazzaracchio, P.; Pifferi, P.; Kindt, M.; Munyaneza, A.; Barbiroli, G. Interactions between anthocyanins and organic food molecules in model systems. *Int. J. Food Sci. Technol.* **2004**, *39*, 53–59. [CrossRef]

52. Buchweitz, M.; Carle, R.; Kammerer, D.R. Bathochromic and stabilising effects of sugar beet pectin and an isolated pectic fraction on anthocyanins exhibiting pyrogallol and catechol moieties. *Food Chem.* **2012**, *135*, 3010–3019. [CrossRef] [PubMed]

53. Araújo, P.; Basílio, N.; Azevedo, J.; Fernandes, A.; Mateus, N.; Pina, F.; de Freitas, V.; Oliveira, J. Colour modulation of blue anthocyanin-derivatives. Lignosulfonates as a tool to improve the water solubility of natural blue dyes. *Dyes Pigment.* **2018**, *153*, 150–159. [CrossRef]

54. Moreira, P.F.; Giestas, L.; Yihwa, C.; Vautier-Giongo, C.; Quina, F.H.; Maçanita, A.L.; Lima, J.C. Ground- and excited-state proton transfer in anthocyanins: From weak acids to superphotoacids. *J. Phys. Chem. A* **2003**, *107*, 4203–4210. [CrossRef]

55. Anouar, E.H.; Gierschner, J.; Duroux, J.-L.; Trouillas, P. UV/Visible spectra of natural polyphenols: A time-dependent density functional theory study. *Food Chem.* **2012**, *131*, 79–89. [CrossRef]

56. Ferreira da Silva, P.; Lima, J.C.; Quina, F.H.; Maçanita, A.L. Excited-state electron transfer in anthocyanins and related flavylium salts. *J. Phys. Chem. A* **2004**, *108*, 10133–10140. [CrossRef]

57. Rodrigues, R.F.; Ferreira da Silva, P.; Shimizu, K.; Freitas, A.A.; Kovalenko, S.A.; Ernsting, N.P.; Quina, F.H.; Maçanita, A. Ultrafast internal conversion in a model anthocyanin-polyphenol complex: Implications for the biological role of anthocyanins in vegetative tissues of plants. *Chem. A Eur. J.* **2009**, *15*, 1397–1402. [CrossRef] [PubMed]

58. Ferreira da Silva, P.; Paulo, L.; Barbafina, A.; Elisei, F.; Quina, F.H.; Maçanita, A.L. Photoprotection and the photophysics of acylated anthocyanins. *Chem. A Eur. J.* **2012**, *18*, 3736–3744. [CrossRef] [PubMed]

59. Wu, J.; Guan, Y.; Zhong, Q. Yeast mannoproteins improve thermal stability of anthocyanins at pH 7.0. *Food Chem.* **2015**, *172*, 121–128. [CrossRef] [PubMed]

60. Chung, C.; Rojanasasithara, T.; Mutilangi, W.; McClements, D.J. Enhancement of colour stability of anthocyanins in model beverages by gum arabic addition. *Food Chem.* **2016**, *201*, 14–22. [CrossRef] [PubMed]

61. Chung, C.; Rojanasasithara, T.; Mutilangi, W.; McClements, D.J. Enhanced stability of anthocyanin-based color in model beverage systems through whey protein isolate complexation. *Food Res. Int.* **2015**, *76*, 761–768. [CrossRef] [PubMed]

62. Arroyo-Maya, I.J.; McClements, D.J. Biopolymer nanoparticles as potential delivery systems for anthocyanins: Fabrication and properties. *Food Res. Int.* **2015**, *69*, 1–8. [CrossRef]

63. Ge, J.; Yue, P.; Chi, J.; Liang, J.; Gao, X. Formation and stability of anthocyanins-loaded nanocomplexes prepared with chitosan hydrochloride and carboxymethyl chitosan. *Food Hydrocoll.* **2018**, *74*, 23–31. [CrossRef]

64. Klimaviciute, R.; Navikaite, V.; Jakstas, V.; Ivanauskas, L. Complexes of dextran sulfate and anthocyanins from *Vaccinium myrtillus*: Formation and stability. *Carbohydr. Polym.* **2015**, *129*, 70–78. [CrossRef] [PubMed]

65. Navikaite, V.; Simanaviciute, D.; Klimaviciute, R.; Jakstas, V.; Ivanauskas, L. Interaction between κ- and ι-carrageenan and anthocyanins from *Vaccinium myrtillus*. *Carbohydr. Polym.* **2016**, *148*, 36–44. [CrossRef] [PubMed]

66. Wang, W.; Jung, J.; Zhao, Y. Chitosan-cellulose nanocrystal microencapsulation to improve encapsulation efficiency and stability of entrapped fruit anthocyanins. *Carbohydr. Polym.* **2017**, *157*, 1246–1253. [CrossRef] [PubMed]

67. Guo, J.; Giusti, M.M.; Kaletunç, G. Encapsulation of purple corn and blueberry extracts in alginate-pectin hydrogel particles: Impact of processing and storage parameters on encapsulation efficiency. *Food Res. Int.* **2018**, *107*, 414–422. [CrossRef] [PubMed]

68. Sivam, A.S.; Sun-Waterhouse, D.; Perera, C.O.; Waterhouse, G.I.N. Exploring the interactions between blackcurrant polyphenols, pectin and wheat biopolymers in model breads; a FTIR and HPLC investigation. *Food Chem.* **2012**, *131*, 802–810. [CrossRef]

69. Del Rio, D.; Rodriguez-Mateos, A.; Spencer, J.P.E.; Tognolini, M.; Borges, G.; Crozier, A. Dietary (Poly) phenolics in human health: Structures, bioavailability, and evidence of protective effects against chronic diseases. *Antioxid. Redox Signal.* **2013**, *18*, 1818–1892. [CrossRef] [PubMed]

70. Kay, C.D. Aspects of anthocyanin absorption, metabolism and pharmacokinetics in humans. *Nutr. Res. Rev.* **2006**, *19*, 137–146. [CrossRef] [PubMed]

71. Manach, C.; Williamson, G.; Morand, C.; Scalbert, A.; Rémésy, C. Bioavailability and bioefficacy of polyphenols in humans. I. Review of 97 bioavailability studies. *Am. J. Clin. Nutr.* **2005**, *81*, 230S–242S. [CrossRef] [PubMed]

72. Passamonti, S.; Vrhovsek, U.; Mattivi, F. The interaction of anthocyanins with bilitranslocase. *Biochem. Biophys. Res. Commun.* **2002**, *296*, 631–636. [CrossRef]

73. Fernandes, I.; Freitas, V.D.; Reis, C.; Mateus, N. A new approach on the gastric absorption of anthocyanins. *Food Funct.* **2012**, *3*, 508–516. [CrossRef] [PubMed]

74. Talavéra, S.; Felgines, C.; Texier, O.; Besson, C.; Lamaison, J.-L.; Rémésy, C. Anthocyanins are efficiently absorbed from the stomach in anesthetized rats. *J. Nutr.* **2003**, *133*, 4178–4182. [CrossRef] [PubMed]

75. Vitaglione, P.; Donnarumma, G.; Napolitano, A.; Galvano, F.; Gallo, A.; Scalfi, L.; Fogliano, V. Protocatechuic acid is the major human metabolite of cyanidin-glucosides. *J. Nutr.* **2007**, *137*, 2043–2048. [CrossRef] [PubMed]

76. Bouayed, J.; Deußer, H.; Hoffmann, L.; Bohn, T. Bioaccessible and dialysable polyphenols in selected apple varieties following in vitro digestion vs. their native patterns. *Food Chem.* **2012**, *131*, 1466–1472. [CrossRef]

77. Tagliazucchi, D.; Verzelloni, E.; Bertolini, D.; Conte, A. In vitro bio-accessibility and antioxidant activity of grape polyphenols. *Food Chem.* **2010**, *120*, 599–606. [CrossRef]

78. Woodward, G.; Kroon, P.; Cassidy, A.; Kay, C. Anthocyanin stability and recovery: Implications for the analysis of clinical and experimental samples. *J. Agric. Food Chem.* **2009**, *57*, 5271–5278. [CrossRef] [PubMed]

79. Mullen, W.; Edwards, C.A.; Serafini, M.; Crozier, A. Bioavailability of pelargonidin-3-*O*-glucoside and its metabolites in humans following the ingestion of strawberries with and without cream. *J. Agric. Food Chem.* **2008**, *56*, 713–719. [CrossRef] [PubMed]

80. McDougall, G.J.; Fyffe, S.; Dobson, P.; Stewart, D. Anthocyanins from red wine—Their stability under simulated gastrointestinal digestion. *Phytochemistry* **2005**, *66*, 2540–2548. [CrossRef] [PubMed]

81. Dangles, O. Antioxidant activity of plant phenols: Chemical mechanisms and biological significance. *Curr. Org. Chem.* **2012**, *16*, 1–23. [CrossRef]

82. Vanzo, A.; Terdoslavich, M.; Brandoni, A.; Torres, A.M.; Vrhovsek, U.; Passamonti, S. Uptake of grape anthocyanins into the rat kidney and the involvement of bilitranslocase. *Mol. Nutr. Food Res.* **2008**, *52*, 1106–1116. [CrossRef] [PubMed]

83. Tsuda, T. Dietary anthocyanin-rich plants: Biochemical basis and recent progress in health benefits studies. *Mol. Nutr. Food Res.* **2012**, *56*, 159–170. [CrossRef] [PubMed]

84. De Pascual-Teresa, S. Molecular mechanisms involved in the cardiovascular and neuroprotective effects of anthocyanins. *Arch. Biochem. Biophys.* **2014**, *559*, 68–74. [CrossRef] [PubMed]

85. Hwang, Y.P.; Choi, J.H.; Yun, H.J.; Han, E.H.; Kim, H.G.; Kim, J.Y.; Park, B.H.; Khanal, T.; Choi, J.M.; Chung, Y.C.; et al. Anthocyanins from purple sweet potato attenuate dimethylnitrosamine-induced liver injury in rats by inducing Nrf2-mediated antioxidant enzymes and reducing COX-2 and iNOS expression. *Food Chem. Toxicol.* **2011**, *49*, 93–99. [CrossRef] [PubMed]

86. Wang, D.; Wei, X.; Yan, X.; Jin, T.; Ling, W. Protocatechuic acid, a metabolite of anthocyanins, inhibits monocyte adhesion and reduces atherosclerosis in apolipoprotein E-deficient mice. *J. Agric. Food Chem.* **2010**, *58*, 12722–12728. [CrossRef] [PubMed]

MDPI

Article

Exogenous 24-Epibrassinolide Interacts with Light to Regulate Anthocyanin and Proanthocyanidin Biosynthesis in Cabernet Sauvignon (*Vitis vinifera* L.)

Yali Zhou [1], Chunlong Yuan [1,2], Shicheng Ruan [3], Zhenwen Zhang [1,2], Jiangfei Meng [1,2,]* and Zhumei Xi [1,2,]*

[1] College of Enology, Northwest A&F University, Yangling 712100, China; zhouyali@nwafu.edu.cn (Y.Z.);
     yuanchl69@nwsuaf.edu.cn (C.Y.); zhangzhw60@nwsuaf.edu.cn (Z.Z.)
[2] Shaanxi Engineering Research Center for Viti-Viniculture, Yangling 712100, China
[3] Chateau Changyu Rena Co., Ltd., Xianyang 712000, China; ruanshicheng@sina.cn
*   Correspondence: mjfwine@nwafu.edu.cn (J.M.); xizhumei@nwsuaf.edu.cn (Z.X.);
    Tel.: +86-29-87092107 (Z.X.)

Received: 2 November 2017; Accepted: 29 December 2017; Published: 9 January 2018

Abstract: Anthocyanins and proanthocyanidins (PAs) are crucial factors that affect the quality of grapes and the making of wine, which were stimulated by various stimuli and environment factors (sugar, hormones, light, and temperature). The aim of the study was to investigate the influence of exogenous 24-Epibrassinolide (EBR) and light on the mechanism of anthocyanins and PAs accumulation in grape berries. Grape clusters were sprayed with EBR (0.4 mg/L) under light and darkness conditions (EBR + L, EBR + D), or sprayed with deionized water under light and darkness conditions as controls (L, D), at the onset of veraison. A large amount of anthocyanins accumulated in the grape skins and was measured under EBR + L and L treatments, whereas EBR + D and D treatments severely suppressed anthocyanin accumulation. This indicated that EBR treatment could produce overlay effects under light, in comparison to that in dark. Real-time quantitative PCR analysis indicated that EBR application up-regulated the expression of genes (*VvCHI1*, *VvCHS2*, *VvCHS3*, *VvDFR*, *VvLDOX*, *VvMYBA1*) under light conditions. Under darkness conditions, only early biosynthetic genes of anthocyanin biosynthesis responded to EBR. Furthermore, we also analyzed the expression levels of the BR-regulated transcription factor VvBZR1 (Brassinazole-resistant 1) and light-regulated transcription factor VvHY5 (Elongated hypocotyl 5). Our results suggested that EBR and light had synergistic effects on the expression of genes in the anthocyanin biosynthesis pathway.

Keywords: grape; anthocyanin biosynthesis; VvBZR1; VvHY5; 24-Epibrassinolide

---

1. Introduction

Anthocyanins and proanthocyanidins (PAs) are crucial factors that affect the quality of grapes and the making of wine [1–3]. The accumulation of anthocyanin is stimulated by various stimuli: sugar, plant hormones and other environmental factors (light, UV irradiation, temperature, wounding, drought) [4–11]. To promote anthocyanin accumulation, application of plant growth regulators has been proposed as an economically viable alternative. The biosynthesis mechanisms of anthocyanins and PAs has been deeply studied [2,12–16]. The anthocyanins and PAs are synthesized via the phenylpropanoid pathway and flavonoids pathway in the cytoplasm as shown in Figure 1 [17]. The biosynthetic pathway leading to anthocyanins and PAs is well known and the key regulatory genes controlling the pathway have been reported [18,19]. In anthocyanin biosynthetic pathway, genes encoding enzymes are divided into two groups: the upstream genes of anthocyanin biosynthetic pathway, such as chalcone synthase (CHS) and chalcone isomerase (CHI), and the downstream genes of anthocyanin biosynthetic pathway including dihydroflavonol reductase (DFR), leucoanthocyandin dioxygenase (LDOX), and UDP-glucose: flavonoid-3-O-glucosyl transferase (UFGT) [15,20–23].

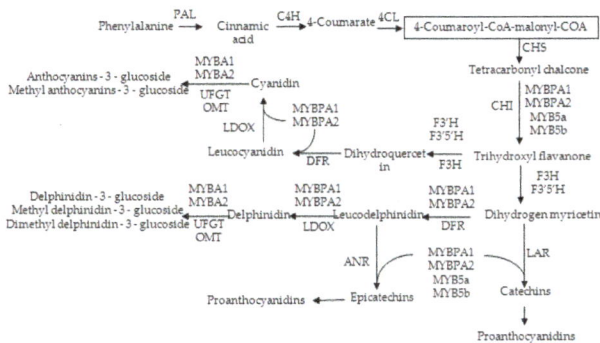

Figure 1. Biosynthetic pathway of anthocyanins and proanthocyanins in grape. Notes: Transcription regulators: VvMYBPA1, VvMYBPA2, VvMYBA1, VvMYBA2, VvMYB5a, VvMYB5b. PAL: phenylalanine ammonia-lyase, CHI: chalcone isomerase, DFR: dihydroflavonol 4-reductase, LAR: leucoanthocyanin reductase, ANR anthocyanin reductase, UFGT UDP-glucose: flavonoid 3-O-glucosyltransferase, LDOX: leucoanthocyandin dioxygenase.

As reported previously, plant hormones including brassinosteroids (BRs), abscisic acid(ABA) and ethylene, etc., also regulate anthocyanin biosynthesis [24,25]. Brassinosteroids (BRs) has been considered the sixth steroidal hormone and has been intensively researched; it regulates a wide range of physiological processes and plays vital roles in plant growth and development [26–30]. In recent years, the effects of BRs on fruit growth and development have been investigated. BRs are used for the development and ripening of fleshy fruit, such as tomatoes [31], strawberries [32] and apples [33]. Exogenous application of BRs significantly promoted grapes ripening and enhanced the anthocyanin content [27,34]. BRs, through BR signaling mainly via BZR1 and BES1 (BRI1-EMS-suppressor 1), regulate a large number of genes involved in plant developmental and physiological processes. Thus, the transcription factors that directly regulate the expression of the structural genes in the development of plants have been identified in many species [35]. However, the molecular mechanism for the regulation of anthocyanin accumulation in grape by VvBZR1 remains unknown.

Another important environmental factor in anthocyanin synthesis is light. Understanding of the light-mediated mechanism involved in the regulation of anthocyanin biosynthesis in fruits has increased markedly [19,36]. Shin et al. (2007) reported that the combination of HY5 (Elongated hypocotyl 5) and PIF3 (Phytochrome-interacting factor 3) bound directly to the promoters of anthocyanin structural genes such as CHS, CHI, F3H, DFR and LDOX in Arabidopsis thaliana [5,37]. Wang et al. (2016) reported that HY5 directly binds the MYBL2 promoter and represses its expression to activate anthocyanin biosynthetic pathway in *Arabidopsis thaliana* [38].

Recent studies suggested that hormones participating in light transduction pathways control anthocyanin accumulation or, alternatively, that hormones and photoreceptors shared common molecular targets regulating the response [35]. Many studies had been performed to determine the impact of BRs on phenolic accumulation in grape berries. However, there are few reports concerning the interrelationship of EBR and light on anthocyanin accumulation in grape skin. Thus, it has not yet been elucidated how flavonoid biosynthesis pathway genes respond to various combinations of EBR and light.

In the experiment presented in this study, the effects of exogenous EBR and light treatment on the anthocyanins and PA content as well as mRNA expression levels of structural genes were measured in grape skins. It clarified the interrelationship of EBR and light effects on anthocyanin accumulation and the expression of flavonoid-related genes. The purpose of this study is to provide physiological support for researching the regulatory mechanism of VvBZR1 and VvHY5 involved in anthocyanin synthesis. These findings provide new information about the relationships between EBR, light and anthocyanin accumulation in grape berry skin.

2. Results

2.1. Physiochemical Parameters

Grapes were treated at E-L35 phenological stage and collected from 1 DAT to 46 DAT. EBR treatment significantly influenced 100-berry weight under the darkness condition (Figure 2). From fruit version to maturity, 100-berry weight was dramatically increased (46 DAT, E-L 38), then increased slightly and reached maximum value at harvest. Bagging without EBR (D) treatment had no significant effects on 100-berry weight, but EBR + D treatments increased the 100-berry weight at 15 and 46 DAT. At harvest (46 DAT, E-L 38), the mean 100-berry weight was 2.04% (EBR + D) higher than that of the bagging without EBR (D). The result suggests that EBR treatment could increase 100-berry weight under darkness condition. Correspondingly, the four treatments had the same effects on reducing levels of sugar and titratable acid, which were significantly increased under the light exposure with EBR (EBR + L) treatments relative to D, L (L: light exposure without EBR) and EBR + D (Figure 2b,c). The EBR treatments (EBR + D, EBR + L) enhanced reducing sugar accumulation and decreased the total acid content. During berry development, the reducing sugar content and total acid content in juice of EBR-treated (EBR + D, EBR + L) berries were significantly higher and lower, respectively, than that of the D and L treatments. In addition, light exposure significantly increased the content of reducing sugar during berry development (Figure 2). The result showed that EBR was more effective in the light than dark in the synthesis of the reducing sugar.

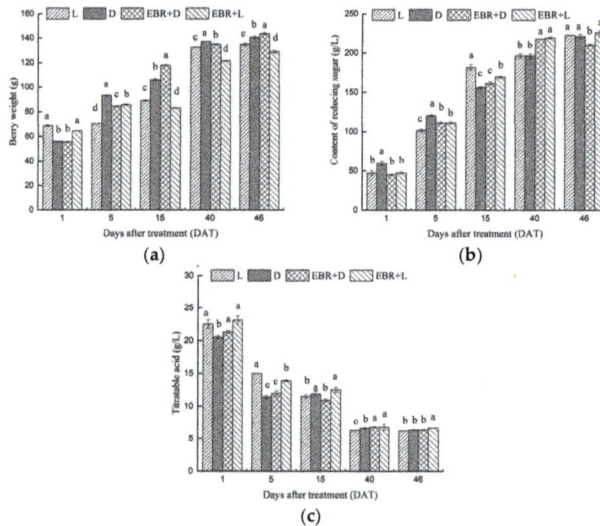

Figure 2. Effects of the four treatments on 100-berry weight (**a**); reducing sugar (**b**) and total acidity (c) in grape berry during fruit development. Data represent the mean of three replicates ± standard deviation (error bars). The different letters (a, b, c, d) indicate significant differences between treatments at $p < 0.05$ (Duncan's multiple range test). DAT: Days after treatment. The same as belows.

2.2. The Content of Anthocyanins and Proanthocyanidins in Grape Skins

The effects of different treatments on anthocyanin content in grape skins at different ripening stages are shown in Figure 3a. Color development of berries started after 5 DAT and sufficient anthocyanins accumulation was observed, whereas D or EBR + D treatment suppressed anthocyanins accumulation severely (Figure 3a). The total anthocyanin content of berry skins were 0.0089, 0.699, 0.024 and 0.862 mg/g fresh weight in the 5 DAT in the D, L, EBR + D and EBR + L treatments, respectively (Figure 3a). The EBR treatment (EBR + D, EBR + L) significantly increased the content of anthocyanins in the skin, under light treatment, EBR treatment significantly increased the content of anthocyanins compared with the treatment of EBR + D. The total anthocyanin content in the EBR + L treatment was significantly higher than other treatments. The total anthocyanin content for each treatment in the 40 and 46 DAT was similar to that in the 5 and 15 DAT. Thus, the anthocyanin content was enhanced under the EBR + D treatment compared with D treatment at 46 DAT. These results suggested that light treatment with exogenous EBR treatment induced anthocyanin accumulation, but D treatment severely reduced it.

The effects of different treatments on PA content in grape skins at different ripening stages are shown in Figure 3b. The result suggested that EBR could increase the PA content significantly under darkness condition at 15 DAT and 46 DAT. Compared to D treatment, the PAs content in the skins was enhanced by 13.74%, 81.64%, 15.8% under the EBR + D treatment at 5 DAT, 15 DAT, 46 DAT, respectively.

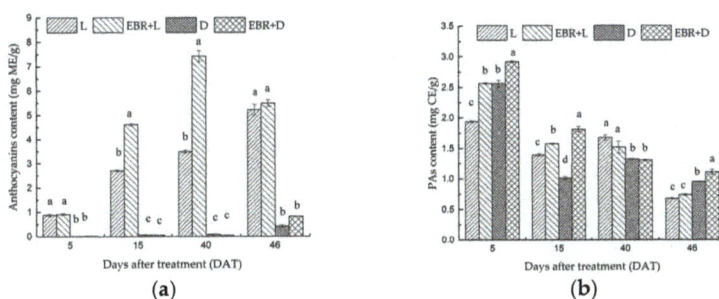

Figure 3. (a) Accumulation of anthocyanins per gram of skins dry weight in berry skins (cv. Cabernet Sauvignon) during development. The amounts are expressed as milligrams of cyanidin-3-monoglucoside equivalence (ME) per gram of dry berry skins (mg ME/g); (b) Accumulation of proanthocyanidins (PAs) per gram of skins dry weight in grape berries (cv. Cabernet Sauvignon) during development. The amounts are expressed as milligrams (+)-catechin equivalence (CE) per gram of dry berry skins (mg CE/g) (mean ± SE; n = 3). The different letters (a, b, c, d) indicate significant differences between treatments at $p < 0.05$ (Duncan's multiple range test).

2.3. Determination Monomer Anthocyanins Content in Grape Skins

Anthocyanins composition of grape skins was determined by LC-20AT HPLC system (Shimadzu, Kyoto, Japan). The anthocyanin compositions at 15 and 46 DAT upon different treatments are shown in Figure 4. The 15 DAT was the mid-maturity stage and the 46 DAT was the maturity stage, which were important stages in the development of grape berries. The results revealed that EBR treatment could significantly increase the content of monomer anthocyanins and the Ma-derivatives were the most abundant components among them (Figure 4). In L and EBR + L treatments, most of the Ma-derivatives were malvidin-3-O-glucoside and its acyl derivatives, the content was 3.55 mg/g fresh weight (FW), 3.85 mg/g FW at 15 DAT; 7.49 mg/g FW, 11.55 mg/g FW at 46 DAT, respectively. Thus, the fact was that malvidin-3-O-glucoside made up the vast majority of the total anthocyanins. The higher content of malvidin-3-O-glucoside and their acyl derivatives were detected in the skins of

L and EBR + L treated berries, but the content of malvidin derivatives decreased in D and EBR + D treatments. Meanwhile, the result revealed that the content of Pn-derivatives, which is the second most abundant component in the skins, was 0.85 mg/g FW, 2.39 mg/g FW in 15 DAT, 46 DAT, respectively. However, both Cy-derivatives and De-derivatives were presented in low concentrations. In the skin of grape, malvidin-3-*O*-glucose was the major pigment, followed by malvidin-3-*O*-(6-*O*-acetyl)-glucoside (Figure 4). The results indicated that EBR and light firstly influenced malvidin-3-*O*-glucoside and its acyl derivatives accumulation and then promoted the biosynthesis of AT. It was apparent that the nine kinds of anthocyanidin derivatives were enhanced in response to EBR treatment (EBR + D, EBR + L). At harvest (46 DAT), the grape treated with L showed increases in total content of 1.91 and 3.64 times for Ma-derivatives and De-derivatives, respectively, in comparison with the D treatment. Likewise, the grape treated with EBR + L showed increases in total content of 5.01 and 7.25 times for Ma-derivatives and De-derivatives, respectively, in comparison with the EBR + D treatment, which demonstrated that EBR and light conditions affected the anthocyanins compositions in grape berries skins.

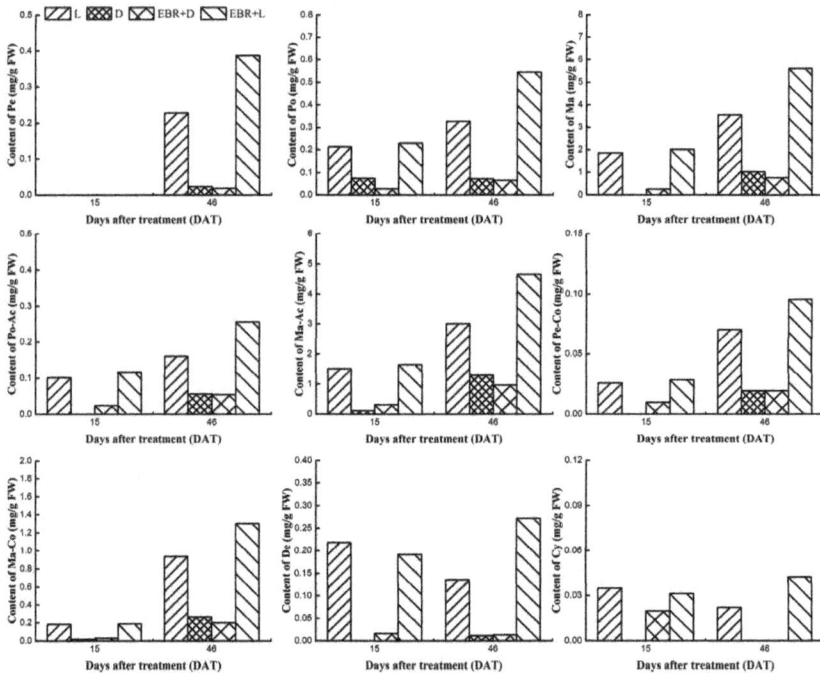

Figure 4. Individual anthocyanins content identified in Cabernet Sauvignon skins in 15 DAT and 46 DAT. Notes: Pe: 3′-Petunidin-3-*O*-glucoside; Po: Peonidin-3-*O*-glucoside; Ma: malvidin-3-*O*-glucoside; Po-Ac: Peonidin-3-*O*-(6-*O*-Acetyl)-glucoside; Ma-Ac: malvidin-3-*O*-(6-*O*-Acetyl)-glucoside; Pe-Co.: Petunidin-3-*O*-(6-*O*-Coumaryl)-glucoside; Ma-Co.: malvidin-3-*O*-(6-*O*-Coumaryl)-glucoside; De: Delphinidin-3-*O*-glucoside; Cy: Cyanidin-3-*O*-glucoside; D: dark; L: light; EBR + D: EBR + dark; EBR + L: EBR + light; Data represent the mean of three replicates ± standard deviation (error bars); FW: fresh weight.

2.4. The Expression Patterns of Anthocyanins Biosynthetic Genes and Transcriptional Regulator Genes in Grape Skins

For further dissecting the EBR modulation of light controlling anthocyanin accumulation in grape, we measured the mRNA levels of *VvCHI1*, *VvCHS2*, *VvCHS3*, *VvF3′5′H*, *VvDFR*, *VvLDOX*,

VvUFGT and *VvMYBA1* by real-time quantitative PCR (RT-qPCR) during berry development which are presented in Figure 5. The expression levels of anthocyanin biosynthesis structural genes were significantly higher under EBR treatment than under that D treatment, but having no clear effects compared to light treatment. Under different conditions of EBR treatments, the expression levels of *VvCHI1*, *VvCHS3*, *VvF3′5′H* and *VvUFGT* were not affected by light, while dark treatment significantly decreased the expression of *VvCHS2*, *VvDFR*, *VvLDOX* and *VvMYBA1*. The expression levels of *VvCHS2*, *VvDFR* and *VvLDOX* were the highest in EBR + L and drastically down-regulated by L or D treatment at 46 DAT. Meanwhile, L treatment improved the transcription levels of *VvCHI1*, *VvF3′5′H* and *VvDFR* by 1.75, 2, 2.82-fold than the D treatment respectively, and EBR + L treatment improved the transcription levels by 0.34, 0.21, 0.31-fold than L treatment respectively, at 46 DAT. For *VvCHI1*, *VvCHS3*, *VvF3′5′H* and *VvUFGT*, D treatment down-regulated gene expression, but the effects of D treatment was particularly dramatic under EBR treatment. The results indicated that both EBR and light are required to induce the expression of these genes. Although the expression levels of *VvCHS2* and *VvLDOX* were the highest in EBR + L, the effects of EBR and light treatment were unclear. Thus, many different expression patterns were observed among the flavonoid biosynthesis pathway genes. Meanwhile, under darkness conditions, several genes also exhibited up-regulated expression after EBR treatment. These genes also included *VvCHI*, *VvCHS3*, *VvF3′5′H*, *VvDFR*, *VvLDOX* and *VvUFGT*. Under darkness conditions, genes of the late-steps of anthocyanins biosynthesis, which played a more direct role in the biosynthesis of anthocyanins, showed significant change in expression with EBR treatment at 46 DAT.

As shown in Figure 5, the treatment of EBR + L also increased the mRNA expression level of the transcriptional regulator gene *VvMYBA1* at each sampling date, but D and EBR + D treatment severely suppressed the expression of this gene. The pattern of *VvMYBA1* expression across the four treatments was similar to that of the anthocyanins content and suggested that both EBR and light were needed to induce its expression. The expression level of *VvMYBA1* was significantly induced by light treatment, indicating that *VvMYBA1* might be a light-responsive gene.

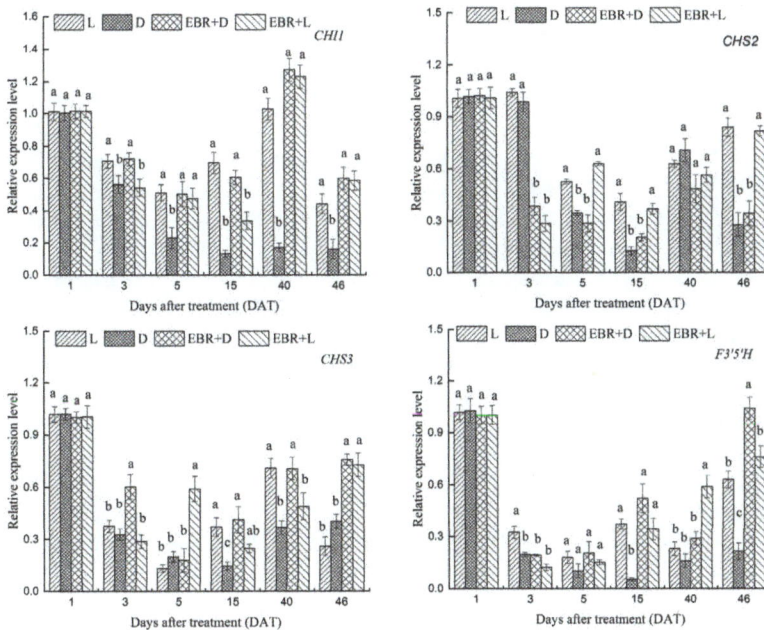

Figure 5. *Cont.*

29

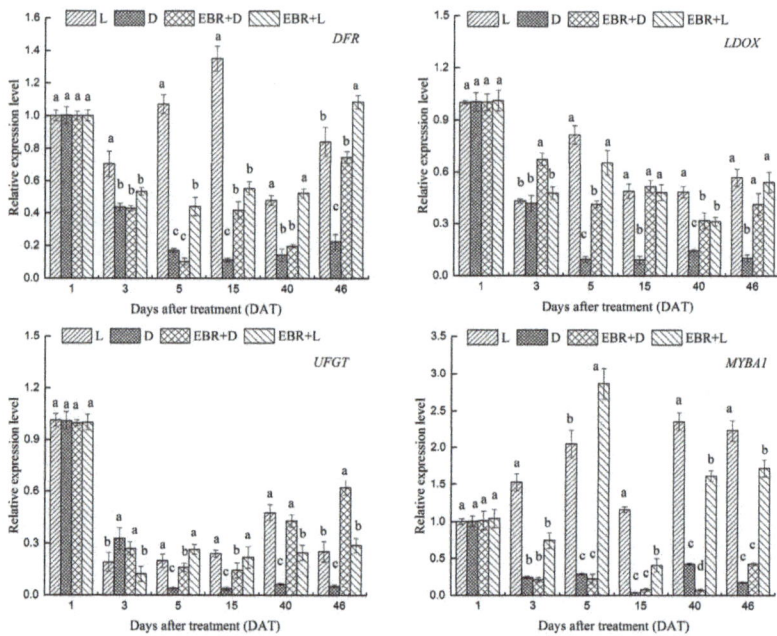

Figure 5. Transcript profiles of *VvCHI1*, *VvCHS2*, *VvCHS3*, *VvF3'5'H*, *VvDFR*, *VvLDOX*, *VvUFGT* and *VvMYBA1* as the molar ratio of the mRNA level to that of *VvGAPDH* in each sample (mean ± SE; *n* = 3). The different letters (a, b, c, d) indicate significant differences between treatments at *p* < 0.05 (Duncan's multiple range test).

2.5. The Expression Patterns of Pas Biosynthetic Genes and Transcriptional Regulator Genes in Grape Skins

We investigated the effects of the four treatments on the expression of PA biosynthesis genes (*VvLAR1*, *VvLAR2*, *VvANR*) and their transcriptional regulators (VvMYB5a, VvMYB5b, VvMYBPA1 and VvMYBPA2) by RT-qPCR in berry skins (Figure 6). The EBR + L treatment induced the expression of *VvMYB5a* and *VvMYB5b*, but D or L treatment severely suppressed the expression of these genes. The pattern of *VvMYB5a* and *VvMYB5b* across the four treatments was also similar to that of the anthocyanin content and suggested that both EBR and L were needed to induce its expression. Expression levels of VvMYBPA1 increased to a maximum at 5 DAT, and then declined to a low level after 15 DAT (Figure 6). As for VvMYBPA1, EBR and light treatments could not significantly induce its expression, in contrast, D treatment induced VvMYBPA1 synthesis from five to 15 DAT. At the beginning of veraison, EBR and light treatments could significantly induce the expression of VvMYBPA2 from three to 15 DAT, and the expression level of VvMYBPA1 nearly declined to zero after 40 DAT. The expression level of *VvANR* was higher in the EBR + D treatment than the D treatment at 46 DAT. Although the PA accumulation in the EBR + D treatment was severely suppressed (Figure 3b), the expression level of *VvANR* in the EBR + D treatment was only moderately suppressed. At 5 DAT, EBR treatments (EBR + D, EBR + L) significantly increased the expression levels of *VvLAR1* and *VvLAR2*. Thus, each of the PAs biosynthesis genes (*VvLAR1*, *VvLAR2*, *VvANR*) and their transcriptional regulators (VvMYB5a, VvMYB5b, VvMYBPA1 and VvMYBPA2) reacted differently to the light and EBR treatments.

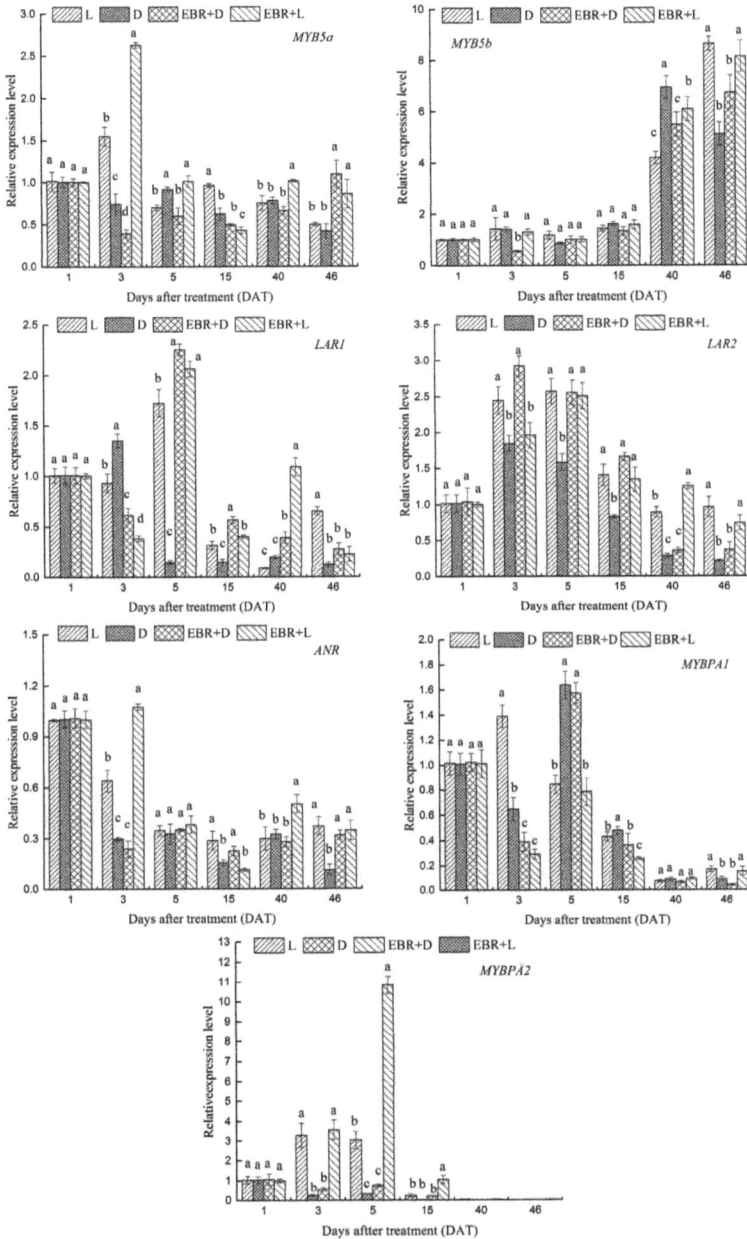

Figure 6. Transcript profiles of *VvMYB5a*, *VvMYB5b*, *VvLAR1*, *VvLAR2*, *VvANR*, *VvMYBPA1* and *VvMYBPA2* as the molar ratio of the mRNA level to that of *VvGAPDH* in each sample (mean ± SE; *n* = 3). The different letters (a, b, c, d) indicate significant differences between treatments at *p* < 0.05 (Duncan's multiple range test).

2.6. Expression Patterns of VvBZR1, VvHY5 and VvBRI1 Transcriptional Regulator Genes in Grape Skins

The effects of different treatments on the mRNA levels of *VvBRI1* (brassinosteroid insensitive 1), *VvBZR1* (brassinazole-resistant 1) and *VvHY5* (elongated hypocotyls 5) genes were determined by RT-qPCR during berry development. The expression level of *VvBZR1* during berry development is presented in Figure 7. As illustrated in Figure 7, the expression level of *VvBZR1* was at a relatively high level enhanced by EBR treatment at 40 and 46 DAT. However, the expression level of *VvBZR1* was at a relatively low level during veraison. Meanwhile, the result displayed that EBR + L treatment significantly increased the mRNA level of *VvHY5* at 5 DAT.

VvBRI1 (brassinosteroid insensitive 1) is a LRR receptor kinase of BR receptor, which exists on the surface of the cell. Usually higher expression of upstream genes leads to higher expression of downstream genes. However, EBR and L treatments (EBR + D, EBR + L, L) have no significant effects on the mRNA levels of *VvBRI1* during berry development as can be observed in Figure 7, but compared to D treatment, EBR treatments (EBR + D, EBR + L) had significant effects on the mRNA levels of *VvBRI1*. This indicated that expression of *VvBRI1* might be up-regulated by light, independent of the treatment of EBR.

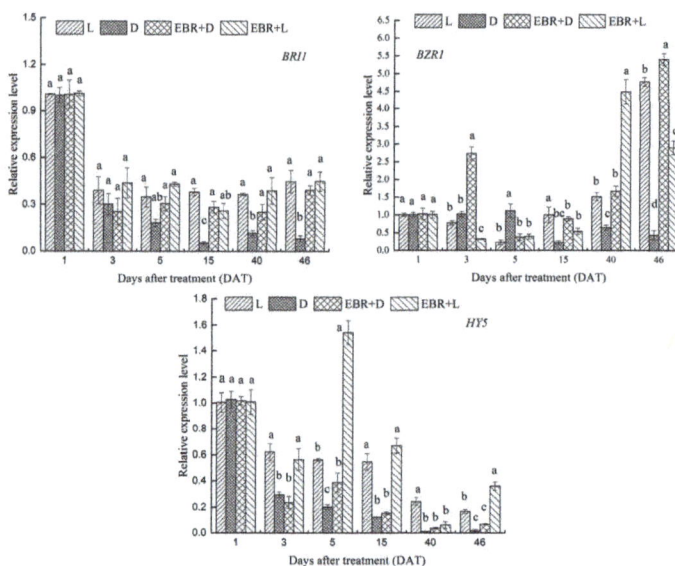

Figure 7. Transcript profiles of *VvBRI1*, *VvBZR1* and *VvHY5* as the molar ratio of the mRNA level to that of *VvGAPDH* in each sample (mean ± SE; *n* = 3). The different letters (a, b, c, d) indicate significant differences between treatments at *p* < 0.05 (Duncan's multiple range test).

2.7. Hierarchical Clustering in the Profiles of All Genes Analyzed

Hierarchical clustering was performed by Multiple Array Viewer to access the similarity of the profiles of gene expression in the four treatments. We identified four clusters (Figure 8). VvBZR1, the key transcription factor in BR signal transcription pathway was clustered with *VvMYB5b*, *VvF3′5′H*, *VvCHS3*, *VvUFGT* and *VvCHI1*. The expression level of *VvMYB5b* was similar to *VvBZR1*, which indicated that *VvMYB5b* had a relation with *VvBZR1*. Furthermore, we could infer that PA synthesis hada relationship with *VvBZR1* at the transcriptional level. The typical anthocyanin structure genes (*VvCHS2*, *VvMYBA1*, *VvDFR*, *VvLDOX*) were clustered in a group with *VvBRI1*. The genes in this group were strongly induced by EBR and light treatment (EBR + D, EBR + L), indicating that the EBR treatment enhanced their expression and signal transduction pathway which is involved in the anthocyanin accumulation and PAs biosynthesis. The PAs biosynthesis genes (*VvLAR1*, *VvLAR2*) and their transcriptional regulators (VvMYBPA1 and VvMYBPA2) were clustered in a group with *VvHY5*. The expression pattern of genes was similar to that of the PAs content in this group. Additional PAs structure genes *VvANR* and *VvMYB5a* were clustered into another group. The genes were up-regulated by EBR + D treatment than EBR + L treatment but were down-regulated under D treatment at 46 DAT, further supporting that EBR was the main stimulus in the biosynthesis of PAs.

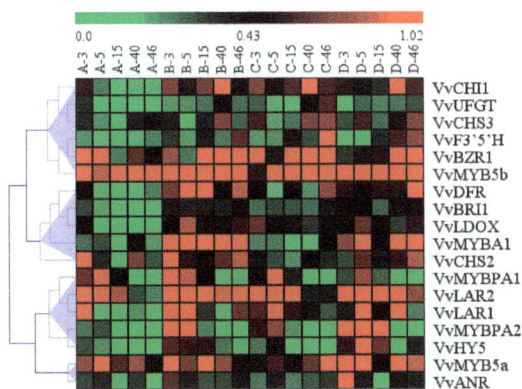

Figure 8. Hierarchical clustering of the transcript profiles of all genes. The relative expression levels of the genes after the four treatments compared to the reference gene, used in the original rank-based algorithm, were used for the hierarchical cluster analysis with Genesis. Red colors represent relatively higher transcript abundances, and green colors represent relatively lower transcript abundances. Sampling times and treatments are indicated at the top; the numbers A, B, C and D represent treatments D, L, EBR + D, EBR + L, respectively, and the numbers 3, 5, 15, 40 and 46 are days after treatment.

3. Discussion

In this study, the content of PAs, anthocyanins and monomer anthocyanins were determined. EBR treatment improved accumulation of phenolic compound in the skins of the grape berries not only when spraying at maturity [27] but also at the stage of veraison. Under light exposure, EBR treatment markedly increased the accumulation of PAs, anthocyanins and monomer anthocyanins. These results, which were in agreement with previous studies, showed that artificial shading of fruits would decrease the content of phenolic compound, and light exposure would increase phenolic compound content [7,39–41]. Before maturity, EBR + L treatment enhanced accumulation of phenolic compound (PAs, anthocyanins) in the skins. At harvest, the content of PAs, anthocyanins and monomer anthocyanins were slightly decreased; this is due to the polyphenols in skins being transferred to the seeds during the growing process, which was corroborated by previous studies [42–44].

It could be observed from Figure 5 that light and EBR were essential in anthocyanin biosynthesis. EBR treatment (EBR + D, EBR + L) was involved in the regulation of *VvCHI1*, *VvCHS3*, *VvF3'5'H*, *VvDFR* and *VvUFGT* expression and reached a peak at the veraison; thereafter, the mRNA levels decreased rapidly then constantly increased until harvest (46 DAT, E-L38) compared with those of the D treatment. In this study, the expression levels of *VvCHI1*, *VvCHS2*, *VvCHS3*, *VvDFR* and *VvLDOX* were up-regulated by light and EBR treatments. However, the expression levels of *VvCHI1*, *VvCHS3*, *VvF3'5'H* and *VvUFGT* were up-regulated by EBR + D higher than EBR + L, which was similar to *VvBRI1* (Figure 8).

We investigated the effects of different treatments on the expression of PAs biosynthetic genes (*VvLAR1*, *VvLAR2*, *VvANR*) and their transcriptional regulators (VvMYB5a, VvMYB5b, VvMYBPA1 and VvMYBPA2). The expression levels of *VvLAR1*, *VvLAR2* were higher in early development berries

(Figure 6) than harvest (E-L38), which was supported by previous research [15,34,45,46]. The research showed that *VvLAR2* mRNA level was always higher than that of *VvLAR1* under the same treatment at harvest, thus, light exposure could increase the *VvLAR2* mRNA level, however, EBR + D treatment significantly increased the *VvLAR1* mRNA level at 5, 15 DAT. Our results predicted that the *VvLAR2* might play more important roles in two LAR isoforms involved in the regulation of the PAs biosynthesis pathway [46,47]. In the PAs biosynthesis pathway, ANR is the other crucial catalyzing enzyme. During berry development, the *VvANR* mRNA level was increased markedly in the EBR treatments (EBR + D, EBR + L), which contributed to the higher levels of *VvLAR1* and *VvLAR2* mRNA under different conditions. These effects of EBR were similar to those of abscisic acid (ABA) and methyl jasmonate (JA) on the synthetic process of proanthocyanidins and anthocyanins [48–50]. In grapes, VvMYB5a and VvMYB5b were transcriptional regulators related to PA metabolism [15,51], and previous studies have already identified VvMYBPA1 and VvMYBPA2 as the transcription factors involved in the regulation of the proanthocyanidin pathway during grape development [52,53]. Bogs et al. found that the expression level of VvMYBPA1 in skins was relatively low before veraison, increased to a maximum two weeks after veraison, and then declined to a low level. Terrier et al. found that VvMYBPA2 expression mostly existed in the skin of very young berries, and the expression level declined to very low levels after veraison. The treatments were carried out at the beginning of veraison, so the expression levels of VvMYBPA1 and VvMYBPA2 were consistent with that of Bogs et al. and Terrier et al., respectively [52,53]. As for VvMYBPA1, D treatment induced its synthesis from five to 15 DAT (Figure 6), maybe due to the maturity processing of grape being delayed by darkness treatment. It is worth noting that the expression of VvMYB5a and VvMYB5b activated *VvLAR*, *VvANR* and *VvCHI1* [1,15,17,54]. Thus, as illustrated by Figure 8, the transcription profiles of transcriptional regulator VvMYB5b was consistent with VvBZR1. Deluc et al. (2008) reported that the transcription factor VvMYB5b contributed to the regulation of anthocyanins and proanthocyanidins biosynthesis in development grape berries [15]. We inferred that VvBZR1 had a relationship with VvMYB5b in regulation of biosynthesis of anthocyanins and proanthocyanidins. VvBZR1 is a key transcription factor associated with the BR signal transduction pathway, which plays an important role in BR-regulated genes [55,56]. The results showed that different treatments made the genes expression different up to 46 DAT.

It is known that the effect of the EBR treatment on the expression of genes and transcription factors is almost nonexistent after 48 hr or even 24 hr treated. In this study, both of the PA biosynthesis genes (*VvLAR1*, *VvLAR2*, *VvANR*) and their transcriptional regulators (VvMYB5a, VvMYB5b, VvMYBPA1 and VvMYBPA2) showed none significant difference at 1 DAT. At 3 DAT, EBR + L treatment significantly increased the expression levels of *VvLAR1* and *VvANR* compared to L treatment. Furthermore, at 5 DAT, EBR + L treatment significantly up-regulated the expression levels of *VvLAR1*, *VvCHS3*, *VvUFGT*, VvMYBA1, VvMYB5a and VvMYBPA2 compared to L treatment. These results indicated that EBR could promote the expression of genes which control anthocyanin and PAs synthesis in the early days after ERB treatment. The differences in the expression of genes and transcription factors in our research up to 46 DAT may be because exogenous EBR promoted the synthesis of endogenous BRs or other endogenous hormones in the grape in later period.

Currently, the molecular mechanism of EBR enhancing the anthocyanin content in the skins is not well understood. We investigated the effects of EBR on the expression of *VvBZR1*, *VvHY5* and *VvBRI1* genes. *VvHY5* was extensively researched as a transcription factor involved in promoting photomorphogenesis. As shown in Figures 5 and 8, light exposure can significantly influence the expression level of *VvHY5* and enhance the content of anthocyanins in grape berries, and photomorphogenic factors, HY5 and PIF3 bind directly to the promoters of anthocyanins structural genes such as *CHS*, *CHI*, *F3H*, *DFR* and *LDOX* in *Arabidopsis thaliana* [37]. Meanwhile, previous research reported that *VvHY5* was involved in flavonoid biosynthesis depending on light [5,57]. Thus, the results shown in Figures 4, 5 and 7, suggested that EBR treatment could increase the content of anthocyanins, up-regulate the expression of *VvHY5* and anthocyanin structural genes. On the basis of these finding, we inferred that *VvBZR1* could enhance the expression of *VvHY5*, which presented opposite trends with *VvBZR1*, further up-regulate the expression of anthocyanins structural genes and increase the content of anthocyanins either at transcription level or protein level.

In the study, we examined whether light participated in the process of BR signal transduction. EBR + L treatment significantly increased the mRNA expression levels of *VvBRI1* and *VvBZR1*, suggesting that light did significantly affect BR signal transduction upstream of *VvBZR1* at harvest and significantly increase the content of anthocyanins. In addition, EBR and light affected the expression level of *VvHY5*. These results supported the interaction between BR signaling, light signaling and anthocyanin biosynthesis pathways. However, how exactly VvBZR1 functions during anthocyanins biosynthesis process is still an open question. Future efforts will be focused on the mechanism by VvBZR1 and VvHY5 regulating anthocyanin biosynthetic pathway to illustrate how BR and light signaling pathways interplay to control the accumulation of anthocyanins.

4. Materials and Methods

4.1. Sample Treatments

Cabernet Sauvignon (*Vitis vinifera* L.) berries were sampled from a commercial vineyard in Xianyang, Shaanxi Province, China (34°650′ N, 108°750′ E). The grapevines were planted in 2009 (Seven-year-old own-rooted grapevines) and employed the single cordon pruning method. The grapevines were planted in North-South oriented rows with spacing of 1.0 m within rows and 2.5 m between rows. The grapevines were trained on a vertical shoot-positioning system with a pair of wires. Three blocks were chosen as biological replicates from one field randomly, and 30 plants of each block received a different spray treatment: light exposure with EBR and without EBR, bagging with and without EBR, once 0.4 mg/L 24-Epibrassinolide (Ruibio, Cologne, Germany) were applied at the onset of veraison. Shading material was 20 cm × 30 cm fruit bag with black double-layer inside.

4.2. Experimental Design and Samples Collection

Stock solutions of EBR were prepared by dissolving EBR in 1 mL of 98% ethanol. The control stock solution contained 1 mL of 98% ethanol without addition of EBR. Each stock solution was mixed with 1 mL of Tween 80 and diluted to 1 L with sterilized deionized water. In this study, the treatments were carried out at the beginning of veraison (softening of 10% of the berries), each solution was sprayed to cover the entire surface area of the berries in the cluster, after the surface was dry, bagging was carried out. 24-Epibrassinolide affects the accumulation of regulatory-gene transcripts at the beginning of veraison [34]. The application dates were 27 July in 2016. All spray applications were carried out at sunset.

Grapes were collected at E-L35 phenological stages (1, 3, 5 days after treatment-DAT), E-L36 (15 DAT), E-L37 (40 DAT) and E-L38 (46 DAT) [58]. Each samples consisted of 300 berries randomly from inside and outside of the cluster, the top, the bottom, and the middle of the cluster. These samples were stored at −20 °C for analysis of the phenolic compounds. In addition, another 60 berries in each treatment were divided into three groups as three replicates, then, the samples were frozen in liquid nitrogen and stored at −80 °C for RNA extraction and quantification of gene expression by real-time quantitative PCR.

4.3. Determination the Physicochemical Indices of Berries

The 100-berry weight for each replicates per treatment was recorded after blotting of residual moisture on the skins surface. Berry juice was collected and used to assay the content of reducing sugars and titratable acids, which were analyzed in accordance with the methods proposed by OIV (2012).

4.4. Extraction of Phenolic Compounds From Grape Skins

Phenolic compounds of grape in skins were extracted according to the methods proposed by Di Stefano and Cravero [59] with minor modifications. Grape skins and seeds of about 90 berries were carefully removed using razor blades. Then the residual water on the surface of grapes was dried and weighed dried skins. Added to 30 mL buffer solution (12% v/v ethanol + 600 mg/L Sodium metabisulfite + 5 g/L Tartaric Acid, 1 M NaOH adjust pH to 3.20), and put in swing bed (100 r/min, 25 °C), extracted for three days, collected the supernatant, finally, placed in −20 °C stored keep away from light before use.

4.5. Determination of Phenolics Content

Total proanthocyanidins content (PAs) were performed as described previously [60,61] with minor modifications. Buffer A was washing buffer containing 200 mM acetic acid and 170 mM sodium chloride, pH adjusted to 4.9 with sodium hydroxide. Buffer B was a model wine (5.0 g/L potassium bitartrate and 12% (v/v) ethanol, pH was adjusted to 3.3 with HCl. Buffer C was a suspending buffer consisting 5% (v/v) triethanolamine and of 5% (w/v) sodium dodecyl sulphate, pH was adjusted to 9.4 with HCl. Ferric chloride reagent was made by 0.01 M HCl and 10 mM ferric chloride.

For PAs determination, a protein solution for tannin precipitation was prepared by dissolving Bovine serum albumin (BSA) in a buffer A, giving a final protein concentration of 1.0 mg/mL. The skins extract was added 200 µL extract sample and 4.0 mL of the protein solution in a 1.5 mL microfuge tube. After incubating for 15 min with slow agitation at room temperature, the mixture was centrifuged at 14,000× g for 5 min at 4 °C. After the precipitate was washed with buffer A three times and then resolubilization in 3.5 mL of buffer C. The absorbance of the solution was read at 510 nm for as tannin background A510. Then, 0.25 mL of ferric chloride reagent was added and shaken for 10 min in the dark. The absorbance of the solution were read at 510 nm as tannin final A510. Buffer C was used as a blank and read at 510 nm for tannin initial A510. After the incubation period the absorbance at 510 nm was determined in Shimadzu 640 spectrophotometer. PAs values are reported in catechin equivalents (C.E.) as described here.

The absorbance for PAs = [(tannin final A510) − (tannin initial A510)] − (tannin background A510) × 0.875.

Total anthocyanins content was estimated using the pH differential method [62] with minor modification. Each grape extract was diluted 20-times with buffers at pH 1.0 and 4.5. The absorbance was measured at 520 and 700 nm in both pH 1.0 and 4.5 buffers. The anthocyanins (expressed in terms of cyanidin-3-glucoside) was calculated using the following formula: A = (A520 − A700) pH 1.0 −(A520 − A700) pH 4.5, the anthocyanins content was expressed as Milligrams of malvidin-3-monoglucoside equivalence per berry (mg ME/berry) and calculated using the equation anthocyanins content = A × DF × MW × 1000/(ε × C), Where: MW is the molecular weight of cyanidin-3-glucoside (449 g/mol), DF is the dilution factor, ε is the molar extinction coefficient of cyanidin-3-O-glucoside (29,600) and C is the concentration of extracted volume.

4.6. Determination the Content of Monomer Anthocyanins

The chromatographic analyses of anthocyanins were performed using LC-20AT HPLC system (Shimadzu, Japan), consisting of Pump LC-20AT, Central controller CBM-20A, Auto sampler SIL-20A, Column oven CTO-20A, Detector (PDA) SPD-M20A, equipped with a reversed phase column (Synergi Hydro-RP C18, 250 × 4.6 mm, 4 µm). The mobile phase was ultrapure water:acetonitrile:methanoic acid (800:100:25) as solvent A, and ultrapure water:acetonitrile:methanoic acid (400:500:25) as solvent B. The elution profile had the following proportions (v/v) of solvent B: 0.00–15 min, 0%–10%; 15–30 min, 10%–20%; 30–45 min, 20%–35%; 45–46.00 min, 35%–100%; 46.00–50.00 min, 100%. The column was held at 35 °C and at a flow rate of 1 mL/min. The injection volume was 20 µL and analyses were detected at 520 nm. Before injection, the extracts were filtered through 0.22 m filters (cellulose acetate and nitrocellulose, CAN).

All phenolic compounds were identified by comparison of their order of elution and retention time with those of standards, the weight of molecular ion and the fragment ion with standards and references. Quantitative determinations were made by the external standard method with the commercial standards. The calibration curves were obtained by injecting of standard solutions under the same conditions as for the samples analyzed, observing the range of concentrations. Anthocyanins, flavan-3-ols, were respectively expressed as micrograms of malvidin-3-*O*-glucoside (ME), catechin equivalence (CE)/L of grape skins.

4.7. RNA Extraction and cDNA Synthesis Real-Time Quantitative PCR Analysis

Expression levels of the anthocyanins and PA biosynthesis genes *VvCHI1*, *VvCHS2*, *VvCHS3*, *VvF3'5'H*, *VvDFR*, *VvLDOX*, *VvUFGT*, *VvMYBA1*, *VvLAR1*, *VvLAR2*, *VvANR* and the transcription factor genes VvMYB5a, VvMYB5b in grape skins were measured by real-time quantitative PCR (RT-qPCR), using the IQ-SYBR Green Supermix on a MyIQTM Single Colour IQ5 Real-Time PCR Detection System (Bio-Rad Laboratories, Berkeley, CA, USA) monitored via the IQ5 Standard Edition Optical System Software 2.0 (Bio-Rad). The primers (Supplementary Materials Table S1) designed by Primer Premier 5. The two-step RT-qPCR Reagent Kit (Vazyme Biotech Co., Ltd., Nanjing, China) was used in accordance with the manufacturer's instructions. The reaction mixture (20 µL) contained 2 µL cDNA (20 times dilution), 0.8 µL of each primer suspension (10 µmol/L) (Supplementary Materials Table S2), 10 µL 2 × Premix (Vazyme), and 7.2 µL ddH$_2$O. The reaction conditions were as follows: 95 °C for 30 s, followed by 40 cycles of 95 °C for 10 s, 58 °C for 30 s. A melting cycle from 60 to 95 °C as the last step was used to check the specificity of each gene product, assisting by gel electrophoresis and sequence analysis. The annealing temperature (58 °C) was determined when designing the primers and by preliminary experiments. Expression levels for each gene were normalized to constitutively expressed transcripts by VvGAPDH, and calculated using the equation $2^{-\Delta\Delta Ct}$, in which $\Delta\Delta Ct$ = (CT, Target CT, VvGAPDH) Time X− (CT, Target CT, VvGAPDH) Time 0 [63]. Time X is any time point, and Time 0 represents E-L35. Three PCR replicates were conducted per sample and the fold change in each target gene of time 0 was set to 1. For each experiment, data were analyzed separately.

4.8. Statistical Analysis

Data were analyzed using SPSS 19.0 software (SPSS, Chicago, IL, USA). The significance of the difference between each treatment was determined by one-way analysis of variance (ANOVA) and Duncan's new multiple range tests at the 0.05 and 0.01 significance levels. The data for physiochemical parameters, AT and PA content, the gene-expression profiles and monomer anthocyanins were analyzed using OriginPro2016. Data were expressed as the mean values of triplicate experiments and different letters (a, b, c, d) indicate a significant difference between treatments and the control.

5. Conclusions

In these experiments, the effects of exogenous EBR and light treatments on anthocyanin and PAs accumulation and the transcript patterns of relative structural genes were studied. The results indicated that EBR interacts with light to induce the synthesis of anthocyanins and PAs by up-regulating biosynthesis genes, which promoted the accumulation of anthocyanins, PAs and monomer anthocyanins in grape skins. In addition, we also found that both EBR and light affected the expression levels of VvBZR1 and VvHY5, which it may help us to illustrate that BR and light signaling pathways interplay to control the accumulation of anthocyanins in our future studies.

Supplementary Materials: The supplementary materials are available online.

Acknowledgments: This study was supported by the National Natural Science Foundation of China (31772258) and National Technology System for Grape Industry (CARS-30-zp-09). We are grateful to Harry Green and Master Liting Wang for providing help in grammar, spelling and punctuation checking to our manuscript.

Author Contributions: Yali Zhou carried out the experiments with the help of Chunlong Yuan and Shicheng Ruan. Zhumei Xi and Zhenwen Zhang provided all of the financial support and critical intellectual input into the study design. Yali Zhou collected the experimental data and drafted the manuscript. Jiangfei Meng provided suggestions for the revision of the manuscript. All authors read and approved the final manuscript.

Conflicts of Interest: The authors declare no conflict of interest.

References

1. González-Neves, G.; Gil, G.; Barreiro, L. Influence of grape variety on the extraction of anthocyanins during the fermentation on skins. *Eur. Food Res. Technol.* **2008**, *226*, 1349–1355. [CrossRef]
2. Lorenzis, G.D.; Rustioni, L.; Parisi, S.G.; Zoli, F.; Brancadoro, L. Anthocyanin biosynthesis during berry development in corvina grape. *Sci. Hortic.* **2016**, *212*, 74–80. [CrossRef]
3. Parpinello, G.P.; Versari, A.; Chinnici, F.; Galassi, S. Relationship among sensory descriptors, consumer preference and color parameters of Italian Novello red wines. *Food Res. Int.* **2010**, *42*, 1389–1395. [CrossRef]
4. Harborne, J.B.; Williams, C.A. Advances in flavonoid research since 1992. *Phytochemistry* **2000**, *55*, 481–504. [CrossRef]
5. Loyola, R.; Herrera, D.; Mas, A.; Wong, D.C.; Höll, J.; Cavallini, E.; Amato, A.; Azuma, A.; Ziegler, T.; Aquea, F. The photomorphogenic factors UV-B RECEPTOR 1, ELONGATED HYPOCOTYL 5, and HY5 HOMOLOGUE are part of the UV-B signalling pathway in grapevine and mediate flavonol accumulation in response to the environment. *J. Exp. Bot.* **2016**, *67*, 5429–5445. [CrossRef] [PubMed]
6. Shinomiya, R.; Fujishima, H.; Muramoto, K.; Shiraishi, M. Impact of temperature and sunlight on the skin coloration of the 'Kyoho' table grape. *Sci. Hortic.* **2015**, *193*, 77–83. [CrossRef]
7. Wang, X.; Wei, Z.; Ma, F. The effects of fruit bagging on levels of phenolic compounds and expression by anthocyanin biosynthetic and regulatory genes in red-fleshed apples. *Process Biochem.* **2015**, *50*, 1774–1782. [CrossRef]
8. Zhang, C.; Jia, H.; Wu, W.; Wang, X.; Fang, J.; Wang, C. Functional conservation analysis and expression modes of grape anthocyanin synthesis genes responsive to low temperature stress. *Gene* **2015**, *574*, 168–177. [CrossRef] [PubMed]
9. Winkelshirley, B. Biosynthesis of flavonoids and effects of stress. *Curr. Opin. Plant Biol.* **2002**, *5*, 218–223. [CrossRef]
10. Dai, Z.W.; Meddar, M.; Renaud, C.; Merlin, I.; Hilbert, G.; Delrot, S.; Gomès, E. Long-term in vitro culture of grape berries and its application to assess the effects of sugar supply on anthocyanin accumulation. *J. Exp. Bot.* **2014**, *65*, 4665–4677. [CrossRef] [PubMed]
11. António, T.; José, E.D.; Castellarin, S.D.; Gerós, H. Berry Phenolics of Grapevine under Challenging Environments. *Int. J. Mol. Sci.* **2013**, *14*, 18711–18739. [CrossRef]
12. Boss, P.K.; Davies, C.; Robinson, S.P. Anthocyanin composition and anthocyanin pathway gene expression in grapevine sports differing in berry skin colour. *Aust. J. Grape Wine Res.* **1996**, *2*, 163–170. [CrossRef]
13. Costantini, L.; Malacarne, G.; Lorenzi, S.; Troggio, M.; Mattivi, F.; Moser, C.; Grando, M.S. New candidate genes for the fine regulation of the colour of grapes. *J. Exp. Bot.* **2015**, *66*, 67–69. [CrossRef] [PubMed]
14. Da, S.F.; Iandolino, A.; Al-Kayal, F.; Bohlmann, M.C.; Cushman, M.A.; Lim, H.; Ergul, A.; Figueroa, R.; Kabuloglu, E.K.; Osborne, C. Characterizing the grape transcriptome. Analysis of expressed sequence tags from multiple Vitis species and development of a compendium of gene expression during berry development. *Plant Physiol.* **2005**, *139*, 574–597.
15. Deluc, L.; Bogs, J.; Walker, A.R.; Ferrier, T.; Decendit, A.; Merillon, J.M.; Robinson, S.P.; Barrieu, F. The transcription factor VvMYB5b contributes to the regulation of anthocyanin and proanthocyanidin biosynthesis in developing grape berries. *Plant Physiol.* **2008**, *147*, 2041–2053. [CrossRef] [PubMed]
16. Fournierlevel, A.; Hugueney, P.; Verriès, C.; This, P.; Ageorges, A. Genetic mechanisms underlying the methylation level of anthocyanins in grape (*Vitis vinifera* L.). *BMC Plant Biol.* **2011**, *11*, 269–287.
17. Ali, M.B.; Howard, S.; Chen, S.; Wang, Y.; Yu, O.; Kovacs, L.G.; Qiu, W. Berry skin development in Norton grape: Distinct patterns of transcriptional regulation and flavonoid biosynthesis. *BMC Plant Biol.* **2011**, *11*, 230–233. [CrossRef] [PubMed]

18. Zoratti, L.; Karppinen, K.; Luengo, E.A.; Häggman, H.; Jaakola, L. Light-controlled flavonoid biosynthesis in fruits. *Front. Plant Sci.* **2014**, *5*, 534. [CrossRef] [PubMed]

19. Wei, Y.Z.; Hu, F.C.; Hu, G.B.; Li, X.J.; Huang, X.M.; Wang, H.C. Differential Expression of Anthocyanin Biosynthetic Genes in Relation to Anthocyanin Accumulation in the Pericarp of Litchi Chinensis Sonn. *PLoS ONE* **2011**, *6*, e19455. [CrossRef] [PubMed]

20. Chen, Y.; Gao, A.; Fu, C.H.; Gui, B.H.; Zhi, C.Z. Functional analysis of the UDP glucose: Flavonoid-3-Oglucosyltransferase (UFGT) promoter from litchi (*Litchi chinesis* Sonn.) and transient expression in onions (Allium cepa Linn.). *Afr. J. Plant Sci.* **2015**, *9*, 244–249. [CrossRef]

21. HE, F.; He, J.J.; Pan, Q.H.; Duan, C.Q. Mass-spectrometry evidence confirming the presence of pelargonidin-3-*O*-glucoside in the berry skins of Cabernet Sauvignon and Pinot Noir (*Vitis vinifera* L.). *Aust. J. Grape Wine Res.* **2010**, *16*, 464–468. [CrossRef]

22. Meng, R.; Qu, D.; Liu, Y.; Gao, Z.; Yang, H.; Shi, X.; Zhao, Z. Anthocyanin accumulation and related gene family expression in the skin of dark-grown red and non-red apples (*Malus domestica* Borkh.) in response to sunlight. *Sci. Hortic.* **2015**, *189*, 66–73. [CrossRef]

23. Zhao, Q.; He, F.; Reeves, M.J.; Pan, Q.H.; Duan, C.Q.; Wang, J. Expression of structural genes related to anthocyanin biosynthesis of Vitis amurensis. *J. For. Res.* **2016**, *27*, 647–657. [CrossRef]

24. Castellarin, S.D.; Gambetta, G.A.; Wada, H.; Krasnow, M.N.; Cramer, G.R.; Peterlunger, E.; Shackel, K.A.; Matthews, M.A. Characterization of major ripening events during softening in grape: Turgor, sugar accumulation, abscisic acid metabolism, colour development, and their relationship with growth. *J. Exp. Bot.* **2016**, *67*, 709–722. [CrossRef] [PubMed]

25. Liu, M.Y.; Song, C.Z.; Chi, M.; Wang, T.M.; Zuo, L.L.; Li, X.L.; Zhang, Z.W.; Xi, Z.M. The effects of light and ethylene and their interaction on the regulation of proanthocyanidin and anthocyanin synthesis in the skins of *Vitis vinifera* berries. *Plant Growth Regul.* **2016**, *79*, 1–14. [CrossRef]

26. Gomes, M.M.A. Physiological effects related to brassinosteroid application in plants. In *Brassinosteroids: A Class of Plant Hormone*; Hayat, S., Ahmad, A., Eds.; Springer: Dordrecht, The Netherlands, 2010; pp. 193–242.

27. Xi, Z.M.; Zhang, Z.W.; Huo, S.S.; Luan, L.Y.; Gao, X.; Ma, L.N.; Fang, Y.L. Regulating the secondary metabolism in grape berry using exogenous 24-epibrassinolide for enhanced phenolics content and antioxidant capacity. *Food Chem.* **2013**, *141*, 3056–3065. [CrossRef] [PubMed]

28. Hayat, S.; Ahmad, A. *Brassinosteroids: A Class of Plant Hormone*; Springer: Dordrecht, The Netherlands, 2011; pp. 1283–1317.

29. Hayat, S.; Ali, B.; Hasan, S.A.; Ahmad, A. Brassinosteroid enhanced the level of antioxidants under cadmium stress in Brassica juncea. *Environ. Exp. Bot.* **2007**, *60*, 33–41. [CrossRef]

30. Wang, Z.; Zheng, P.; Meng, J.; Xi, Z. Effect of exogenous 24-epibrassinolide on chlorophyll fluorescence, leaf surface morphology and cellular ultrastructure of grape seedlings (*Vitis vinifera* L.) under water stress. *Acta Physiol. Plant.* **2015**, *37*, 1–12. [CrossRef]

31. Cui, L.; Zou, Z.; Zhang, J.; Zhao, Y.; Yan, F. 24-Epibrassinoslide enhances plant tolerance to stress from low temperatures and poor light intensities in tomato (*Lycopersicon esculentum* Mill.). *Funct. Integr. Genom.* **2016**, *16*, 1–7. [CrossRef] [PubMed]

32. Asghari, M.; Zahedipour, P. 24-Epibrassinolide Acts as a Growth-Promoting and Resistance-Mediating Factor in Strawberry Plants. *J. Plant Growth Regul.* **2016**, *35*, 1–8. [CrossRef]

33. Liu, Q.; Xi, Z.; Gao, J.; Meng, Y.; Lin, S.; Zhang, Z. Effects of exogenous 24-epibrassinolide to control grey mould and maintain postharvest quality of table grapes. *Int. J. Food Sci. Technol.* **2016**, *51*, 1236–1243. [CrossRef]

34. Xu, F.; Gao, X.; Xi, Z.M.; Zhang, H.; Peng, X.Q.; Wang, Z.Z.; Wang, T.M.; Meng, Y. Application of exogenous 24-epibrassinolide enhances proanthocyanidin biosynthesis in *Vitis vinifera* 'Cabernet Sauvignon' berry skin. *Plant Growth Regul.* **2015**, *75*, 741–750. [CrossRef]

35. Wang, Z.Y.; Bai, M.Y.; Oh, E.; Zhu, J.Y. Brassinosteroid Signaling Network and Regulation of Photomorphogenesis. *Annu. Rev. Genet.* **2012**, *46*, 701–724. [CrossRef] [PubMed]

36. Lee, J. Light exclusion influence on grape anthocyanin. *Heliyon* **2017**, *3*, e00243. [CrossRef] [PubMed]

37. Shin, J.; Park, E.; Choi, G. PIF3 regulates anthocyanin biosynthesis in an HY5-dependent manner with both factors directly binding anthocyanin biosynthetic gene promoters in Arabidopsis. *Plant J.* **2007**, *49*, 981–994. [CrossRef] [PubMed]

38. Wang, Y.; Wang, Y.; Song, Z.; Zhang, H. Repression of MYBL2 by both microRNA858a and HY5 Leads to the Activation of Anthocyanin Biosynthetic Pathway in Arabidopsis. *Mol. Plant* **2016**, *9*, 1395–1405. [CrossRef] [PubMed]

39. Bureau, S.M.; Baumes, R.L.; Razungles, A.J. Effects of vine or bunch shading on the glycosylated flavor precursors in grapes of *Vitis vinifera* L. Cv. syrah. *J. Agric. Food Chem.* **2000**, *48*, 1290–1297. [CrossRef] [PubMed]

40. Friedel, M. Impact of light exposure on fruit composition of white 'Riesling' grape berries ("*Vitis vinifera*" L.). *J. Grapevine Res.* **2015**, *54*, 107–116.

41. Sharma, R.R.; Reddy, S.V.R.; Jhalegar, M.J. Pre-harvest fruit bagging: A useful approach for plant protection and improved post-harvest fruit quality—A review. *J. Hortic. Sci. Biotechnol.* **2014**, *89*, 101–113. [CrossRef]

42. Lorrain, B.; Chira, K.; Teissedre, P.L. Phenolic composition of Merlot and Cabernet-Sauvignon grapes from Bordeaux vineyard for the 2009-vintage: Comparison to 2006, 2007 and 2008 vintages. *Food Chem.* **2011**, *126*, 1991–1999. [CrossRef] [PubMed]

43. Bautista-Ortín, A.B.; Rodríguez-Rodríguez, P.; Gil-Muñoz, R.; Jiménez-Pascual, E.; Busse-Valverde, N.; Martínez-Cutillas, A.; López-Roca, J.M.; Gómez-Plaza, E. Influence of berry ripeness on concentration, qualitative composition and extractability of grape seed tannins. *Aust. J. Grape Wine Res.* **2012**, *18*, 123–130. [CrossRef]

44. Bucchetti, B.; Matthews, M.A.; Falginella, L.; Peterlunger, E.; Castellarin, S.D. Effect of water deficit on Merlot grape tannins and anthocyanins across four seasons. *Sci. Hortic.* **2011**, *128*, 297–305. [CrossRef]

45. Bogs, J.; Downey, M.O.; Harvey, J.S.; Ashton, A.R.; Tanner, G.J.; Robinson, S.P. Proanthocyanidin Synthesis and Expression of Genes Encoding Leucoanthocyanidin Reductase and Anthocyanidin Reductase in Developing Grape Berries and Grapevine Leaves. *Plant Physiol.* **2005**, *139*, 652–663. [CrossRef] [PubMed]

46. Gagné, S.; Lacampagne, S.; Claisse, O.; Gény, L. Leucoanthocyanidin reductase and anthocyanidin reductase gene expression and activity in flowers, young berries and skins of *Vitis vinifera* L. cv. Cabernet-Sauvignon during development. *Plant Physiol. Biochem.* **2009**, *47*, 282–290. [CrossRef] [PubMed]

47. Wen, P.F.; Ji, W.; Gao, M.Y.; Niu, T.Q.; Xing, Y.F.; Niu, X.Y. Accumulation of flavanols and expression of leucoanthocyanidin reductase induced by postharvest UV-C irradiation in grape berry. *Genet. Mol. Res.* **2015**, *14*, 7687–7695. [CrossRef] [PubMed]

48. Ju, Y.L.; Liu, M.; Zhao, H.; Meng, J.F.; Fang, Y.L. Effect of Exogenous Abscisic Acid and Methyl Jasmonate on Anthocyanin Composition, Fatty Acids, and Volatile Compounds of Cabernet Sauvignon (*Vitis vinifera* L.). Grape Berries. *Molecules* **2016**, *21*, E1354. [CrossRef] [PubMed]

49. Luan, L.Y.; Zhang, Z.W.; Xi, Z.M.; Huo, S.S.; Ma, L.N. Comparing the effects of exogenous abscisic acid on the phenolic composition of Yan 73 and Cabernet Sauvignon (*Vitis vinifera* L.) wines. *Eur. Food Res. Technol.* **2014**, *239*, 203–213. [CrossRef]

50. Villalobos-González, L.; Peña-Neira, A.; Ibáñez, F.; Pastenes, C. Long-term effects of abscisic acid (ABA) on the grape berry phenylpropanoid pathway: Gene expression and metabolite content. *Plant Physiol. Biochem.* **2016**, *105*, 213–223. [CrossRef] [PubMed]

51. Zhu, Y.; Peng, Q.Z.; Du, C.; Li, K.G.; Xie, D.Y. Characterization of Flavan-3-ols and Expression of MYB and Late Pathway Genes Involved in Proanthocyanidin Biosynthesis in Foliage of Vitis bellula. *Metabolites* **2013**, *3*, 185–203. [CrossRef] [PubMed]

52. Bogs, J.; Jaffé, F.W.; Takos, A.M.; Walker, A.R.; Robinson, S.P. The grapevine transcription factor VvMYBPA1 regulates proanthocyanidin synthesis during fruit development. *Plant Physiol.* **2007**, *143*, 1347–1361. [CrossRef] [PubMed]

53. Terrier, N.; Torregrosa, L.; Ageorges, A.; Vialet, S.; Verries, C.; Cheynier, V.; Romieu, C. Ectopic expression of VvMybPA2 promotes proanthocyanidin biosynthesis in grapevine and suggests additional targets in the pathway. *Plant Physiol.* **2009**, *149*, 1028–1041. [CrossRef] [PubMed]

54. Koyama, K.; Numata, M.; Nakajima, I.; Goto-Yamamoto, N.; Matsumura, H.; Tanaka, N. Functional characterization of a new grapevine MYB transcription factor and regulation of proanthocyanidin biosynthesis in grapes. *J. Exp. Bot.* **2014**, *65*, 4433–4449. [CrossRef] [PubMed]

55. Tang, W.; Deng, Z.; Wang, Z.Y. Proteomics shed light on the brassinosteroid signaling mechanisms. *Curr. Opin. Plant Biol.* **2009**, *13*, 27–33. [CrossRef] [PubMed]

56. Li, J. Regulation of the Nuclear Activities of Brassinosteroid Signaling. *Curr. Opin. Plant Biol.* **2010**, *13*, 540–547. [CrossRef] [PubMed]

57. Azuma, A.; Fujii, H.; Shimada, T.; Yakushiji, H. Microarray Analysis for the Screening of Genes Inducible by Light or Low Temperature in Post-veraison Grape Berries. *Hortic. J.* **2015**, *84*, 214–216. [CrossRef]
58. Coombe, B.G. Growth Stages of the Grapevine: Adoption of a system for identifying grapevine growth stages. *Aust. J. Grape Wine Res.* **2008**, *1*, 104–110. [CrossRef]
59. Stefano, R.D.; Cravero, M.C. The grape phenolic determination. *Riv. Vitic. Enol.* **1991**, *xliv(58)*, 37–45.
60. Harbertson, J.F.; Picciotto, E.A.; Adams, D.O. Measurement of polymeric pigments in grape berry extracts and wines using a protein precipitation assay combined with bisulfite bleaching. *Am. J. Enol. Vitic.* **2003**, *54*, 301–306.
61. Jayaprakasha, G.K.; Singh, R.P.; Sakariah, K.K. Antioxidant activity of grape seed (*Vitis vinifera*) extracts on peroxidation models in vitro. *Food Chem.* **2001**, *73*, 285–290. [CrossRef]
62. Lee, J.; Durst, R.W.; Wrolstad, R.E. Determination of total monomeric anthocyanin pigment content of fruit juices, beverages, natural colorants, and wines by the pH differential method: Collaborative study. *J. AOAC Int.* **2005**, *88*, 1269–1278. [PubMed]
63. Livak, K.J.; Schmittgen, T.D. Analysis of relative gene expression data using real-time quantitative PCR and the 2(-Delta Delta C(T)) Method. *Methods* **2001**, *25*, 402–408. [CrossRef] [PubMed]

Sample Availability: Samples of the compounds are not available from the authors.

MDPI

Article

Vermicompost Supplementation Improves the Stability of Bioactive Anthocyanin and Phenolic Compounds in *Clinacanthus nutans* Lindau

Zuhaili Yusof [1], Sujatha Ramasamy [1], Noor Zalina Mahmood [1] and Jamilah Syafawati Yaacob [1,2,*]

[1] Institute of Biological Sciences, Faculty of Science, University of Malaya, Kuala Lumpur 50603, Malaysia; zuhailiyusof@yahoo.com (Z.Y.); sujatha@um.edu.my (S.R.); alin@um.edu.my (N.Z.M.)

[2] Centre for Research in Biotechnology for Agriculture (CEBAR), Institute of Biological Sciences, Faculty of Science, University of Malaya, Kuala Lumpur 50603, Malaysia

* Correspondence: jamilahsyafawati@um.edu.my; Tel.: +60-3-7967-4090

Academic Editor: M. Monica Giusti
Received: 25 April 2018; Accepted: 30 May 2018; Published: 4 June 2018

Abstract: This project studied the effect of vermicompost application on the composition of bioactive anthocyanin and phenolic compounds, and the antioxidant activity of *Clinacanthus nutans*. The correlation between the bioactive constituents and antioxidant capacity was also evaluated. In this project, a field study was conducted using a randomized complete block design (RCBD) with four treatment groups, including control plants (CC), plants supplied with chemical fertilizer (CF), plants supplied with vermicompost (VC), and plants supplied with mixed fertilizer (MF). The leaves of *C. nutans* from all treatment groups were harvested, subjected to solvent extraction, and used for quantification of total anthocyanin content (TAC), total phenolic content (TPC), and total flavonoid content (TFC). The initial antioxidant activity of the extracts was evaluated using 2,2-Diphenyl-1-picrylhydrazyl (DPPH) and 2,2′-azinobis(3-ethylbenzothiazoline-6-sulfonic acid) (ABTS) assays, as well as after two and four weeks of storage at $-20\,^{\circ}\text{C}$ and $4\,^{\circ}\text{C}$. Data analysis showed that CC plants contained the highest TAC ($2180.14 \pm 338.43\ \mu\text{g/g}$ dry weight) and TFC (276.25 ± 3.09 mg QE/g dry weight). On the other hand, CF plants showed the highest TPC (181.53 ± 35.58 mg GAE/g dry weight). Moreover, we found that CC plants had the highest antioxidant potential against DPPH radicals whereas MF plants showed the lowest antioxidant potential. After four weeks of extract storage at $-20\,^{\circ}\text{C}$ and $4\,^{\circ}\text{C}$, the TPC, TFC, TAC, and antioxidant potential of the extracts decreased. Extracts from VC showed the lowest percentage of total phenolic and total flavonoid loss after extract storage at $-20\,^{\circ}\text{C}$ and $4\,^{\circ}\text{C}$ compared with other plant extracts. At this juncture, it could be deduced that the application of vermicompost had little effect on the expression of phenolics, flavonoids, or anthocyanin in *C. nutans*. However, the extract from plants treated with vermicompost (VC and MF) showed better stability compared with CC and CF after extract storage at different temperatures.

Keywords: total phenolic content; anthocyanin; flavonoid; antioxidant; extract storage; vermicompost; *Clinacanthus nutans*

1. Introduction

Clinacanthus nutans Lindau is an important medicinal plant from the family Acanthaceae. A tall, erect herbaceous perennial shrub, it grows up to one meter in height and is distributed throughout tropical regions, including Southeast Asia and China. The plant is locally known in Malaysia as "Belalai Gajah" due to the slightly curved stem supporting the leaves that resembles the curve of an elephant's trunk [1]. In Malaysia and Thailand, the leaves of *C. nutans* are used as a remedy against venomous snake bites, scorpion, and insect stings [2,3], resulting in the vernacular name, Sabah snake grass. *C. nutans* is classified as one of Malaysia's high-value herbal products in the Entry Point Project 1 (EPP1) in the agriculture sector, which is one of the identified National Key Economic Areas (NKEAs). *C. nutans* has been reported to contain various phytochemicals with biological activities including antioxidant [4,5], antimicrobial [6], anti-inflammatory [5,7], antivenom [2], and anticancer activities [3,8]. This plant has also been widely used for the treatment of various diseases such as cancer, herpes simplex virus (HSV), varicella-zoster virus (VZV) lesions, skin rashes, and kidney problems [1,3].

Researchers are increasingly embracing green technology and the usage of sustainable and environmentally friendly approaches to enhance the production of valuable secondary metabolites in plants. Plants produce essential secondary metabolites for their adaptation and defense against environmental stressors. For example, water-soluble flavonoids play an important role in protecting plant's cellular processes against ultraviolet-B (UVB) radiation [9]. Flavonoids are used in various industries, such as in the food industry as a food additive. In addition, flavonoids are used in the production of pharmaceutical products [10] and as a source of environmentally friendly colorants in the textile industry [11]. Among the pigments synthesized via the flavonoid pathway, anthocyanins are one of the most important. Anthocyanins can appear red, blue, or purple and have the potential to be a natural food colorant [12]. Anthocyanins have been reported to help in improving cardiovascular disease and cancer [13,14]. The vast benefits of anthocyanins have rendered them a healthier alternative to synthetic dyes and food colorants in the growing food and beverage industry, as the color property is seen as an important factor in influencing consumer acceptance and to reflect the quality of the product [15,16]. Moreover, advances in the use of natural dyes as functional colorants have been made in various sectors, such as in dye-sensitized solar cells, development of biomedical sensors, and optical data storage [17–20]. The applications of plant-derived natural colorants or dyes for green technology development appears to be limitless as demands for sustainable production of these valuable pigments continue to increase. Thus, "green" practices in the synthesis of these valuable pigments, such as through organic production, are key to increase consumer acceptance, especially for its application in the food industry.

In this study, the effects of organic growth supplements, including vermicompost in accumulation of bioactive phenolic, anthocyanin, and flavonoid compounds in *C. nutans*, as well as their antioxidant potential were evaluated. Vermicompost is an organic materials broken down by the interactions between microorganisms and earthworms in a mesophilic process to produce fully stabilized and organic soil amendments with low carbon to nitrogen (C:N) ratios [21]. Vermicompost is rich in NPK (nitrogen 2–3%, phosphorus 1.55–2.25%, and potassium 1.85–2.25%), micronutrients, and beneficial soil microbes, so it is garnering attention as a greener replacement for chemical fertilizers to maintain and further improve soil quality. The application of vermicompost as an alternative to chemical fertilizer not only produces healthier plants, but it also increases plant resistance toward pests and diseases. Moreover, vermicomposting is a quicker and more cost-effective technique for composting as it helps in diverting organic waste from landfills. Vermicompost has been reported to significantly stimulate the growth of a wide range of plant species including several medicinal plants [22], horticultural crops [23–26], fruit crops [27,28], ornamentals [29,30], and forestry species [31–33]. However, few publications were found on the effect of vermicompost supplementation on the availability of phytoconstituents and bioactivity of *C. nutans*. Previous works conducted on the application of vermicompost only focused on the effects of its supplementation on plant growth. The outcomes of this study provide additional knowledge and understanding about the effects of vermicompost supplementation on the availability of bioactive phenolic and anthocyanin compounds in *C. nutans*. Thus, this study adds to the knowledge that can be used for ensuring sustainable production of these bioactive pigments from natural sources, while reducing the impact on the environment.

2. Materials and Methods

2.1. Sample Preparation

In this project, a field study using a randomized complete block design (RCBD) with four treatment groups was conducted at Glami Lemi Biotechnology Research Centre (PPBGL), Jelebu, Malaysia. The treatment groups included control plants (CC), plants supplied with 10 t/ha of NPK fertilizer (CF), plants supplied with vermicompost at 15 t/ha (VC), and plants supplied with mixed fertilizer (containing both 15 t/ha of vermicompost and 5 t/ha of NPK fertilizer; MF). All treatments were applied on the plots as a soil conditioner two weeks before the planting process. The leaves from three *C. nutans* plants of each of the three different blocks per treatment were harvested in February 2017, resulting in a total of nine samples for each treatment. *C. nutans* was identified by comparing the plant with a herbarium specimen and a voucher specimen (KLU 49509) was deposited at the Herbarium of Rimba Ilmu, Institute of Biological Sciences, University of Malaya, Kuala Lumpur, Malaysia. After harvesting the leaves, the methanolic extract of *C. nutans* was prepared, in which the fresh leaves were freeze-dried and subjected to solvent extraction using methanol. Briefly, 3 g of freeze-dried leaves were soaked in 90 mL 70% methanol and ground using a mortar and pestle. Then, the sample mixture was incubated in solvent at 4 °C for 24 h followed by filtration using filter paper. The extraction process was repeated using the residue obtained from the filtration. The filtrates were pooled and evaporated to dryness using a rotary evaporator at 45 °C to yield the methanolic extracts. The concentrated extract was adjusted to a concentration of 20 mg/mL using 70% methanol before it was stored at either 4 °C or −20 °C until further analysis.

2.2. Phytochemical Screening of Bioactive Compounds in C. nutans

The presence of various phytochemicals such as alkaloid, tannin, phenol, flavonoid, and saponin in the samples was evaluated based on standard methods [34].

2.2.1. Measurement of Total Anthocyanin Content

The total anthocyanin content was determined using the pH differential method [35]. The methanolic extracts of *C. nutans* were diluted separately with two types of buffer: potassium chloride (0.025 M) at pH 1.0 and sodium acetate (0.4 M) at pH 4.5 using the ratio 1:4 (1 part test portion and 4 parts buffer). The absorbance was measured at wavelengths of 520 nm and 700 nm using UV-200-RS spectrophotometer (MRC Ltd., Holon, Israel). The concentration of anthocyanin pigment was measured using the following formula:

$$\text{Anthocyanin pigment content (mg/L)} = \frac{(A \times MW \times DF \times 1000)}{(\varepsilon \times 1)}$$

$$\text{where } A = (Abs_{520} - Abs_{700})pH1.0 - (Abs_{520} - Abs_{700})pH4.5$$

$$MW(\text{cyanidin} - 3 - \text{glucoside}) = 449.2 \text{ g/mol}$$

$$DF = \text{dilution factor}$$

$$\varepsilon = 26,900$$

2.2.2. Measurement of Total Phenolic Content

Folin-Ciocalteu's method was used to quantify the total phenolic content of *C. nutans* according to the method described by Sun, et al. [36] with minor changes. Briefly, Folin-Ciocalteu reagent was diluted 10-fold with deionised water. A total of 0.1 mL of *C. nutans* methanolic extract was mixed with 0.75 mL of diluted Folin-Ciocalteu reagent and incubated for 10 min at room temperature. Then, 0.75 mL of 2% sodium carbonate (Na_2CO_3) solution was added. The mixture was allowed to stand in the dark for 45 min before measuring the absorbance at 765 nm using a UV-200-RS spectrophotometer

(MRC Ltd., Holon, Israel) against a blank, containing the solvent (70% methanol). The TPC of the samples was determined from a calibration curve prepared with a series of gallic acid standards (0.01, 0.02, 0.03, 0.04, 0.05, and 0.06 mg/mL). Results were expressed as mg of gallic acid equivalents/g dry weight (mg GAE/g DW) of extract.

2.2.3. Measurement of Total Flavonoid Content

The aluminum chloride colorimetric method [37] was used to quantify the total flavonoid content in the methanolic extracts of *C. nutans*. First, 0.5 mL of each extract was mixed with 1.5 mL of 70% methanol, 0.10 mL of 10% aluminum chloride, $AlCl_3$ ($AlCl_3 \cdot 6H_2O$), 0.10 mL of sodium acetate ($NaC_2H_3O_2 \cdot 3H_2O$) (1 M), and 2.80 mL of distilled water. Then, the absorbance was measured at 415 nm using a UV-200-RS spectrophotometer (MRC Ltd., Holon, Israel) after 40 min of incubation. The flavonoid concentrations were calculated by preparing a calibration curve using quercetin as the standard (0.15–0.4 mg/mL). The flavonoid concentration was expressed as quercetin equivalents in mg per gram of dry weight (mg/g DW) of extract. All assays were performed in triplicate.

2.3. DPPH (2,2-Diphenyl-1-picrylhydrazyl) Radical Scavenging Activity Assay

DPPH free radical scavenging activity of the *C. nutans* methanolic extracts was analyzed following standard procedure [38]. First, 50 µL of extract at six different concentrations (0.5, 1.0, 1.5, 2.0, 2.5, and 3.0 mg/mL) were added to 150 µL of DPPH solution (60 mM) in each well of a 96-well plate. The mixture was then incubated at room temperature for 30 min. Then, 50 µL of methanol was added to the DPPH solution as a blank. At the end of the incubation period, a Multiskan Go plate reader (Thermo Scientific, Waltham, MA, USA) was used to measure the absorbance at 515 nm. All the extracts were assayed in triplicate. The obtained data were then used to determine the concentration of the sample required to scavenge 50% of the DPPH free radicals (IC_{50}). The percentage of inhibition was plotted against the concentration and the IC_{50} was obtained from the fitted linear curve. A lower IC_{50} denotes a more potent antioxidant.

2.4. ABTS (2,2'-Azinobis(3-ethylbenzothiazoline-6-sulfonic acid)) Radical Scavenging Activity Assay

The colorimetric method described by Shao et al. [39] was used to perform the ABTS scavenging activity assay, with slight modifications. First, the preparation of the ABTS radical cation was performed by mixing 10 mL of 2.6 mM $K_2S_2O_4$ solution with 10 mL of 7.4 mM ABTS solution. Then, the mixture was stored for 12 h at room temperature in a dark room before further use. After that, double distilled water (ddH_2O) was used to dilute the mixture and it was adjusted to produce an absorbance reading of 0.70 ± 0.2 at 734 nm. A 20 µL of sample at six different concentrations (0.5, 1, 1.5, 2.0, 2.5, and 3.0 mg/mL) was then added to 200 µL ABTS solution and the mixture was incubated at room temperature for 30 min. The absorbance reading was measured at 734 nm using a Multiskan Go plate reader (Thermo Scientific, Waltham, MA, USA) and the assay was performed in triplicate. The percentage of inhibition was calculated according to the following formula:

$$\% \text{ inhibition} = \frac{\left(\text{Abs}_{\text{blank}} - \text{Abs}_{\text{sample}}\right)}{\left(\text{Abs}_{\text{blank}}\right)} \times 100$$

The antioxidant capacity of the test extracts is expressed as IC_{50}, which is the concentration necessary for a 50% reduction in ABTS radicals.

2.5. Extract Stability after Storage at Different Temperatures

In this study, the effect of the growth supplements on the stability of the methanolic extracts following extract storage at different temperatures was analyzed. For this, the extracts were stored at 4 °C or −20 °C for 2 and 4 weeks and then used in the determination of phenolic, anthocyanin and flavonoid contents. The effect of extract storage on antioxidant properties of the extracts was also determined.

2.6. Statistical Analysis

The data obtained in this study were subjected to statistical analysis using analysis of variance (ANOVA) and the mean values were compared using Duncan's Multiple Range Test (DMRT) in SPSS software version 24. Pearson correlation analysis was also conducted to determine the relationship between the antioxidant potential with the amount of bioactive phenolic, anthocyanin, and flavonoid present in the extracts.

3. Results and Discussion

3.1. Phytochemical Screening

Qualitative screening of *C. nutans* methanolic extract was performed using standard protocols [34]. Results of the present study revealed the presence of phenols, flavonoid, and saponin in the methanolic extracts of *C. nutans* leaves (Table 1) but were negative for alkaloid and tannin (Table 1). These results are in agreement with previous studies [25,40]. However, the methanolic extract of *C. nutans* has also been reported to contain steroids and triterpenes [40].

Table 1. Effect of plant growth supplements on availability of phytoconstituents in methanolic extracts of *Clinacanthus nutans* leaves.

Phytochemical	Treatment			
	No Fertilizer (CC)	NPK Fertilizer (CF)	Vermicompost (VC)	Mixed Fertilizer (MF)
Alkaloid	−	−	−	−
Tannin	−	−	−	−
Phenol	+	+	+	+
Flavonoid	+	+	+	+
Saponin	+	+	+	+

+ Present; − Absent.

3.2. Determination of Pigments Content (Total Anthocyanin, Phenolic and Flavonoid)

The total anthocyanin (TAC), total phenolic (TPC), and total flavonoid (TFC) contents in the methanolic extracts of *C. nutans* were measured. The TPC was expressed as mg gallic acid (GAE) per g dry weight of the sample, whereas the TFC was expressed as mg quercetin (QE) per g dry weight of the sample. Based on Table 2, the highest TAC was obtained in CC plants (2180.14 ± 338.43 µg/g dry weight), followed by CF, VC, and MF (Table 2). However, data analysis revealed that the differences observed among the TAC of all samples were not statistically significant. Conversely, significantly higher TFC was found in the methanolic extracts of CC (276.25 ± 3.09 mg QE/g dry weight) and CF (256.66 ± 45.43 mg QE/g dry weight) plants, compared with VC and MF. CF plant extract was also observed to contain the highest TPC (181.53 ± 35.58 mg GAE/g dry weight), followed by CC, VC, and MF (Table 2).

Table 2. Effect of plant growth supplementation on total phenolic, anthocyanin, and flavonoid contents in methanolic extracts of *C. nutans* leaves.

Treatment	Sample ID	Total Anthocyanin Content (µg/g DW)	Total Phenolic Content (mg GAE/g DW)	Total Flavonoid Content (mg QE/g DW)
No fertilizer	CC	2180.14 ± 338.43 [a]	120.48 ± 6.70 [a]	276.25 ± 3.09 [b]
NPK fertilizer	CF	1933.52 ± 66.06 [a]	181.53 ± 35.58 [b]	256.66 ± 45.43 [b]
Vermicompost	VC	1742.86 ± 62.30 [a]	98.06 ± 2.27 [a]	170.42 ± 7.55 [a]
Mixed fertilizer	MF	1669.91 ± 122.12 [a]	97.47 ± 18.73 [a]	157.30 ± 26.42 [a]

Note: Means with different letters within the same column are significantly different at $p \leq 0.05$ according to Duncan's multiple range test (DMRT).

3.3. Antioxidant Potential of C. Nutans Methanolic Extracts against DPPH and ABTS Radicals

The DPPH assay is typically based on the scavenging of free radicals and converting them to colorless products. When the methanolic extract of *C. nutans* reacts with DPPH solution, the free radicals are reduced by hydrogen donation to produce the reduced form of 1,1-diphenyl-2-picryl hydrazine (non-radical), indicated by the color change from violet to pale yellow [41]. The ABTS (2,2'-azinobis(3-ethylbenzothiazoline-6-sulfonic acid)) assay is a decolourization assay in which the stable radical is generated directly through the reaction of ABTS with potassium persulfate, which resulted in the production of blue or green ABTS chromophore, prior to reaction with the antioxidants. The free radical scavenging activity of the extracts against DPPH and ABTS is expressed as IC_{50} (mg per mL of extract), which is the concentration of antioxidant necessary to decrease the initial DPPH and ABTS concentration by 50%. The antioxidant activities of the extracts against both DPPH and ABTS radicals are displayed in Table 3. Data analysis revealed that the methanolic extract from control plants (CC) exhibited the highest free radical scavenging activity against DPPH and ABTS radicals, with an IC_{50} of 1.18 ± 0.05 mg/mL and 0.98 ± 0.05 mg/mL, respectively (Table 3). The antioxidant activity (denoted by IC_{50}) of the methanolic extracts against both DPPH and ABTS radicals, in decreasing order, is CC, CF, VC, and MF.

Table 3. Effect of plant growth supplements on antioxidant potential of methanolic extracts of *C. nutans* leaves.

Treatment	Sample ID	DPPH IC_{50} (mg/mL)	ABTS IC_{50} (mg/mL)
No fertilizer	CC	1.18 ± 0.05 [a]	0.98 ± 0.05 [a]
NPK fertilizer	CF	1.41 ± 0.02 [b]	1.15 ± 0.16 [a,b]
Vermicompost	VC	1.67 ± 0.04 [c]	1.41 ± 0.01 [b,c]
Mixed fertilizer	MF	2.23 ± 0.02 [d]	1.67 ± 0.17 [c]

Note: Means with different letters within the same column are significantly different at $p \leq 0.05$ according to Duncan's multiple range test (DMRT).

Based on the results of the present study, the antioxidant capacity of the methanolic extracts of *C. nutans* evaluated by the ABTS method was higher than that evaluated with the DPPH method. The DPPH assay underestimated the antioxidant capacity of the methanolic extracts of *C. nutans* by 18.44–33.53% compared to the ABTS assay. This observation is in agreement with a previous study by Almeida et al. [42] that showed that a higher antioxidant capacity was obtained for fresh Brazilian exotic fruits when evaluated by ABTS assay compared with evaluation using the DPPH method. This observation can be attributed to several factors, such as the wavelengths used during the spectrophotometric measurement in both assays, where 515 nm was used in DPPH assays and 734 nm was used in ABTS assays. Arnao [43] reported that the underestimation by DPPH assays could be due to the pigments contained in the colored extracts (such as carotenoids and anthocyanins) having overlapping absorbance spectra in the visible region with that of DPPH at 515 nm, thus interfering with the absorbance readings.

This underestimate could be due to the structural conformation of the antioxidants that influenced the reaction mechanism of free radical scavengers and DPPH. Larger molecules that have less access to the radical site have a lower antioxidant activity for a particular test compared with smaller molecules [44]. Otherwise, the underestimation could be due to the reactions of certain phenols such as eugenol and its derivatives that are reversible when reacted with DPPH, resulting in low readings of the antioxidant capacity [45]. However, DPPH assays also have an advantage over ABTS assays, as the DPPH free radicals can be directly acquired without preparation so they are ready to dissolve, whereas ABTS radical cations must be produced through enzymatic (peroxidase and myoglobin) or chemical (manganese dioxide and potassium persulfate) reactions [43]. Nevertheless, both the ABTS and DPPH assays have been the most popular spectrophotometric methods for determination of antioxidant capacity of foods and chemical compounds [46].

3.4. Effect of C. Nutans Extracts Storage (Duration and Temperature) on Stability of Pigments and Antioxidant Activity

In this study, the TPC, TAC, and TFC of the methanolic extracts after two and four weeks of storage at −20 °C and 4 °C were also evaluated. As shown in Figure 1, the TAC of all plant extracts decreased after four weeks of storage at −20 °C and 4 °C. More than 50% of the total anthocyanin loss for CC and CF plant extracts occurred after four weeks of storage at −20 °C and 4 °C compared to VC and MF. VC plant extract showed the lowest percentage of total anthocyanin loss (21.0%) after four weeks of storage at 4 °C, compared to other extracts that exhibited a TAC loss of more than 50%.

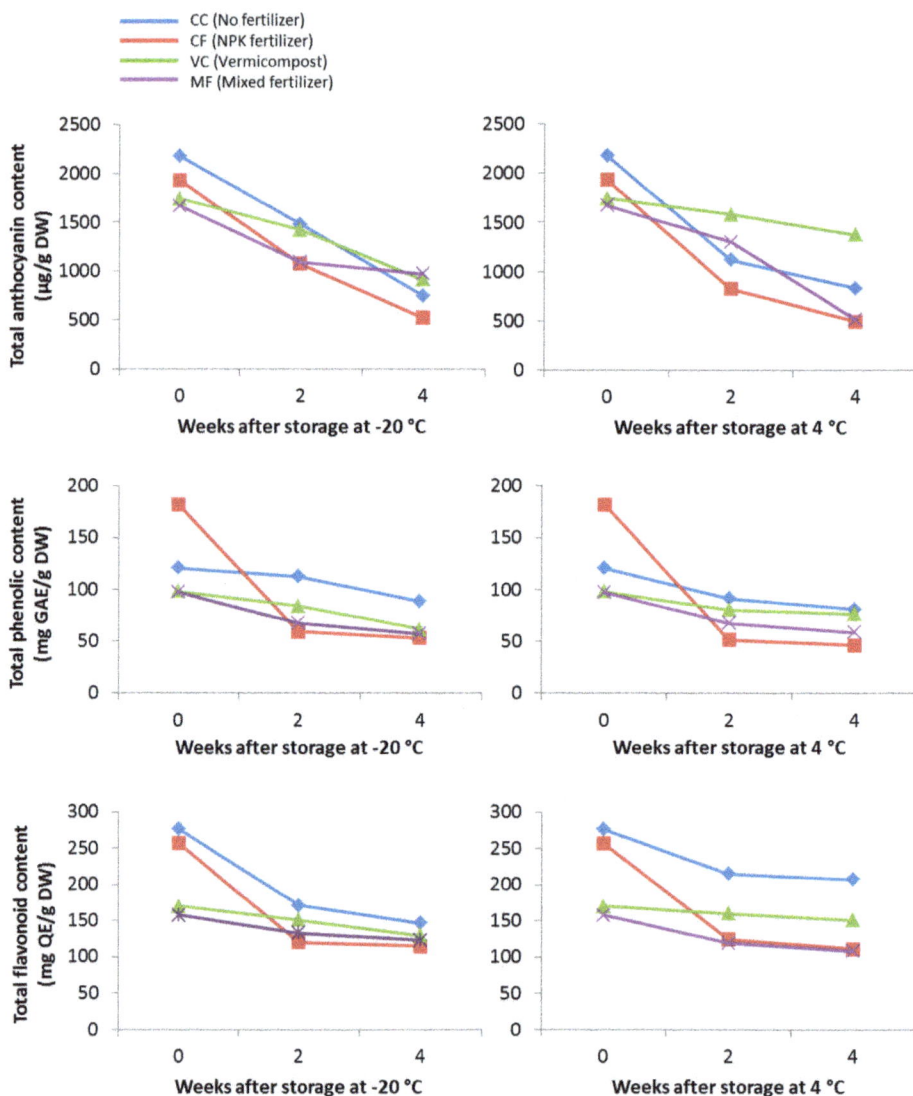

Figure 1. Effect of extract storage at −20 °C and 4 °C on total anthocyanin (TAC), total phenolic (TPC), and total flavonoid (TFC) contents in the methanolic extracts of *C. nutans* supplemented with different fertilizers.

Data analysis showed that TPC of the extracts decreased gradually with extract storage (Figure 1). After 4 weeks of storage at −20 °C, the TPC was observed to decrease by 26.62%, 70.75%, 37.04% and 41.19% for CC, CF, VC, and MF extracts, respectively. Plants supplemented with vermicompost (VC) showed the least percentage of TPC loss after extract storage at 4 °C, compared to other plant extracts (Figure 1). It was also observed that plants supplemented with only NPK fertilizer (CF) showed the highest TPC loss after extract storage at both −20 °C and 4 °C (Figure 1). Similar results were observed for TFC, where CF plant extracts were found to exhibit the highest TFC loss after extract storage (Figure 1). Interestingly, data analysis again revealed that plants supplemented with vermicompost (VC) exhibited the lowest percentage of TFC loss, with a decrease of 23.96% and 11.30% after 4 weeks of storage at −20 °C and 4 °C, respectively.

Furthermore, the antioxidant potential of the plant extracts following extract storage at −20 °C and 4 °C were also monitored. Data analysis revealed that after 4 weeks of storage, the antioxidant potential of the extracts was significantly reduced, as denoted by the increase in IC_{50} values. As observed in Figure 2, plant extracts from CF showed the highest loss of antioxidant potential against DPPH and ABTS radicals, when stored at both −20 °C and 4 °C. It was also evident that supplementation with vermicompost (VC and MF) improved the stability of antioxidants present in the extracts (Figure 2). Based on correlation analysis (Table 4), the antioxidant potential exhibited by the plant extracts could be due to their pigments content (anthocyanin, phenolic and flavonoid), as shown by the negative correlations observed. The significant negative correlations indicate that the DPPH and ABTS IC_{50} values decreased with increasing anthocyanin, phenolic and flavonoid content. However, these correlations were only moderate, except for TAC, which was shown to be strongly correlated with DPPH IC_{50}.

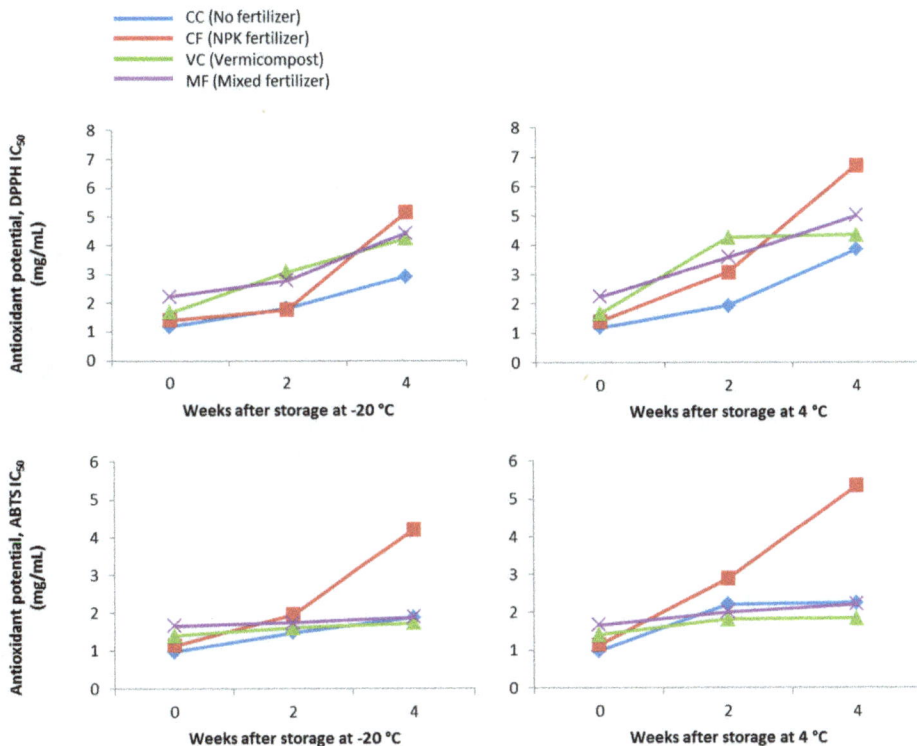

Figure 2. Effect of extract storage at −20 °C and 4 °C on antioxidant potential of the methanolic extracts of *Clinacanthus nutans* supplemented with different fertilizers.

Table 4. Pearson's correlation coefficients between the variables.

Variables	TFC	TPC	TAC	DPPH	ABTS
TFC	1				
TPC	0.834 **	1			
TAC	0.399 **	0.429 **	1		
DPPH	−0.465 **	−0.464 **	−0.518 **	1	
ABTS	−0.376 **	−0.401 **	−0.427 **	0.545 **	1

** Correlation is significant at $p < 0.01$.

Based on the results obtained, vermicompost does not exert any significant effect on the expression of bioactive compounds, as evident by the lower values (TAC, TPC, and TFC) obtained for all vermicompost-treated plants. This finding is in agreement with a previous study conducted on lettuce (*Lactuca sativa* L.), which showed that vermicompost fertilization significantly increased the crop yield, but reduced the levels of phenolic compounds and antioxidant activity [47]. Similarly, pak choi (*Brassica rapa* cv Bonsai, Chinensis group) and chincuya (*Annona purpurea* Moc and Sesse ex Dunal) plants grown with vermicompost were also reported to contain lower amounts of phenolic compounds [48,49]. These observations indicate that some factors may be present in vermicompost that negatively interferes with the synthesis of phenolic compounds through the phenylpropanoid pathway [47]. The accumulation of these various secondary metabolites has been shown to be affected by the interactions between plant genotype (species and variety within species) and environmental factors, including cultivation technique, season, abiotic and biotic stresses, and nutrient status [50,51]. The nutritional status of plants plays an important role in supporting plant growth, and various studies have demonstrated that supplementation of vermicompost provides nutrients to the plants in readily available forms such as nitrogen and enhances plant nutrient-uptake [52]. Thus, the production of the resulting phenolics and flavonoids are low when plants are not nutrient-limited.

The degradation of the compounds following extract storage that was observed in this study was consistent with previous findings [53], where the total phenolic content in Cornelian cherries extracts was found to decrease after being stored at 2 °C for up to 60 days. A drastic decrease was observed in the total phenolic and flavonoid content of *Anemopsis californica* extracts after being stored at 4 °C [54]. The decrease in total phenolics and flavonoids after prolonged storage was caused by various factors including oxidation by enzymes such as glycosidase, phenolase, and polyphenol oxidase (PPO), degradation of compounds, the polymerization of these compounds with proteins, and the conversion between free and bound phenolic substances [55–57]. PPO is an enzyme that causes the oxidation of phenolic compounds to quinones [58] by using polyphenols as the substrate [59]. Additionally, sample processing could cause disruption of the cell structure, which results in the loss of total phenolics and flavonoids [60]. Conversely, the decrease in total phenolic content may be due to non-enzymatic reactions that can contribute to the loss of total phenolic and flavonoid contents after storage. For example, the hydrogen peroxide produced as the by-product from the oxidation of ascorbic acid to dehydroascorbic acid can adversely affect the production of phenolic compounds [61], and the production of ascorbic acid and dehydroascorbic acid has been reported to increase due to chilling stress [62].

Results of the current study indicate that organic fertilization through supplementation with vermicompost contributed to the stability of the bioactive compounds during storage. At present, no published scientific literature was found on the exact mechanisms behind this observation. However, we postulate that the chemical compounds present in the plants supplemented with vermicompost (VC and MF) may undergo less oxidation, less degradation, and less polymerization compared with the compounds present in plants supplemented with chemical fertilizer (CF) and control plants (CC). Moreover, soil and fertilizer application has been found to affect protein composition and concentration in plants [63], which can interact with plant polyphenols. Various scientific literature has reported that the covalent interaction between plant polyphenols and proteins can further influence the content of free polyphenols and their antioxidant capacity during processing, transportation, and storage [64–66]. This implies that further studies involving multi-omics platforms, such as through metabolomics and proteomics, are essential for better understanding of the dynamics of protein and compound synthesis in relation to fertilization practices, as well as their interactions.

4. Conclusions

Vermicompost does not exert any significant effect on the plant antioxidant activity and the composition of bioactive compounds. However, the usage of vermicompost produced plants with the same content of bioactive compounds as the control and as the plants supplied with chemical fertilizer. Different storage conditions affected the stability of the bioactive compounds present in the methanolic extracts; however, plants supplemented with vermicompost were found to exhibit better stability when stored at different storage conditions.

Author Contributions: N.Z.M. and J.S.Y. conceived and designed the experiments; Z.Y. performed the experiments; Z.Y., S.R. and J.S.Y. analyzed the data; S.R., N.Z.M. and J.S.Y. contributed reagents/materials/analysis tools; Z.Y., S.R. and J.S.Y. wrote the paper.

Funding: This research was funded by University of Malaya, Malaysia (grant number RP022A-16SUS, RP015B-14AFR and CEBAR RU006-2017).

Acknowledgments: The authors thank the University of Malaya, Malaysia for experimental facilities and financial support (Grant No. RP022A-16SUS, RP015B-14AFR and CEBAR RU006-2017) provided.

Conflicts of Interest: The authors declare no conflict of interest.

References

1. Zakaria, Z.A.; Rahim, M.H.A.; Mohtarrudin, N.; Kadir, A.A.; Cheema, M.S.; Ahmad, Z.; Mooi, C.S.; Tohid, S.F.M. Acute and sub-chronic oral toxicity studies of methanol extract of *Clinacanthus nutans* in mice. *Afr. J. Tradit. Complement. Altern. Med.* **2016**, *13*, 210–222. [CrossRef]

2. Uawonggul, N.; Chaveerach, A.; Thammasirirak, S.; Arkaravichien, T.; Chuachan, C.; Daduang, S. Screening of plants acting against *Heterometrus laoticus* scorpion venom activity on fibroblast cell lysis. *J. Ethnopharmacol.* **2006**, *103*, 201–207. [CrossRef] [PubMed]

3. Kamarudin, M.N.A.; Sarker, M.M.R.; Kadir, H.A.; Ming, L.C. Ethnopharmacological uses, phytochemistry, biological activities, and therapeutic applications of *Clinacanthus nutans* (Burm. f.) Lindau: A comprehensive review. *J. Ethnopharmacol.* **2017**, *206*, 245–266. [CrossRef] [PubMed]

4. Yong, Y.K.; Tan, J.J.; Teh, S.S.; Mah, S.H.; Ee, G.C.L.; Chiong, H.S.; Ahmad, Z. *Clinacanthus nutans* extracts are antioxidant with antiproliferative effect on cultured human cancer cell lines. *Evid.-Based Complement. Altern. Med.* **2013**, *2013*, 462751. [CrossRef] [PubMed]

5. Alam, A.; Ferdosh, S.; Ghafoor, K.; Hakim, A.; Juraimi, A.S.; Khatib, A.; Sarker, Z.I. *Clinacanthus nutans*: A review of the medicinal uses, pharmacology and phytochemistry. *Asian Pac. J. Trop. Med.* **2016**, *9*, 402–409. [CrossRef] [PubMed]

6. Arullappan, S.; Rajamanickam, P.; Thevar, N.; Kodimani, C.C. In vitro screening of cytotoxic, antimicrobial and antioxidant activities of *Clinacanthus nutans* (Acanthaceae) leaf extracts. *Trop. J. Pharm. Res.* **2014**, *13*, 1455–1461. [CrossRef]

7. Wanikiat, P.; Panthong, A.; Sujayanon, P.; Yoosook, C.; Rossi, A.G.; Reutrakul, V. The anti-inflammatory effects and the inhibition of neutrophil responsiveness by *Barleria lupulina* and *Clinacanthus nutans* extracts. *J. Ethnopharmacol.* **2008**, *116*, 234–244. [CrossRef] [PubMed]

8. Farooqui, M.; Hassali, M.A.; Shatar, A.K.A.; Farooqui, M.A.; Saleem, F.; ul Haq, N.; Othman, C.N. Use of complementary and alternative medicines among Malaysian cancer patients: A descriptive study. *J. Tradit. Complement. Med.* **2016**, *6*, 321–326. [CrossRef] [PubMed]

9. Jordan, B.R. Plant pigments and protection against UV-B radiation. In *Plant Pigments Their Manipulation*; Blackwell: Oxford, UK, 2004; pp. 275–292.

10. Azmir, J.; Zaidul, I.; Rahman, M.; Sharif, K.; Mohamed, A.; Sahena, F.; Jahurul, M.; Ghafoor, K.; Norulaini, N.; Omar, A. Techniques for extraction of bioactive compounds from plant materials: A review. *J. Food Eng.* **2013**, *117*, 426–436. [CrossRef]

11. Gürses, A.; Açıkyıldız, M.; Güneş, K.; Gürses, M.S. Colorants in Health and Environmental Aspects. In *Dyes and Pigments*; Springer International Publishing: Cham, Switzerland, 2016; pp. 69–83.

12. Khoo, H.E.; Azlan, A.; Tang, S.T.; Lim, S.M. Anthocyanidins and anthocyanins: Colored pigments as food, pharmaceutical ingredients, and the potential health benefits. *Food Nutr. Res.* **2017**, *61*, 1361779. [CrossRef] [PubMed]

13. Ananga, A.; Georgiev, V.; Ochieng, J.; Phills, B.; Tsolova, V. Production of anthocyanins in grape cell cultures: A potential source of raw material for pharmaceutical, food, and cosmetic industries. In *The Mediterranean Genetic Code-Grapevine and Olive*; In Tech: Vienna, Austria, 2013.

14. Jacobo-Herrera, N.J.; Jacobo-Herrera, F.E.; Zentella-Dehesa, A.; Andrade-Cetto, A.; Heinrich, M.; Pérez-Plasencia, C. Medicinal plants used in Mexican traditional medicine for the treatment of colorectal cancer. *J. Ethnopharmacol.* **2016**, *179*, 391–402. [CrossRef] [PubMed]

15. Giusti, M.M.; Wallace, T.C. Flavonoids as Natural Pigments. In *Handbook of Natural Colorants*; John Wiley & Sons, Ltd.: Chichester, UK, 2009; pp. 255–275.

16. Mateus, N.; de Freitas, V. Anthocyanins as Food Colorants. In *Anthocyanins: Biosynthesis, Functions, and Applications*; Winefield, C., Davies, K., Gould, K., Eds.; Springer: New York, NY, USA, 2009; pp. 284–304.

17. Kim, S.H. *Functional Dyes*; Elsevier Science: Amsterdam, The Netherlands, 2006.

18. Ayalew, W.A.; Ayele, D.W. Dye-sensitized solar cells using natural dye as light-harvesting materials extracted from *Acanthus sennii* chiovenda flower and *Euphorbia cotinifolia* leaf. *J. Sci. Adv. Mater. Devices* **2016**, *1*, 488–494. [CrossRef]

19. Mustroph, H.; Stollenwerk, M.; Bressau, V. Current Developments in Optical Data Storage with Organic Dyes. *Angew. Chem. Int. Ed.* **2006**, *45*, 2016–2035. [CrossRef] [PubMed]

20. El-Shishtawy, R.M. Functional Dyes, and Some Hi-Tech Applications. *Int. J. Photoenergy* **2009**, *2009*, 434897. [CrossRef]

21. Ramasamy, P.K.; Suresh, S.N. Effect of vermicompost on root numbers and length of sunflower plant (*Helianthus annuus* L.). *J. Pure Appl. Microbiol.* **2010**, *4*, 297–302.

22. Chiluvuru, N.; Tartte, V.; Kalla, C.M.; Kommalapati, R. Plant bioassay for assessing the effects of vermicompost on growth and yield of *Vigna radiata* and *Centella asiatica*, two important medicinal plants. *J. Dev. Sustain. Agric.* **2009**, *4*, 160–164.

23. Sundararasu, K.; Neelanarayanan, P. Effect of vermicompost and inorganic fertilizer on the growth and yield of tomato, *Lycorpersium esculentum* L. *Int. J. Curr. Res.* **2012**, *4*, 049–051.

24. Kashem, M.A.; Sarker, A.; Hossain, I.; Islam, M.S. Comparison of the effect of vermicompost and inorganic fertilizers on vegetative growth and fruit production of tomato (*Solanum lycopersicum* L.). *Open J. Soil Sci.* **2015**, *5*, 53. [CrossRef]

25. Yang, H.S.; Peng, T.W.; Madhavan, P.; Abdul, M.S.S.; Akowuah, G.A. Phytochemical Analysis and Antibacterial activity of Methanolic extract of *Clinacanthus nutans* Leaf. *Int. J. Drug Dev. Res.* **2013**, *5*, 349–355.

26. Zucco, M.A.; Walters, S.A.; Chong, S.-K.; Klubek, B.P.; Masabni, J.G. Effect of soil type and vermicompost applications on tomato growth. *Int. J. Recycl. Org. Waste Agric.* **2015**, *4*, 135–141. [CrossRef]

27. Cabanas-Echevarría, M.; Torres–García, A.; Díaz-Rodríguez, B.; Ardisana, E.F.H.; Creme-Ramos, Y. Influence of three bioproducts of organic origin on the production of two banana clones (*Musa* spp. AAB.) obtained by tissue cultures. *Alimentaria* **2005**, *369*, 111–116.

28. Acevedo, I.C.; Pire, R. Effects of vermicompost as substrate amendment on the growth of papaya (*Carica papaya* L.). *Interciencia* **2004**, *29*, 274–279.

29. Sardoei, A.S.; Roien, A.; Sadeghi, T.; Shahadadi, F.; Mokhtari, T.S. Effect of Vermicompost on the Growth and Flowering of African Marigold (*Tagetes erecta*). *Am.-Euras. J. Agric. Environ. Sci.* **2014**, *14*, 631–635.

30. Chattopadhyay, A. Effect of Vermiwash and Vermicompost on an Ornamental Flower, *Zinnia* sp. *J. Hortic.* **2014**, *1*, 112. [CrossRef]

31. Donald, D.G.M.; Visser, L.B. Vermicompost as a possible growth medium for the production of commercial forest nursery stock. *Appl. Plant Sci.* **1989**, *3*, 110–113.

32. Lazcano, C.; Sampedro, L.; Zas, R.; Domínguez, J. Vermicompost enhances germination of the maritime pine (*Pinus pinaster* Ait.). *New For.* **2010**, *39*, 387–400. [CrossRef]

33. Lazcano, C.; Sampedro, L.; Zas, R.; Domínguez, J. Assessment of plant growth promotion by vermicompost in different progenies of maritime pine (*Pinus pinaster* Ait.). *Compost Sci. Util.* **2010**, *18*, 111–118. [CrossRef]

34. Solihah, M.; Wan Rosli, W.; Nurhanan, A. Phytochemicals screening and total phenolic content of Malaysian *Zea mays* hair extracts. *Int. Food Res. J.* **2012**, *19*, 1533–1538.

35. Giusti, M.M.; Wrolstad, R.E. Characterization and measurement of anthocyanins by UV-visible spectroscopy. In *Current Protocols in Food Analytical Chemistry*; John Wiley and Sons, Inc.: Hoboken, NJ, USA, 2001.

36. Sun, T.; Powers, J.R.; Tang, J. Effect of enzymatic macerate treatment on rutin content, antioxidant activity, yield, and physical properties of asparagus juice. *J. Food Sci.* **2007**, *72*, S267–S271. [CrossRef] [PubMed]

37. Lin, J.-Y.; Tang, C.-Y. Determination of total phenolic and flavonoid contents in selected fruits and vegetables, as well as their stimulatory effects on mouse splenocyte proliferation. *Food Chem.* **2007**, *101*, 140–147. [CrossRef]

38. Sulaiman, S.F.; Ooi, K.L. Polyphenolic and vitamin C contents and antioxidant activities of aqueous extracts from mature-green and ripe fruit fleshes of *Mangifera* sp. *J. Agric. Food Chem.* **2012**, *60*, 11832–11838. [CrossRef] [PubMed]

39. Shao, P.; Chen, X.; Sun, P. Chemical characterization, antioxidant and antitumor activity of sulfated polysaccharide from *Sargassum horneri. Carbohydr. Polym.* **2014**, *105*, 260–269. [CrossRef] [PubMed]

40. Abdul Rahim, M.H.; Zakaria, Z.A.; Mohd Sani, M.H.; Omar, M.H.; Yakob, Y.; Cheema, M.S.; Ching, S.M.; Ahmad, Z.; Abdul Kadir, A. Methanolic extract of *Clinacanthus nutans* exerts antinociceptive activity via the opioid/nitric oxide-mediated, but cGMP-independent, pathways. *Evid.-Based Complement. Altern. Med.* **2016**, *2016*, 1494981. [CrossRef] [PubMed]

41. Molyneux, P. The use of the stable free radical diphenylpicrylhydrazyl (DPPH) for estimating antioxidant activity. *Songklanakarin J. Sci. Technol.* **2004**, *26*, 211–219.

42. Almeida, M.M.B.; de Sousa, P.H.M.; Arriaga, Â.M.C.; do Prado, G.M.; de Carvalho Magalhães, C.E.; Maia, G.A.; de Lemos, T.L.G. Bioactive compounds and antioxidant activity of fresh exotic fruits from northeastern Brazil. *Food Res. Int.* **2011**, *44*, 2155–2159. [CrossRef]

43. Arnao, M.B. Some methodological problems in the determination of antioxidant activity using chromogen radicals: A practical case. *Trends Food Sci. Technol.* **2000**, *11*, 419–421. [CrossRef]

44. Prior, R.L.; Wu, X.; Schaich, K. Standardized methods for the determination of antioxidant capacity and phenolics in foods and dietary supplements. *J. Agric. Food Chem.* **2005**, *53*, 4290–4302. [CrossRef] [PubMed]

45. Bondet, V.; Brand-Williams, W.; Berset, C. Kinetics and mechanisms of antioxidant activity using the DPPH. free radical method. *LWT-Food Sci. Technol.* **1997**, *30*, 609–615. [CrossRef]

46. Kim, D.-O.; Lee, K.W.; Lee, H.J.; Lee, C.Y. Vitamin C equivalent antioxidant capacity (VCEAC) of phenolic phytochemicals. *J. Agric. Food Chem.* **2002**, *50*, 3713–3717. [CrossRef] [PubMed]

47. Coria-Cayupán, Y.S.; Sánchez de Pinto, M.A.I.; Nazareno, M.N.A. Variations in bioactive substance contents and crop yields of lettuce (*Lactuca sativa* L.) cultivated in soils with different fertilization treatments. *J. Agric. Food Chem.* **2009**, *57*, 10122–10129. [CrossRef] [PubMed]

48. Pant, A.; Radovich, T.J.K.; Hue, N.V.; Arancon, N.Q. Effects of vermicompost tea (aqueous extract) on pak choi yield, quality, and on soil biological properties. *Compost Sci. Util.* **2011**, *19*, 279–292. [CrossRef]

49. Luján-Hidalgo, M.C.; Pérez-Gómez, L.E.; Abud-Archila, M.; Meza-Gordillo, R.; Ruiz-Valdiviezo, V.M.; Dendooven, L.; Gutiérrez-Miceli, F.A. Growth, phenolic content and antioxidant activity in Chincuya (*Annona purpurea* Moc & Sesse ex Dunal) cultivated with vermicompost and phosphate rock. *Compost Sci. Util.* **2015**, *23*, 276–283.

50. Ksouri, R.; Megdiche, W.; Debez, A.; Falleh, H.; Grignon, C.; Abdelly, C. Salinity effects on polyphenol content and antioxidant activities in leaves of the halophyte *Cakile maritima. Plant Physiol. Biochem.* **2007**, *45*, 244–249. [CrossRef] [PubMed]

51. Downey, P.J.; Levine, L.H.; Musgrave, M.E.; McKeon-Bennett, M.; Moane, S. Effect of hypergravity and phytohormones on isoflavonoid accumulation in soybean (*Glycine max* L.) callus. *Micrograv. Sci. Technol.* **2013**, *25*, 9–15. [CrossRef]

52. Adhikary, S. Vermicompost, the story of organic gold: A review. *Agric. Sci.* **2012**, *3*, 905. [CrossRef]

53. Moldovan, B.; Popa, A.; David, L. Effects of storage temperature on the total phenolic content of Cornelian Cherry (*Cornus mas* L.) fruits extracts. *J. Appl. Bot. Food Qual.* **2016**, *89*. [CrossRef]

54. Del-Toro-Sánchez, C.L.; Gutiérrez-Lomelí, M.; Lugo-Cervantes, E.; Zurita, F.; Robles-García, M.A.; Ruiz-Cruz, S.; Aguilar, J.A.; Rio, M.-D.; Alfredo, J.; Guerrero-Medina, P.J. Storage effect on phenols and on the antioxidant activity of extracts from *Anemopsis californica* and inhibition of elastase enzyme. *J. Chem.* **2015**, *2015*, 602136. [CrossRef]

55. Ferrante, A.; Maggiore, T. Chlorophyll a fluorescence measurements to evaluate storage time and temperature of Valeriana leafy vegetables. *Postharvest Biol. Technol.* **2007**, *45*, 73–80. [CrossRef]

56. Varela-Santos, E.; Ochoa-Martinez, A.; Tabilo-Munizaga, G.; Reyes, J.E.; Pérez-Won, M.; Briones-Labarca, V.; Morales-Castro, J. Effect of high hydrostatic pressure (HHP) processing on physicochemical properties, bioactive compounds and shelf-life of pomegranate juice. *Innov. Food Sci. Emerg. Technol.* **2012**, *13*, 13–22. [CrossRef]

57. Serea, C.; Barna, O.; Manley, M.; Kidd, M. Effect of storage temperature on the ascorbic acid content, total phenolic content and antioxidant activity in lettuce (*Lactuca sativa* L.). *J. Anim. Plant Sci.* **2014**, *24*, 1173–1177.

58. Lee, C.Y.; Kagan, V.; Jaworski, A.W.; Brown, S.K. Enzymic browning in relation to phenolic compounds and polyphenoloxidase activity among various peach cultivars. *J. Agric. Food Chem.* **1990**, *38*, 99–101. [CrossRef]

59. Janovitz-Klapp, A.H.; Richard, F.C.; Goupy, P.M.; Nicolas, J.J. Kinetic studies on apple polyphenol oxidase. *J. Agric. Food Chem.* **1990**, *38*, 1437–1441. [CrossRef]

60. Kim, D.O.; Padilla-Zakour, O.I. Jam processing effect on phenolics and antioxidant capacity in anthocyanin-rich fruits: Cherry, plum, and raspberry. *J. Food Sci.* **2004**, *69*, S395–S400. [CrossRef]

61. He, J. Isolation of Anthocyanin Mixtures from Fruits and Vegetables and Evaluation of Their Stability, Availability and Biotransformation in the Gastrointestinal Tract. Ph.D. Thesis, The Ohio State University, Columbus, OH, USA, 2008.

62. Shen, W.; Nada, K.; Tachibana, S. Involvement of polyamines in the chilling tolerance of cucumber cultivars. *Plant Physiol.* **2000**, *124*, 431–440. [CrossRef] [PubMed]

63. Malik, A.; Holm, L.; Johansson, E. Soil and starter fertilizer and its effect on yield and protein composition of malting barley. *J. Soil Sci. Plant Nutr.* **2012**, *12*, 835–849. [CrossRef]

64. Trombley, J.D.; Loegel, T.N.; Danielson, N.D.; Hagerman, A.E. Capillary electrophoresis methods for the determination of covalent polyphenol–protein complexes. *Anal. Bioanal. Chem.* **2011**, *401*, 1523–1529. [CrossRef] [PubMed]

65. Le Bourvellec, C.; Renard, C. Interactions between polyphenols and macromolecules: Quantification methods and mechanisms. *Crit. Rev. Food Sci. Nutr.* **2012**, *52*, 213–248. [CrossRef] [PubMed]

66. Ozdal, T.; Capanoglu, E.; Altay, F. A review on protein–phenolic interactions and associated changes. *Food Res. Int.* **2013**, *51*, 954–970. [CrossRef]

Sample Availability: Samples of the compounds are not available from the authors.

molecules

MDPI

Article

Anthocyanins of Coloured Wheat Genotypes in Specific Response to SalStress

Sonia Mbarki [1,2,†], Oksana Sytar [3,4,*,†], Marek Zivcak [4], Chedly Abdelly [2], Artemio Cerda [5] and Marian Brestic [4]

1   National Research Institute of Rural Engineering, Water and Forests (INRGREF), BP 10, Aryanah 2080, Tunisia; mbarkisonia14@gmail.com
2   Laboratory of Plant Extremophiles, Biotechnology Center at the Technopark of Borj-Cedria Tunisia, BP 901, Hammam Lif 2050, Tunisia; chedly.abdelly@cbbc.rnrt.tn
3   Department of Plant Biology, Institute of Biology, Kiev National University of Taras Shevchenko, Volodymyrska St, 64, 02000 Kyiv, Ukraine; oksana.sytar@gmail.com
4   Department of Plant Physiology, Slovak University of Agriculture, Nitra, Tr. A. Hlinku 2, 949 01 Nitra, Slovak Republic; marek.zivcak@uniag.sk (M.Z.); marian.brestic@uniag.sk (M.B.)
5   Departament of Geografy, University of València, Blasco Ibàñez, 28, 46010 Valencia, Spain; artemio.cerda@uv.es
*   Correspondence: oksana.sytar@gmail.com; Tel.: +421-37-3414-822
†   Sonia Mbarki and Oksana Sytar equally contributed to this work.

Received: 16 May 2018; Accepted: 21 June 2018; Published: 23 June 2018

Abstract: The present study investigated the effect of salt stress on the development of adaptive responses and growth parameters of different coloured wheat genotypes. The different coloured wheat genotypes have revealed variation in the anthocyanin content, which may affect the development of adaptive responses under increasing salinity stress. In the early stage of treatment with salt at a lower NaCl concentration (100 mM), anthocyanins and proline accumulate, which shows rapid development of the stress reaction. A dose-dependent increase in flavonol content was observed for wheat genotypes with more intense purple-blue pigmentation after treatment with 150 mM and 200 mM NaCl. The content of Na^+ and K^+ obtained at different levels of salinity based on dry weight (DW) was more than 3 times greater than the control, with a significant increase of both ions under salt stress. Overall, our results demonstrated that coloured wheat genotypes with high anthocyanin content are able to maintain significantly higher dry matter production after salt stress treatment.

Keywords: salinity; anthocyanins; proline; MDA; flavonol; wheat

1. Introduction

Land degradation has become currently a strong issue at a global level as its spatial distribution varies from 1 billion ha to over 6 billion ha [1]. Climate change, high salt levels in soils and irrigation waters are major environmental concerns, and a problem for agriculture in many arid and semi-arid regions [2]. Approximately 15% of the cultivated land has an excess of salt [3], and large quantities of water are of very poor quality. Globally, no less than 10 million hectares of agricultural land are abandoned annually because of salinity. Salt-induced land degradation is a very important issue affecting the status of food productivity worldwide [2–4].

Salinity can increase the toxic levels of some ions and cause water stress and malnutrition in plants. The difference in salt tolerance may be connected with plant ontogenetic stages and plant species. It is usually visible in the reduction of biomass and yield or in decreasing survival rates [5,6]. In plants, osmotic stress and ionic toxicity are two stresses that follow salinity and can be destructive under prolonged treatment [5]. Plants respond to these variations in soil salinity by provoking resistance mechanisms [7,8]. Among known mechanisms, osmotic adjustment plays a major role in plant tolerance to stress [9]. A high level of Na^+ compromises carbon fixation, which supports an over-reduction of light-harvesting complexes, reducing photosynthesis and the production of reactive oxygen (ROS) [10]. This process generates a disturbance in ionic homeostasis and/or nutritional imbalance [11].

Under stress conditions, secondary metabolite accumulation often occurs in plants [12]. Secondary metabolites may participate as signal molecules and some specific elicitors. Ionic and osmotic stresses, which are part of salt stress in plants, stimulate the accumulation or decrease of specific secondary metabolites in plants [13]. Anthocyanins are increased during the salt stress response [14]. At the same time, salt stress is able to decrease the level of anthocyanins in salt-sensitive plant species [14]. Flavonoids are one of the largest groups of plant phenolics that participate in plant defence [15] via an effective strategy against ROS [16]. Under salt stress, *Azospirillum brasilense* showed higher secretion of nod-gene-inducing flavonoid species and improved root branching in the seedling roots of bean plants [17]. Other salt-sensitive crops like potato, eggplant, pepper, cabbage, lettuce, rise and maize shows their sensitivity to hyperosmotic stresses by the rise of phenolics, as well [18–21]. Under drought and salinity stresses, wheat plants raise antioxidant defence mechanisms under abiotic stresses to alleviate oxidative damage [22,23].

It was suggested that accession-dependent capacity to stimulate salt-stress response antioxidative mechanisms may affect a corresponding changeability for the growth sustainability of some plants. The high leaf antioxidant activity and total phenolic content have been estimated in the halophyte *Cakile maritima* leaves [24]. The wheat genotypes demonstrated a wide range of responses to high salinity stress during the growth stages [25]. The effect of salt stress on wheat plants is one of the most serious problems in arid and semi-arid regions, which reduces both the quantity and quality of the production of this cereal [26].

The response to the salt stress of plant species depends on the plant species and varieties, salt concentration, growing conditions and stage of development of the plant. The identification of salt-tolerant varieties can positively develop the production of areas at risk or irrigated with salinized water and would be of obvious interest in helping to improve varieties.

Therefore, the purpose of this work was to study the effect of salt stress on the development of adaptive responses and growth parameters of different wheat cultivars (coloured and not coloured) to estimate parameters that can help to choose salt-tolerant wheat cultivars.

2. Results

2.1. SFR, ANTH, FLAV and MFI

The results of the presented experimental work showed that the SFR values at the first stage of treatment with a concentration of 100 mM increased in two genotypes, Citrus yellow and PS Karkulka. Under 150 mM NaCl, increasing SFR values have been observed in the non-coloured genotypes (Citrus yellow) at the first stage of treatment and continue in the second stage of treatment. For the two coloured genotypes, KM 178-14 purple and Skorpion Blue aleurone under the first stage of treatment with 100 mM and 150 mM NaCl showed no significant changes compared to the control. At the second stage of treatment, with increasing salt concentration (150 mM and 200 mM NaCl), the SFR value decreased (Figure 1).

Flavonol accumulation, which is also part of the adaptive reaction to salt stress, was observed slightly after the first stage of treatment with NaCl concentrations of 100 mM and 150 mM. At the second stage of treatment with 150 mM and 200 mM NaCl, there was a significant difference in the flavonol reaction between the non-coloured and coloured wheat genotypes (Figure 2A). A dose-dependent increase in flavonol content was observed for the KM 178-14 purple and Skorpion Blue aleurone wheat genotypes after treatment with 150 mM and 200 mM NaCl, respectively (Figure 2B).

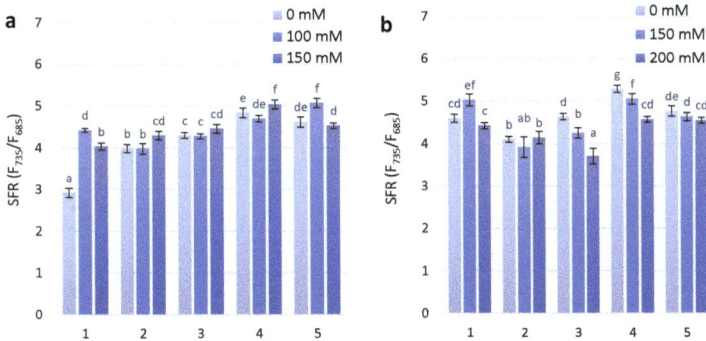

Figure 1. Values of SFR (simple fluorescence ratio) in the leaves of wheat plant genotypes exposed to salt stress during 30 days after seedlings (**a**)—first stage of treatment; (**b**)—second stage of treatment. The numbers indicate individual cultivars of wheat as follow: 1—Citrus yellow, 2—KM 53-14 Blue, 3—KM 178-14 purple, 4—Skorpion Blue aleurone, 5—PS Karkulka purple. The columns represent the mean values ± S.E. for six replicates. Statistically significant differences among treatments at each time are indicated by different small letters (Duncan test, $p < 0.05$).

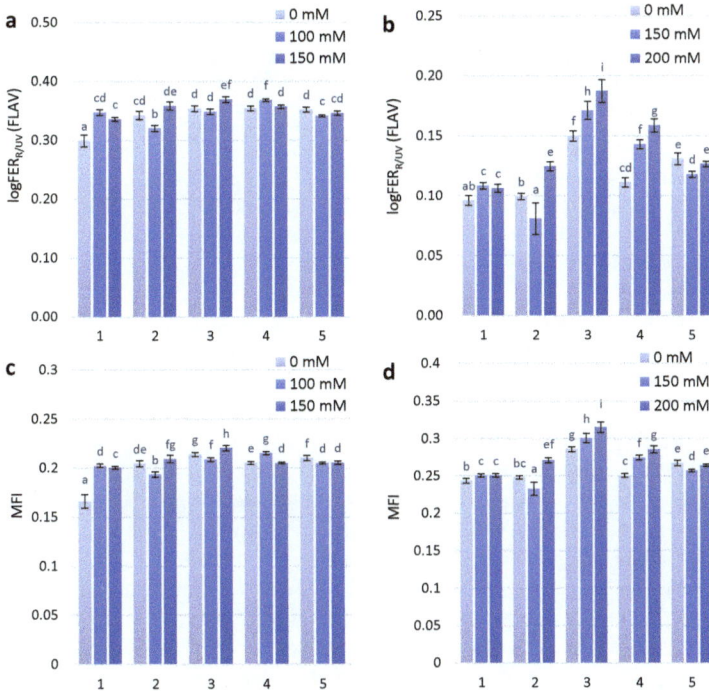

Figure 2. Flavonols accumulation (logFERR/UV-FLAV) in the leaves of investigated wheat plant genotypes exposed to salt stress during 30 days after seedlings (**a**)—first stage of treatment; (**b**)—second stage of treatment. MFI parameter in the leaves of investigated wheat plant genotypes exposed to salt stress during 30 days after seedlings (**c**)—first stage of treatment; (**d**)—second stage of treatment. The numbers shown individual cultivars of wheat: 1—Citrus yellow, 2—KM 53-14 Blue, 3—KM 178-14 purple, 4—Skorpion Blue aleurone, 5—PS Karkulka purple The columns represent the mean values ± S.E. for six replicates. Statistically significant differences among treatments at each time are indicated by different small letters (Duncam test, $p < 0.05$).

The modified flavonoid index (MFI), which takes into consideration the accumulation of both flavonols and anthocyanins, is a better estimate of flavonoids than FLAV. The results of statistical analyses using MFI are highly similar to those that use biochemical analysis data [27]. The MFI index showed a similar dose-dependent tendency as the FLAV index after the second stage of treatment (Figure 2D).

2.2. Anthocyanins Content

The experimental wheat genotypes have been characterized by different anthocyanin contents (Figure 3), which could affect the specific development of adaptive responses to increasing salinity stress [25].

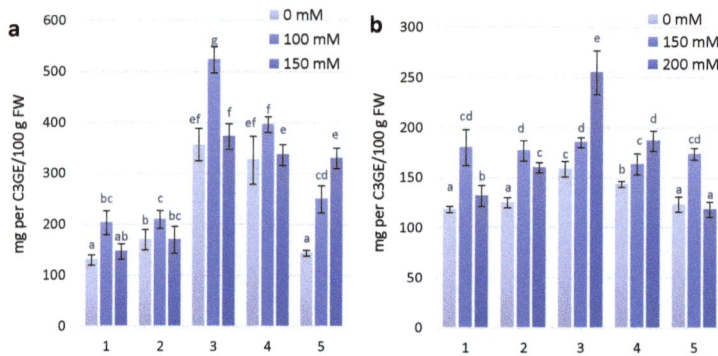

Figure 3. Anthocyanins content in wheat leaves under effect of different NaCl concentrations (**a**)—first stage of treatment; (**b**)—second stage of treatment. Wheat genotypes: 1—Citrus yellow, 2—KM 53-14 Blue, 3—KM 178-14 purple, 4—Skorpion Blue aleurone, 5—PS Karkulka purple. The columns represent the mean values ± S.E. for six replicates. Statistically significant differences among treatments at each time are indicated by different small letters (Duncan test, $p < 0.05$).

Coloured plant genotypes that have high anthocyanin content are a good object to study abiotic tolerance, especially salt-tolerant traits. Two wheat genotypes, KM 178-14 purple and Skorpion Blue aleurone, had the highest anthocyanin content in the leaves among all experimental variants. At the first stage of the experimental treatment with 100 mM NaCl, increasing anthocyanin content was observed in almost all experimental variants. Under treatment with 150 mM NaCl, anthocyanin content was similar to the control level in all experimental variants, except for the wheat genotype PS Karkulka. The higher content of anthocyanins did not result in development of the same adaptation to salt stress in the genotypes. At the second stage of the experimental treatment, genotypes (KM 178-14 purple and Skorpion Blue aleurone) with higher anthocyanin content at 100 mM NaCl had no changes in anthocyanin content. The genotype KM 178-14 purple showed a significant increase in anthocyanin content of 30% at 200 mM NaCl compared to the control. Other genotypes (Citrus yellow, KM 53-14 Blue and PS Karkulka) have been characterized by an increase in anthocyanins at 150 mM NaCl, but at 200 mM NaCl, they were similar to the level of the control.

2.3. Proline Content

The results of the presented experimental work found that wheat genotypes characterized by increased anthocyanin content in the first and second stages of salt treatment show increasing proline content. In the first stage, a significant increase in proline content was observed for the wheat genotype KM 178-14 purple (Figure 4A). Under the second stage of treatment with 100 mM NaCl, a significant increase in proline content was observed for the KM 53-14 Blue and KM 178-14 purple wheat genotypes. However, under the 200 mM NaCl treatment in the second stage of treatment, all experimental genotypes showed increased proline content, especially the coloured cultivars KM 178-14 purple and Skorpion Blue aleurone (Figure 4B).

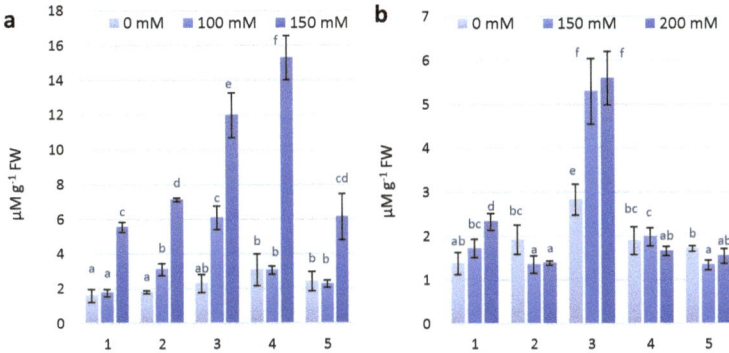

Figure 4. Proline content in wheat leaves under effect of different NaCl concentrations (**a**)—first stage of treatment; (**b**)—second stage of treatment. Wheat genotypes: 1—Citrus yellow, 2—KM 53-14 Blue, 3—KM 178-14 purple, 4—Skorpion Blue aleurone, 5—PS Karkulka purple. The columns represent the mean values ± S.E. for six replicates. Statistically significant differences among treatments at each time are indicated by different small letters (Duncan test, $p < 0.05$).

2.4. Lipid Peroxidation Assay

Lipid peroxidation (LP) is a biochemical marker for ROS-mediated damage. MDA is one of the better known secondary metabolite of LP [28,29]. The results of the experimental work consisting of two stages of NaCl treatment have shown the tendency to increase the MDA level in KM 53-14 Blue and KM 178-14 purple wheat genotypes, whereas other wheat genotypes were almost on a control level after the first treatment (Figure 5A). The significant increase of MDA accumulation was observed in the second treatment with concentration of 200 mM NaCl in all experimental variants (Figure 5B). Under the concentration 100 mM, almost all experimental variants have been characterized by MDA content equal to the control level, except of the genotypes Skorpion Blue aleurone and PS Karkula purple, in which we observed the decreasing tendency of MDA.

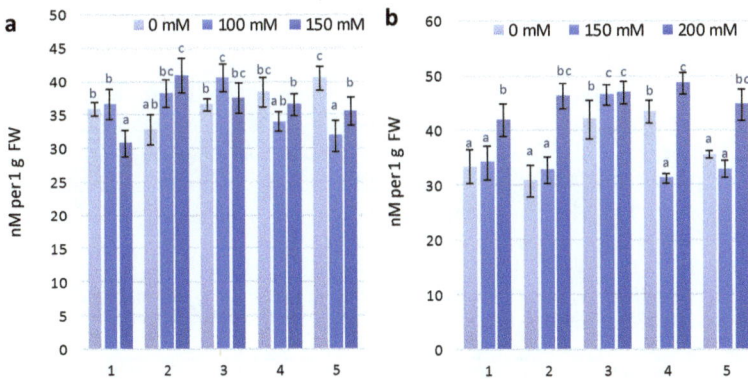

Figure 5. MDA content in wheat leaves under effect of different NaCl concentrations (**a**)—first stage of treatment; (**b**)—second stage of treatment. Wheat genotypes: 1—Citrus yellow, 2—KM 53-14 Blue, 3—KM 178-14 purple, 4—Skorpion Blue aleurone, 5—PS Karkulka purple. The columns represent the mean values ± S.E. for six replicates. Statistically significant differences among treatments at each time are indicated by different small letters (Duncan test, $p < 0.05$).

2.5. Plant Biomass

It has been suggested that stimulation of antioxidative responses under salt stress may result in comparable variability in growth sustainability [23]. An increase in the dry weight of roots under both concentrations of NaCl treatment was observed for the wheat genotype KM 178-14 purple. At a concentration of 200 mM NaCl, a higher dry weight was observed more than two times compared to the control for genotypes KM 178-14 purple, Skorpion blue aleurone, PS Karkulka (Table 1). These genotypes were also characterized by a high anthocyanin content in the control variants and different anthocyanin changes during the development of salt stress (Figure 2). The presence of anthocyanins as antioxidants can partly affect the development of stress adaptation responses under salt stress.

Table 1. Fresh and dry mass after salt stress influence.

Wheat genotypes	Fresh Weight [1]			Dry Weight [1]		
	Roots					
	Control	150 mM	200 mM	Control	150 mM	200 mM
Citrus Yellow	0.048 ± 0.019 [ab]	$0.069 + 0.023$ [a]	0.23 ± 0.07 [b]	0.028 ± 0.009 [a]	0.019 ± 0.003 [ab]	0.017 ± 0.009 [b]
KM 53-14 Blue	0.056 ± 0.003 [b]	0.042 ± 0.004 [a]	0.086 ± 0.02 [c]	0.011 ± 0.005 [a]	0.015 ± 0.002 [a]	0.016 ± 0.008 [a]
KM 178-14 Purple	0.042 ± 0.001 [a]	0.050 ± 0.016 [a]	0.13 ± 0.05 [b]	0.006 ± 0.001 [a]	0.010 ± 0.004 [a]	0.013 ± 0.006 [a]
Skorpion Blue Aleur.	0.091 ± 0.022 [a]	0.038 ± 0.014 [b]	0.079 ± 0.025 [a]	0.007 ± 0.003 [a]	0.007 ± 0.001 [a]	0.020 ± 0.012 [a]
PS Karkulka	0.046 ± 0.012 [b]	0.039 ± 0.014 [b]	0.228 ± 0.091 [a]	0.016 ± 0.006 [ab]	0.011 ± 0.002 [b]	0.023 ± 0.007 [a]
	Shoots					
	Control	150 mM	200 mM	Control	150 mM	200 mM
Citrus yellow	0.278 ± 0.076 [a]	0.297 ± 0.085 [a]	0.233 ± 0.073 [a]	0.038 ± 0.011 [ab]	0.043 ± 0.014 [a]	0.019 ± 0.008 [b]
KM 53-14 Blue	0.194 ± 0.117 [ab]	0.187 ± 0.028 [a]	0.086 ± 0.032 [b]	0.028 ± 0.014 [a]	0.028 ± 0.024 [a]	0.018 ± 0.004 [a]
KM 178-14 Purple	0.252 ± 0.062 [a]	0.188 ± 0.06 [a]	0.127 ± 0.047 [b]	0.032 ± 0.009 [a]	0.032 ± 0.012 [ab]	0.015 ± 0.007 [b]
Skorpion Blue Aleur.	0.556 ± 0.085 [a]	0.283 ± 0.071 [b]	0.079 ± 0.035 [c]	0.025 ± 0.009 [a]	0.042 ± 0.009 [a]	0.074 ± 0.013 [b]
PS Karkulka	0.209 ± 0.069 [a]	0.241 ± 0.063 [a]	0.228 ± 0.052 [a]	0.030 ± 0.011 [a]	0.042 ± 0.009 [a]	0.046 ± 0.015 [a]

[1] Values are mean $\pm$ S.E. for six replicates. Statistically significant differences among treatments at each time within one genotype are indicated by different small letters (Duncan test; $p < 0.05$).

The dry weight of shoot biomass after salt stress influence increased again in Skorpion blue aleurone and PS Karkulka, but for genotype KM 178-14, the purple dry weight of shoots was on the control level. The genotype Citrus yellow, with a lower anthocyanin content, was characterized by a decrease in dry weight under both NaCl concentrations (150 mM and 200 mM). The genotype KM, 53-14 Blue demonstrated decreased shoot dry weight by 45% compared to the control after treatment with 200 mM NaCl.

2.6. Na+ and K+ Level

The content of Na+ and K+ obtained at different levels of salinity on the basis of DW was more than 3 times greater than the control, with a significant increase of both ions under salt stress. For some wheat genotypes, a significant increase in Na+ and K+ was observed under treatment with 150 mM NaCl (KM 53-14 Blue and KM 178-14 purple) and under treatment with 200 mM NaCl (Skorpion Blue aleurone and PS Karkulka). The Na+/K+ ratio was decreased in all experimental wheat genotypes (Table 2).

Table 2. Effects of different levels of salinity on total dry weight, shoot water content after 30 days of sowing of investigated wheat cultivars.

Wheat Cultivars [1]	Salinity Level (mM NaCl)	Water Content (g H$_2$O g^{-1} DW)	Total Dry Weight (g/plant)	Na+ mmol/g DW	K+ mmol/g DW	K+/Na+ Ratio
	0 mM	6.30 ± 0.650 [b]	0.066 ± 0.016 [a]	0.692 ± 0.007 [b]	0.179 ± 0.002 [b]	0.259
Citrus Yellow	150 mM	5.99 ± 0.588 [b]	0.062 ± 0.017 [a]	1.602 ± 0.097 [a]	0.333 ± 0.020 [a]	0.208
	200 mM	12.15 ± 5.24 [a]	0.036 ± 0.017 [a]	1.663 ± 0.129 [a]	0.345 ± 0.027 [a]	0.208

Table 2. *Cont.*

Wheat Cultivars [1]	Salinity Level (mM NaCl)	Water Content (g H$_2$O g^{-1} DW)	Total Dry Weight (g/plant)	Na$^+$ mmol/g DW	K$^+$ mmol/g DW	K$^+$/Na$^+$ Ratio
KM 53-14 Blue	0 mM	5.96 ± 1.50 ^{a}	0.039 ± 0.019 ^{a}	0.918 ± 0.085 ^{b}	0.238 ± 0.022 ^{b}	0.259
	150 mM	6.92 ± 0.341 ^{a}	0.043 ± 0.016 ^{a}	2.582 ± 0.080 ^{a}	0.508 ± 0.016 ^{a}	0.197
	200 mM	6.47 ± 5.04 ^{a}	0.034 ± 0.012 ^{a}	2.272 ± 0.162 ^{a}	0.447 ± 0.032 ^{a}	0.197
KM 178-14 Purple	0 mM	6.86 ± 0.91 ^{b}	0.038 ± 0.010 ^{a}	0.643 ± 0.054 ^{c}	0.148 ± 0.012 ^{c}	0.231
	150 mM	4.65 ± 1.66 ^{b}	0.042 ± 0.011 ^{a}	2.606 ± 0.138 ^{a}	0.455 ± 0.024 ^{a}	0.175
	200 mM	9.92 ± 2.79 ^{a}	0.056 ± 0.009 ^{a}	1.597 ± 0.093 ^{b}	0.295 ± 0.017 ^{b}	0.185
Skorpion Blue Aleurone	0 mM	22.11 ± 5.88 ^{a}	0.027 ± 0.012 ^{b}	0.664 ± 0.074 ^{b}	0.153 ± 0.017 ^{b}	0.185
	150 mM	5.63 ± 0.38 ^{c}	0.049 ± 0.010 ^{b}	1.636 ± 0.093 ^{a}	0.302 ± 0.017 ^{a}	0.185
	200 mM	3.98 ± 0.93 ^{b}	0.094 ± 0.025 ^{a}	1.791 ± 0.090 ^{a}	0.331 ± 0.017 ^{a}	0.231
PS Karkulka	0 mM	6.62 ± 1.73 ^{b}	0.046 ± 0.017 ^{a}	0.633 ± 0.023 ^{c}	0.146 ± 0.005 ^{b}	0.231
	150 mM	4.69 ± 0.46 ^{b}	0.053 ± 0.011 ^{a}	1.634 ± 0.093 ^{b}	0.301 ± 0.017 ^{a}	0.185
	200 mM	10.10 ± 3.67 ^{a}	0.069 ± 0.020 ^{a}	1.972 ± 0.063 ^{a}	0.345 ± 0.011 ^{a}	0.175

[1] Values are mean ± S.E. for six replicates. Statistically significant differences among treatments at each time within one genotype are indicated by different small letters ($p < 0.05$).

3. Discussion

In the present research, to discover the development of adaptive responses of different wheat genotypes under salt stress, non-destructive chlorophyll fluorescence techniques were used for the screening of biologically active compounds of a phenolic nature and some photosynthesis parameters. The majority of published vegetation indices for non-invasive remote sensing techniques are not responsive to fast changes in the status of plant photosynthesis under the effects of common environmental stressors [30]. The SFR index is connected with chlorophyll concentration in the leaves [27]. It was shown that under salt stress the SFR values at the first stage of treatment with a concentration of 100 mM in the two coloured genotypes, KM 178-14 purple and Skorpion Blue aleurone found no significant changes compared to the control. However, at the second stage of treatment, with increasing salt concentration, the SFR value decreased.

The level of photosynthetic pigments can be reduced under the effect of different stressful environments [30]. Plus the kinetics of pigments accumulation can be different at the two investigated developmental stages like 2-d-old and 10-d-old leaves [31] with differences between control variants at two investigated stages. The difference in salt tolerance may be connected with plant ontogenetic stages and plant species. Photosynthetic ability is significantly decreased in plants under salinity stress, which affects plant growth and development [32]. Decreasing photosynthetic ability by increased salt stress was connected to the inhibition of carbon uptake via specific metabolic processes, changes in photochemical capacity, lower stomatal conductance or a combination of these factors [33–35].

Flavonol accumulation, which is also a part of the adaptive reaction to salt stress, was observed after the first stage of treatment with NaCl (100 mM and 150 mM). After the second stage of treatment with 150 mM and 200 mM NaCl, we found a significant difference in the flavonol reaction between the coloured wheat genotypes (with higher anthocyanins content in grains) and non-coloured (with higher carotenoids content in grains). The FLAV and MFI indices have shown a similar dose-dependent tendency of flavonol content increase after the second stage of treatment in the coloured wheat genotypes. The grains of these wheat genotypes have been characterized by the presence of anthocyanins as well [36].

Intensive detoxification of reactive oxygen species (ROS) is a key factor in improving the tolerance of plants to salinity effects. An analysis of phenylpropanoid metabolism at the gene and enzyme levels showed that oxidative damage was lower when flavonols accumulated to a higher degree [37]. Therefore, the higher flavonol accumulation in coloured wheat genotypes (KM 178-14 purple and Skorpion Blue aleurone) showed faster development of adaptive responses to salt stress.

It was estimated that high contents of proline and anthocyanin support an active protective response to salinity [28]. At the first and second stages of salt treatment the coloured wheat genotypes characterized by higher anthocyanin content show increasing proline content. After first treatment the wheat genotypes KM 178-14 purple shows the highest increase in proline content. At the second stage of treatment with a concentration of 200 mM NaCl all experimental genotypes showed increased proline content.

Some studies have cited that the build-up of amino acid proline is an essential regulatory factor under salt stress and may be one of the mechanisms underlying their higher salt tolerance [38–40]. Matysik et al. (2002) showed that proline mitigates salt stress-induced increases in the carboxylase and oxygenase activities of Rubisco [41]. It protects plants from ROS damage by suppressing singlet oxygen. The early responses to salt and water stress have been found to be mostly equal [42]. Slama et al. (2015) found that in plants grown under salt stress, the elevation of the synthesis of proline can be regarded as an adaptation strategy in this species when exposed to salt stress [10]. Organic osmolytes can indirectly participate in osmotic adjustment by changing K^+ flow across membranes, thus reducing sodium-induced efflux of such inorganic osmolytes [42].

Increased lipid peroxidation (LP) and proline contents were observed to significantly accumulate in barley and *Catharanthus roseus* seedlings under salinity effects [43,44]. LP process changes depend on plant sensitivity and the degree of influence stressor. The effect of increasing NaCl doses on the accumulation of proline and LP can result in their high correlation with the tolerance capacity of plant cultivars [45,46]. Low MDA values, as an important indicator of a high oxidative damage-limiting capacity, were shown to be important for tolerating the salinity [46,47].

The results of our experiment showed the different development of an LP process in different wheat genotypes. It was found that in salt-stressed seedlings, MDA contents were negatively correlated with the accumulation of proline and total anthocyanins [27,47]. The results of the experimental work on seedings have shown that Citrus yellow and KM 53-14 Blue cultivars were salt-stress sensitive genotypes compared to other tested varieties. Different changes in MDA, proline and anthocyanin accumulations under salt stress have been observed for coloured and non-coloured wheat genotypes. It can provide evidence for the development of adaptive responses under salt stress the wheat genotypes tested in the experiment.

The tolerant genotypes of wheat showed a better capacity to maintain the low accumulation of Na^+, higher shoot K^+ concentrations, stable osmotic potential, increased values of PSII activity, lower non-photochemical quenching (NPQ) and maximal photochemical efficiency derived in the significantly greater dry weight production detected under salt stress [25,26]. Low Na^+ accumulation in leaf can be used as the best screening criteria, employing a large set of genotypes in a breeding program [48]. The dry weight of shoot biomass after salt stress influence increased again in Skorpion blue aleurone and PS Karkulka. The genotype Citrus yellow, with lower anthocyanin content, was characterized by a decrease in dry weight under both NaCl concentrations.

The content of Na^+ and K^+ obtained at different levels of salinity on the basis of DW was more than three times greater than the control, with a significant increase of both ions under salt stress. Cramer et al. (1985) demonstrated that in the presence of high NaCl concentrations, Na^+ moves Ca from the plasmalemma of root cells, resulting in increased membrane permeability that causes K efflux and alteration of the selectivity ratio K^+/Na^+ [49]. This suggests that plants that are successfully grown in saline conditions are those that maintain a higher K^+/Na^+ ratio in their cytoplasm than in the rhizosphere [50].

The toxicity of excessive salt on the cell level corresponds to the Na^+ level accumulated in the cells [51]. In plants, K^+ is engaged in cell metabolism, photosynthesis, and protein synthesis and is also essential for regulating enzyme activation and stomatal movement [52]. The regulation of Na^+ uptake and support of the relative Na^+/K^+ ratio can be recognized as an important cellular mechanism that assists plant adaptation to saline environments [53].

Siddiqui et al. (2017) obtained similar results regarding Na^+ ions. In the wheat leaves of 10 cultivars, significant increases in Na^+ content, but decreases in K^+ content were found under salinity stress conditions [54]. The ratio of Na^+/K^+ was significantly increased at all NaCl concentrations. The different plant genotypes reveal different salt tolerance mechanisms. These mechanisms are all based on the function and regulation of K^+ and Na^+ transporters and H^+ pumps, which create the driving force for K^+ and Na^+ transport [55,56]. It can be concluded that the contribution of Na^+ and K^+/H^+ antiporter proteins to plant salt stress tolerance is involved in K^+ homeostasis support rather than Na^+ sequestration into the vacuole.

The water content (WC) increased at high doses of NaCl under both treatment levels. During the growth phase and after salt application, the WC was effectively protected. This content is one of the key indicators of plant water status. Relative turgidity was less affected by stress and reflects good efficiency in saving water. This improvement can be explained, in part, by the effective accumulation of organic osmolytes, which suggests that the existence of mechanisms of osmotic adjustment lead to the protection of the structural and functional integrity of the tissue [10]. On the other hand, the improvement of the water status of plants may be, in part, a consequence of the lack of a significant reduction in stomatal conductance. It was found that salinity significantly reduced root hydraulic conductance [57]. In contrast, in cotton and common bean, WC decreased with salt stress at low rates of salt (50 mM) due to osmotic adjustment and a decrease in transpiration [58].

4. Materials and Methods

4.1. Plant Material

Seeds of five wheat genotypes (*Triticum* sp.) with different pigments (Citrus yellow, KM 53-14 Blue, KM 178-14 purple, Skorpion Blue aleurone, and PS Karkulka purple) were provided by the Agricultural Research Institute Kromeriz, Kromeriz, Czech Republic, respectively. The characteristics of coloured grains were based on the visual pink, blue and yellow colour assessment (Figure 6).

Figure 6. Coloured wheat seeds of the genotypes used in experiments 1—Citrus yellow, 2—KM 53-14 Blue, 3—KM 178-14 purple, 4—Skorpion Blue aleurone, 5—PS Karkulka purple.

Coloured grains have blue aleurone (Ba genes), purple pericarp (Pp genes) and yellow endosperm (Psy genes), which are determined by the presence of anthocyanins and carotenoids [59]. The blue aleurone (Ba genes) pericarp got grains of KM 53-14 Blue and Skorpion Blue aleurone wheat genotype. The purple pericarp (Pp genes) got grains of KM 178-14 purple and PS Karkulka purple wheat genotype. The yellow endosperm (Psy genes) have grains of Citrus yellow wheat genotype.

4.2. Growth Conditions

Sterile sandy soil of fine texture was filled in plastic pots (12 cm diameter) to conduct culture under controlled conditions. Grains were washed extensively with distilled water and sterilized with 5% sodium hypochlorite for 5 min. Then, seeds were sown in sandy soil directly watered with 3 mL of 1/4 strength Hoagland's nutrient solution [60]. The germination process was under controlled conditions of 25 ± 2 °C. The controlled growing conditions in a growing chamber had the following parameters: relative humidity of 60–70% and light/dark regime of 16/8 h at 25/20 °C. Four pots per treatment with 6 plants per pot were harvested after 30 days of the experiment. Pots were irrigated every three days, alternatively with filtrated water and a nutrient solution, according to Etherton (1963) [61], until they were 4 weeks old. It was used different concentrations of salinity treatments (0 (Control), 100, 150 and 250 mM NaCl). The control plants only received nutrient solution. To avoid osmotic shock due to high concentrations, plants were started on lower concentrations 100 and 150 mM NaCl (first stage of treatment), then the concentration was increased after 5 days with 50 mM NaCl, until wheat genotypes reached the concentration 150 and 200 mM NaCl (second stage of treatment). Tap water was used daily to balance the water presence after evaporative loss in plant seedlings.

4.3. Growth Parameters

After harvest, the shoot fresh weight (FW) was measured. The dry weight (DW) was estimated after putting plant samples in a forced-draft oven at 60 °C for 48 h or until a fixed weight was obtained.

$$\text{Water content (g } H_2O \text{ g}^{-1} \text{ DW)} = (FW - DW)/DW$$

4.4. Chlorophyll Fluorescence Records and Analyses Using the Fluorescence Excitation Ratio Method

Chlorophyll fluorescence analysis was performed using the portable optical fluorescence sensor Multiplex-3® (Force-A, Paris, France). Multiplex-3® is a multi-parametric, hand-operated sensor. The determination principle is based on light-emitting-diode excitation and a filtered photodiode. The portable optical fluorescence sensor Multiplex-3® is arranged to work in the laboratory, greenhouse and field conditions. The Multiplex-3® sensor has three red-blue-green LED matrices emitting light at 153 and 635 nm (red), 470 nm (blue) and 516 nm (green). There are three integrated photodiode detectors for fluorescence recording: far-red, red and yellow [29]. The values of fluorescence measured at UV (375 nm), green light (516 nm), red light (635 nm) and far-red light (735 nm) have been used.

The evaluation of phenolic compound contents in plants was performed via the calculation of fluorescence values detected after excitation by light of the defined wavelengths (details are below). Similar to the spectrophotometric method for assessing leaf absorbance, the parameters were based on Beer-Lambert's law and calculated as the logarithm of the fluorescence ratio values.

The UV absorbing compound (mostly flavonol) content described by the flavonoid (FLAV) index [62] was estimated using the modified formula [27] as the logarithm of the ratio of the red-light induced far-red fluorescence (FRF_R) and the UV-induced far-red fluorescence (FRF_{UV}):

$$FLAV = \log[FRF_R/(k_{UV} \times FRF_{UV})]$$

Similarly, the ANTH Index that provides estimates of green-light absorbing components ($\log FER_{R/G}$), mostly red-coloured anthocyanins and flavonoids, was determined as the logarithm of the ratio of the red-light induced fluorescence (FRF_R) and the green light-induced fluorescence (FRF_G):

$$ANTH = \log[FRF_R/(k_G \times FRF_G)]$$

The correction coefficients k_{UV} or k_G were applied to measurements of fluorescence to avoid negative values [27]. The constant values of the coefficients were used as the minimum values of the FRF_{UV}/FRF_R and FRF_G/FRF_R ratios found in the database that contains several thousand records from over three hundred plant species grown in diverse environments. The same constants have been used when processing data across all experiments and cultivars. We also calculated the modified flavonoid index (MFI), which provides a better estimate of total flavonoid content when plants with different colours are compared [27]. The logarithm of the ratio of the red-light induced fluorescence (FRF_R) and the green light-induced fluorescence (FRF_G) was used to calculate the MFI parameter.

$$MFI = \log[2 \times FRF_R/(kG \times FRF_G + k_{UV} \times FRFU_V)]$$

The values of correction coefficients (k_G, k_{UV}) for MFI were the same as for ANTH and FLAV.

Chlorophyll content was estimated from values of fluorescence measured at 735 nm (FRF) and 685 nm (RF) after excitation by red light (635 nm). The simple fluorescence ratio (SFR) was calculated:

$$SFR = FRF_R / RF_R$$

Because the diameter of the measuring area was only 50 mm, 6–7 measurements were taken on each plant in different positions to account for heterogeneity in leaf colour and structure. This number of measurements from the top view provides sufficient data to characterize the entire plant. The number of measurements for each plant was 30.

4.5. Proline Assay

The extraction was made by the method of Troll and Lindsley (1955) and simplified based on Wittmer (1987) [63,64]. One hundred milligrams of fresh material was placed with 2 mL of 40% methanol. The reaction mixture was heated at 85 °C in a water bath for 1 h. After cooling, 1 mL was removed from the extract to which 1 mL of acetic acid and 1 mL of the mixture containing 120 mL distilled water + 300 mL of ortho-phosphoric acid was added. The resulting solution was boiled for 30 min. After cooling, 5 mL of toluene was added. Two phases were separated and the upper phase (organic phase) was recovered. Absorbance measurements were determined using a Jenway UV/Vis. 6405 spectrophotometer (Jenway, Chelmsford, UK) at 528 nm. Proline concentration was estimated from a standard curve.

4.6. Lipid Peroxidation (LP) Assay

LP in leaves was identified by estimating the malondialdehyde (MDA) concentration [65]. Plant material (0.2 g) was homogenized in 3 mL of 0.1 mol kg^{-1} Tris buffer containing 0.3 mol kg^{-1} NaCl. Two millilitres of 20% trichloroacetic acid (TCA) containing 0.5% thiobarbituric acid and 2 mL of 20% TCA were added. The mixture was heated at 95 °C for 30 min. Then, the homogenate was centrifuged at $10,000 \times g$ for 5 min. The absorbance of the supernatant was measured at 532 nm with use of spectrophotometer Jenway UV/Vis. 6405 (Jenway, Chelmsford, UK).

4.7. Anthocyanin Estimation

Plant material (0.1 g) was soaked in 3 mL of acidified methanol (1% *v/v* HCl) for 12 h in darkness at 4 °C with occasional shaking. The mixture was centrifuged for 10 min at 14 000 rpm at 4 °C. Absorption of the extracts was estimated spectrophotometrically at 530 and 657 nm wavelengths on spectrophotometer Jenway UV/Vis. 6405 (Jenway, Chelmsford, UK). The blank was acidified methanol. The anthocyanin concentration was revealed as mg·g^{-1} DW and was calculated by the following formula:

$$Anthocyanins = [OD530 - 0.25\ OD657] \times TV/[d\ wt \times 1000] \tag{1}$$

OD = optical density; TV = total volume of the extract (mL); d wt = weight of the dry leaf tissue (g).

4.8. Na$^+$ and K$^+$ Accumulation

Measurements of Na$^+$ and K$^+$ accumulation in the flag leaf were carried out 3 days after flowering. Oven-dried simple flag leaves were finely ground before passing through a 2-mm sieve. Samples (0.5 g) were mixed with 10 mL of concentrated nitric acid and 3 mL of perchlorate acid in digesting tubes for 12 h and then dissolved at 300 °C for 6 h. The concentration of K$^+$ and Na$^+$ was detected using an atomic absorption spectrophotometer (Sherwood 410, Cambridge, UK).

4.9. Statistical Analysis

Statistical analyses were performed using two factor (genotype x salt concentration) analysis of variance (ANOVA) and Duncan's multiple range test performed at $p = 0.05$ (STATISTICA 10, StatSoft, Tulsa, OK, USA). Two terms of measurements were analysed separately. Mean values were calculated from six plants per cultivar in each experimental variant. Data are presented as mean ± standard error from six replicates (SE).

5. Conclusions

In the presented work has been presented the role of anthocyanins in the development salt tolerant stress responses in wheat plants. The coloured wheat genotypes, which were used as models to study the role of anthocyanins in the development of adaptive response to salt stress, may be used in the farther agricultural cultivation practices. It was found that higher flavonol and anthocyanin accumulation in coloured wheat genotypes showed a faster development of adaptive responses of wheat genotypes to salt stress. Furthermore, the KM 178-14 purple, Skorpion blue aleurone, and PS Karkulka, genotypes demonstrated an increased content of anthocyanins, proline and flavonol in the presence of NaCl and are more salt tolerant than Citrus yellow and KM 53-14 Blue, as most of the parameters (especially growth) were rarely influenced or affected by the use of 150 mM NaCl.

Author Contributions: S.M. and O.S. conceived and designed the experiments; M.Z. and A.C. performed the experiments; S.M. and O.S. wrote the paper; M.B. and A.C. reviewed and corrected the document.

Funding: This research was funded by Slovak Research and Development Agency under the project APVV-15-0721 and the Scientific Grant Agency of Ministry of Education, Slovak Republic under the project 1, VEGA 1-0923-16.

Acknowledgments: We thank Mark Kovar for his help and support during experiment.

Conflicts of Interest: The authors declare no conflict of interest. The founding sponsors had no role in the design of the study; in the collection, analyses, or interpretation of data; in the writing of the manuscript, and in the decision to publish the results.

References

1. Gibbs, H.K.; Salmon, J.M. Mapping the world's degraded lands. *Appl. Geogr.* **2015**, *57*, 12–21. [CrossRef]
2. Geist, H. *The Causes and Progression of Desertification*; Routledge: London, UK, 2017.
3. Lachhab, I.; Louahlia, S.; Laamarti, M.; Hammani, K. Effet d'un stress salin sur la germination et l'activité enzymatique chez deux génotypes de *Medicago sativa*. *Int. J. Innov. Appl. Stud.* **2013**, *3*, 511–516.
4. FAO. FAOSTAT. Online Statistical Database. 2016. Available online: http://faostat.fao.org/ (accessed on 30 July 2016).
5. Minhas, P.S. Edaphic stresses: Concerns and opportunities for management. In *Abiotic Stress Management for Resilient Agriculture*; Minhas, P.S., Rane, J., Pasala, R.K., Eds.; Springer: Singapore, 2017.
6. Munns, R.; Tester, M. Mechanisms of salinity tolerance. *Ann. Rev. Plant Biol.* **2008**, *59*, 651–681. [CrossRef] [PubMed]
7. Wang, Z.S.; Li, X.N.; Zhu, X.C.; Liu, S.Q.; Song, F.B.; Liu, F.L.; Wang, Y.; Qi, X.N.; Wang, F.H.; Zuo, Z.Y.; et al. Salt acclimation induced salt tolerance is enhancedby abscisic acid priming in wheat. *Plant Soil Environ.* **2017**, *63*, 307–314.
8. Mbarki, S.; Cerdà, A.; Zivcak, M.; Brestic, M.; Rabhi, M.; Mezni, M.; Abdelly, C.; Pascual, J.A. Alfalfa crops amended with MSW compost can compensate the effect of salty water irrigation depending on the soil texture. *Process Saf. Environ. Prot.* **2018**, *115*, 8–16. [CrossRef]
9. Sharifi, R.S.; Khalilzadeh, R.; Jalilian, J. Effects of biofertilizers and cycocel on some physiological and biochemical traits of wheat (*Triticum aestivum* L.) under salinity stress. *Arch. Agron. Soil Sci.* **2016**, *63*, 308–318. [CrossRef]
10. Slama, I.; Abdelly, C.; Bouchereau, A.; Flowers, T.; Savoure, A. Diversity, distribution and roles of osmoprotective compounds accumulated in halophytes under abiotic stress. *Ann. Bot.* **2015**, *115*, 433–447. [CrossRef] [PubMed]

11. El Midaoui, M.; Benbella, M.; Aït Houssa, A.; Ibriz, M.; Talouizte, A. Contribution to the study of some mechanisms of adaptation to salinity in cultivated sunflower (*Helianthus annuus* L.). *Revue HTE* **2007**, *136*, 29–34.

12. Hamed, K.B.; Chibani, F.; Abdelly, C.; Magne, C. Growth, sodium uptake and antioxidant responses of coastal plants differing in their ecological status under increasing salinity. *Biologia* **2014**, *69*, 193–201.

13. Zhu, J.K. Regulation of ion homeostasis under salt stress. *Curr. Opin. Plant Biol.* **2003**, *6*, 441–445. [CrossRef]

14. Liang, W.; Ma, X.; Wan, P.; Liu, L. Plant salt-tolerance mechanism: A review. *Biochem. Biophys. Res. Commun.* **2018**, *495*, 286–291. [CrossRef] [PubMed]

15. Mahajan, S.; Tuteja, N. Cold, salinity and drought stresses: An overview. *Arch. Biochem. Biophys.* **2005**, *444*, 139–158. [CrossRef] [PubMed]

16. Parida, A.K.; Das, A.B. Salt tolerance and salinity effects on plants: A review. *Ecotoxicol. Environ. Saf.* **2005**, *60*, 324–349. [CrossRef] [PubMed]

17. Dardanelli, M.S.; Fernández de Córdoba, F.J.; Espuny, M.R.; Rodríguez Carvajal, M.A.; Soria Díaz, M.E.; Gil Serrano, A.M.; Yaacov, O.; Megías, M. Effect of *Azospirillum brasilense* coinoculated with *Rhizobium* on *Phaseolus vulgaris* flavonoids and Nod factor production under salt stress. *Soil Biol. Biochem.* **2008**, *40*, 2713–2721. [CrossRef]

18. Daneshmand, F.; Arvin, M.J.; Kalantari, K.M. Physiological responses to NaCl stress in three wild species of potato in vitro. *Acta Physiol. Plant* **2010**, *32*, 91–101. [CrossRef]

19. Posmyk, M.M.; Kontek, R.; Janas, K.M. Antioxidant enzymes activity and phenolic compounds content in red cabbage seedlings exposed to copper stress. *Ecotoxicol. Environ. Saf.* **2009**, *72*, 596–602. [CrossRef] [PubMed]

20. Kondo, T.; Yoshida, K.; Nakagawa, A.; Kawai, T.; Tamura, H.; Goto, T. Commelinin, a highly associated metalloanthocyanin present in the blue flower petals of *Commelina communis*. *Nature* **1992**, *358*, 515–517.

21. Shannon, M.C.; Grieve, C.M. Tolerance of vegetable crops to salinity. *Scientia Hortic.* **1999**, *78*, 5–38. [CrossRef]

22. Ngara, R.; Ndimba, R.; Jensen, J.B.; Jensen, O.N.; Ndimb, B. Identification and profiling of salinity stress-responsive proteins in *Sorghum bicolor* seedlings. *J. Proteomics* **2012**, *75*, 4139–4150. [CrossRef] [PubMed]

23. Caverzan, A.; Casassola, A.; Brammer, S.P. Antioxidant responses of wheat plants under stress. *Genet. Mol. Biol.* **2016**, *39*, 1–6. [CrossRef] [PubMed]

24. Ksouri, R.; Megdiche, W.; Debez, A.; Falleh, H.; Grignon, C.; Abdelly, C. Salinity effects on polyphenol content and antioxidant activities in leaves of the halophyte *Cakile maritima*. *Plant Physiol. Biochem.* **2007**, *45*, 244–249. [CrossRef] [PubMed]

25. Oyiga, B.C.; Sharma, R.C.; Shen, J.; Baum, M.; Ogbonnaya, F.C.; Léon, J.; Ballvora, A. Identification and characterization of salt tolerance of wheat germplasm using a multivariable screening approach. *J. Agron. Crop Sci.* **2016**, *202*, 472–485. [CrossRef]

26. Kalhoro, N.A.; Rajpar, I.; Kalhoro, S.A.; Ali, A.; Raza, S.; Ahmed, M.; Kalhoro, F.A.; Ramzan, M.; Wahid, F. Effect of salts stress on the growth and yield of wheat (*Triticum aestivum* L.). *Am. J. Plant Sci.* **2016**, *7*, 2257–2271. [CrossRef]

27. Zivcak, M.; Brückova, K.; Sytar, O.; Brestic, M.; Olsovska, K.; Allakhverdiev, S.I. Lettuce flavonoids screening and phenotyping by chlorophyll fluorescence excitation ratio. *Planta* **2017**, *245*, 1215–1229. [CrossRef] [PubMed]

28. Chutipaijit, S.; Chaum, S.; Sompornpailin, K. High contents of proline and anthocyanin increase protective response to salinity in *Oryza sativa* L. spp. indica. *Aust. J. Crop Sci.* **2011**, *5*, 1191–1198.

29. Gaweł, S.; Wardas, M.; Niedworok, E.; Wardas, P. Malondialdehyde (MDA) as a lipid peroxidation marker. *Wiad Lek.* **2004**, *57*, 453–455. [PubMed]

30. Ghozlen, N.B.; Cerovic, Z.G.; Germain, C.; Toutain, S.; Latouche, G. Non-destructive optical monitoring of grape maturation by proximal sensing. *Sensors* **2010**, *10*, 10040–10068. [CrossRef] [PubMed]

31. Schoefs, B.; Bertrand, M.; Lemoine, Y. Changes in the photosynthetic pigments in bean leaves during the first photoperiod of greening and the subsequent dark-phase. Comparison between old (10-d-old) leaves and young (2-d-old) leaves. *Photosynth. Res.* **1998**, *57*, 203. [CrossRef]

32. Ashraf, M.; Harris, P.J.C. Photosynthesis under stressful environments: An overview. *Photosynthetica* **2013**, *51*, 163–190. [CrossRef]

33. Ashraf, M. Some important physiological selection criteria for salt tolerance in plants. *Flora* **2004**, *199*, 361–376. [CrossRef]

34. Dubey, R.S. Photosynthesis in plants under stressful conditions. In *Hand Book Photosynthesis*, 2nd ed.; Pessarakli, M., Ed.; C.R.C. Press: New York, NY, USA, 2005; pp. 717–718.

35. Ashraf, M.; Foolad, M.R. Roles of glycinebetaine and proline in improving plant abiotic stress resistance. *Environ. Exp. Bot.* **2007**, *59*, 206–216. [CrossRef]

36. Abdel-Aal, E.S.M.; Hucl, P. Composition and stability of anthocyanins in blue-grained wheat. *J. Agric. Food Chem.* **2003**, *51*, 2174–2180. [CrossRef] [PubMed]

37. Knievel, D.C.; Abdel-Aal, E.S.M.; Rabalski, I.; Nakamura, T.; Hucl, P. Grain color development and the inheritance of high anthocyanin blue aleurone and purple pericarp in spring wheat (*Triticum aestivum* L.). *J. Cereal Sci.* **2009**, *50*, 113–120. [CrossRef]

38. Martinez, V.; Mestre, T.C.; Rubio, F.; Girones-Vilaplana, A.; Moreno, D.A.; Mittler, R.; Rivero, R.M. Accumulation of flavonols over hydroxycinnamic acids favors oxidative damage protection under abiotic stress. *Front. Plant Sci.* **2016**, *7*, 838. [CrossRef] [PubMed]

39. Huang, Z.; Zhao, L.; Chen, D.; Liang, M.; Liu, Z.; Shao, H.; Long, X. Salt stress encourages proline accumulation by regulating proline biosynthesis and degradation in Jerusalem artichoke plantlets. *PLoS ONE* **2013**, *8*, e62085. [CrossRef] [PubMed]

40. Shirazi, M.; Khan, M.; Mahboob, W.; Khan, M.A.; Shereen, A.; Mujtaba, S.; Asad, A. Inconsistency in salt tolerance of some wheat (*Triticum aestivium* L.) genotypes evaluated under various growing environments. *Pak. J. Bot.* **2018**, *50*, 471–479.

41. Matysik, J.; Alia, T.A.; Bhalu, B.A.; Mohanty, P. Molecular mechanisms of quenching of reactive oxygen species by proline under stress in plants. *Curr. Sci.* **2002**, *82*, 525–532.

42. Munns, R. Comparative physiology of salt and water stress. *Plant Cell Environ.* **2002**, *25*, 239–250. [CrossRef] [PubMed]

43. Cuin, T.A.; Shabala, S. Amino acids regulate salinity-induced potassium efflux in barley root epidermis. *Planta* **2007**, *225*, 753–761. [CrossRef] [PubMed]

44. Jaleel, C.A.; Gopi, R.; Sankar, B.; Manivannan, P.; Kishorekumar, A.; Sridharan, R.; Panneerselvam, R. Studies on germination, seedling vigour, lipid peroxidation and proline metabolism in *Catharanthus roseus* seedlings under salt stress. *S. Afr. J. Bot.* **2007**, *73*, 190–195. [CrossRef]

45. Shabala, S.; Cuin, T.A. Potassium transport and plant salt tolerance. *Physiol. Plant.* **2008**, *133*, 651–669. [CrossRef] [PubMed]

46. Koca, H.; Melike, B.; Özdemir, F.; Türkan, İ. The effect of salt stress on lipid peroxidation, antioxidative enzymes and proline content of sesame cultivars. *Environ. Exp. Bot.* **2007**, *60*, 344–351. [CrossRef]

47. Feki, K.; Tounsi, S.; Brini, F. Comparison of an antioxidant system in tolerant and susceptible wheat seedlings in response to salt stress. *Span. J. Agric. Res.* **2018**, *15*, 0805. [CrossRef]

48. Saddiq, M.S.; Afzal, I.; Basra, S.M.; Ali, Z.; Ibrahim, A.M. Sodium exclusion is a reliable trait for the improvement of salinity tolerance in bread wheat. *Arch. Agron. Soil Sci.* **2018**, *64*, 272–284. [CrossRef]

49. Cramer, G.R.; Läuchli, A.; Polito, V.S. Displacement of Ca^{2+} by Na^+ from the plasmalemma of root cells: A primary response to salt stress? *Plant Physiol.* **1985**, *79*, 207–211. [CrossRef] [PubMed]

50. El-Iklil, Y.; Karrou, M.; Benichou, M. Salt stress effect on epinasty in relation to ethylene production and water relations in tomato. *Agronomy* **2000**, *20*, 399–406. [CrossRef]

51. Zafar, S.A.; Shokat, S.; Ahmed, H.G.M.; Khan, A.; Ali, M.Z.; Atif, R.M. Assessment of salinity tolerance in rice using seedling-based morpho-physiological indices. *Adv. Life Sci.* **2015**, *2*, 142–149.

52. Nievens-Cordones, M.; Al Shiblawi, F.R.; Sentenac, H. Role and transport of sodium and potassium in plants. *Met. Ions Life Sci.* **2016**, *16*, 291–324.

53. Munns, R.; James, R.A.; Islam, A.K.M.R.; Colmer, T.D. Hordeum marinum-wheat amphiploids maintain higher leaf K^+:Na^+ and suffer less leaf injury than wheat parents. *Plant Soil* **2011**, *348*, 365–377. [CrossRef]

54. Siddiqui, M.N.; Mostofa, M.G.; Akter, M.M.; Srivastava, A.K.; Sayed, M.A.; Hasan, M.S.; Tran, L.P. Impact of salt-induced toxicity on growth and yield-potential of local wheat cultivars: Oxidative stress and ion toxicity are among the major determinants of salt-tolerant capacity. *Chemosphere* **2017**, *187*, 385–394. [CrossRef] [PubMed]

55. Almeida, D.M.; Oliveira, M.M.; Saibo, N.J.M. Regulation of Na^+ and K^+ homeostasis in plants: Towards improved salt stress tolerance in crop plants. *Genet. Mol. Biol.* **2017**, *40*, 326–345. [CrossRef] [PubMed]

56. Maathuis, F.J.; Ahmad, I.; Patishtan, J. Regulation of $Na^{(+)}$ fluxes in plants. *Front. Plant Sci.* **2014**, *5*, 467. [CrossRef] [PubMed]

57. Silva, C.; Martinez, V.; Carvajal, M. Osmotic versus toxic effects of NaCl on pepper plants. *Biol. Plant* **2008**, *52*, 72–79. [CrossRef]

58. Kadri, A.; Madiouni, N. Effet du stress salin sur quelques paramètres biochimiques de la luzerne cultivée (Medicago sativa L.) Mémoire En vue de l'obtention du diplôme de Master Academique. Master's Thesis, University of Ouargla, Ouargla, Algeria, 4 June 2015.

59. Martinek, P.; Jirsa, O.; Vaculová, K.; Chrpová, J.; Watanabe, N.; Burešová, V.; Kopecký, D.; Štiasna, K.; Vyhnánek, T.; Trojan, V. Use of wheat gene resources with different grain colour in breeding. In Proceedings of the 64 Tagung der Vereinigung der Pflanzenzüchter und Saatgutkaufleute Österreichs, Raumber-Gumpenstein, Austria, 25–26 November 2013; pp. 75–78.

60. Epstein, E. *Mineral Nutrition of Plants: Principles and Perspectives*; John Wiley and Sons, Inc.: New York, NY, USA, 1972; 412p.

61. Etherton, B. Relationship of cell transmembrane electropotential to potassium and sodium accumulation ratios in oat and pea seedlings. *Plant Physiol.* **1963**, *38*, 581–585. [CrossRef] [PubMed]

62. Agati, G.; Cerovic, Z.G.; Pinelli, P.; Tattini, M. Light-induced accumulation of ortho-dihydroxylated flavonoids as non-destructively monitored by chlorophyll fluorescence excitation techniques. *Environ. Exp. Bot.* **2011**, *73*, 3–9. [CrossRef]

63. Troll, W.; Lindsley, J. A photometric method for the determination of proline. *J. Biol. Chem.* **1955**, *215*, 655–660. [PubMed]

64. Wittmer, G. Osmotic and elastic adjustment of durum wheat leaves under drought stress conditions. *Genet. Agrar.* **1987**, *41*, 427–436.

65. Heath, R.L.; Packer, L. Photoperoxidation in isolated chloroplasts 1. Kinetics and stoichiometry of fatty acid peroxidatoin. *Arch. Biochem. Biophys.* **1968**, *125*, 189–198. [CrossRef]

Sample Availability: Samples of the compounds are not available from the authors.

molecules

MDPI

Article

Improved Cold Tolerance of Mango Fruit with Enhanced Anthocyanin and Flavonoid Contents

Pradeep Kumar Sudheeran, Oleg Feygenberg, Dalia Maurer and Noam Alkan *

Department of Postharvest Science of Fresh Produce, Agricultural Research Organization, Volcani Center, P.O. Box 15159, HaMaccabim Road 68, Rishon LeZion 7505101, Israel; pradeepkumar2k@gmail.com (P.K.S.); fgboleg@volcani.agri.gov.il (O.F.); daliam@volcani.agri.gov.il (D.M.)
* Correspondence: noamal@agri.gov.il

Received: 21 June 2018; Accepted: 20 July 2018; Published: 23 July 2018

Abstract: Red fruits were suggested to be tolerant to cold. To understand cold-storage tolerance of red mango fruit that were subjected to sunlight at the orchard, mango cv. Shelly from inside (green fruit) or outside (red fruit) the tree canopy was stored for 3 weeks at 5, 8 or 12 °C and examined for flavonoids, antioxidant, volatiles and tolerance to biotic and abiotic stress. Red fruit from the outer canopy showed significant increases in total anthocyanin and flavonoids, and antioxidant activity. Ripening parameters for red and green mango fruit were similar at harvest and during storage. However, red fruit with high anthocyanin and flavonoid contents were more tolerant to biotic and abiotic stresses. After 3 weeks of suboptimal cold storage, green fruit showed significantly more lipid peroxidation and developed significantly more chilling-injury symptoms—black spots and pitting—than red fruit. Volatiles of red and green peels revealed significant modulations in response to cold-storage. Moreover, red fruit were more tolerant to biotic stress and had reduced general decay incidence. However, during long storage at 10 °C for 4, 5 or 6 weeks, red fruit showed a non-significant reduction in decay and chilling injuries. These results suggest new approaches to avoiding chilling injury during cold storage.

Keywords: anthocyanin; antioxidant activity; cold storage; ethylene; GC-MS; luminescence; multiplex; ROS

---

1. Introduction

Mango (*Mangifera indica* L.), is an economically important fruit, distributed worldwide, which belongs to the family *Anacardiaceae*. Fully ripe mango is known for its aroma, peel color, good taste and nutritional value [1,2]. It is also highly perishable; soon after ripening, the fruit starts decaying and quickly becomes unfit for consumption. Cold storage is probably the best postharvest technology to maintain high-quality fruit [3]. However, as a tropical fruit which was not exposed to extreme cold temperature during its evolution, mango is highly susceptible to cold storage. Storage of mango fruit at temperatures lower than 12 °C leads to the development of chilling injuries, which are expressed as physiological and biochemical alterations and cellular dysfunctions. These alterations include stimulated ethylene production, increased respiratory rate, enzyme inactivation, membrane dysfunction, production of reactive oxygen species (ROS) and changes in cellular structure, which lead to the development of chilling-injury symptoms such as pitting, black spots, peel discoloration, water-soaked appearance, internal breakdown and browning, uneven ripening, off-flavor and decay [4,5]. This limits the application of cold storage to extend mango's lifetime. Recently, mango fruit transcriptome was characterized in response to storage at suboptimal temperature and showed activation of defense response signal transduction, lipid peroxidation and a significant activation of the phenylpropanoids biosynthetic pathway [6].

Anthocyanins are phenylpropanoids that are widely distributed in vascular plants where they serve as inducible sunshields [7]. Polyphenols, like anthocyanins and flavonols, are one of the mango fruit's antioxidant compounds [8]. In mango fruit peel, anthocyanins are responsible for the red color, while carotenoids are responsible for the yellow to orange color [9].

Anthocyanins' antioxidant capacity makes them important phytonutrients in a healthy diet, along with their antitumor, anti-inflammatory and antineurodegenerative properties [10,11]. Both anthocyanins and flavonols are products of the phenylpropanoid biosynthetic pathway and are involved in plant protection against pathogens [12]. Indeed, when 83 mango cultivars were infected with *Colletotrichum gloeosporioides* or stored at suboptimal temperature of 6 °C, the red mango cultivars showed a significant increased tolerance to both anthracnose caused by *C. gloeosporioides* and to chilling injuries after storage in comparison to green cultivars [13].

Anthocyanin production can be induced by abiotic stress. Anthocyanins and flavonols are known to increase in response to cold [14,15]. Another environmental cue regulating anthocyanins is light. Indeed, mango fruit that are exposed to sunlight in the orchard accumulate anthocyanins and a red skin color [12].

This manuscript's aim is to characterize and study the cold tolerance mechanism of red mango fruit that contain high anthocyanin and flavonol contents in order to store red mango fruit at lower temperature and to extend its storage period.

2. Results

2.1. Evaluation of Mango Fruit Color, and Anthocyanin and Flavonoid Contents

Mango (*Mangifera indica* L., cv. Shelly) fruit were harvested from two different positions on the tree canopy: red-colored fruit from the exterior position with direct exposure to sunlight, and green-colored fruit from the interior position, in the shaded part of the canopy. Uniform, unblemished fruit were selected on the basis of skin color: red fruit with more than 60%, and green fruit with less than 10%, of the fruit peel colored in red. The fruit were stored in the cold at 5, 8 or 12 °C for 3 weeks and transferred to 1 week of shelf life at 20 °C.

Mango skin color was measured after cold storage and presented as hue. Both the green fruit and the red fruit on their green side had similar values (100–120) correlated to green color, while the red side of the red fruit presented hue values of 13–28, correlated to red color (Figure 1). Using a Multiplex III fluorescence detector, we measured the fruit fluorescence and assessed different chemical groups as flavonoids (FLAV) and anthocyanin (FER_RG) [15]. The FER_RG and FLAV ratio measurements, representing anthocyanin and flavonoids, revealed that the red side of the red fruit had about 3-fold higher flavonoid content about 10-fold higher anthocyanin content than the red side of the green fruit after cold storage (Figure 1). Similar results were found for extended storage (see Section 2.6) at 10 °C (Figure S1). Interestingly, after storage of 'Shelly' mango fruit at 8 °C, a minor change of color was observed on both the green side of the red fruit and in the green fruit (Figure 1C), correlated to an increase in anthocyanin content in the peel of the green fruit after cold storage (Figure 1A).

Total amount of anthocyanins and flavonoids was estimated chemically for red and green fruit stored at 5, 8 or 12 °C for 3 weeks and an additional 7 days of shelf life (Figure 2). A significant 2-fold increase in the amount of anthocyanins was seen on the red side of the red fruit compared to both red and green sides of the green fruit (Figure 2A–C). While the green side of the red fruit showed only a moderate level of anthocyanin, it was still significantly higher than that in the green fruit. In all, the different cold-storage conditions showed similar amounts of anthocyanin (Figure 2A–C). In addition, contents of both anthocyanin and flavonols declined moderately, by about 20%, during storage and as the fruit ripened. Similarly, the amount of flavonoids was significantly higher on the red side of the red fruit compared to both sides of the green fruit (Figure 2D–F), whereas storage at different temperatures did not affect the flavonoid content in the mango peel (Figure 2D–F).

Thus, both anthocyanins and flavonoids were accumulated mostly at the red side of the red fruit and gradually decreased during storage.

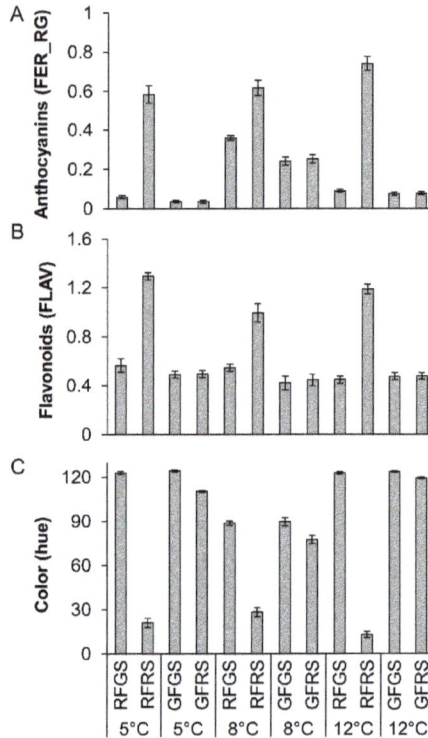

Figure 1. Quantification of color, anthocyanins and flavonols by fluorescence and light-based methods in the peel of red mango fruit (RF) or green fruit (GF) on their red side (RS) or green side (GS) after cold storage at 5 °C, 8 °C or 12 °C. (**A**)Total anthocyanin content. (**B**) Total flavonolcontent. (**C**) Fruit color displayed as hue.

2.2. Antioxidants and ROS

To measure the antioxidant activity of flavonols and anthocyanins in red and green peels of mango fruit, radical-scavenging activity of DPPH was assessed in peels of fruit stored at 5, 8 or 12 °C for 3 weeks and an additional 7 days of shelf life. A significant 2-time increase in the amount of scavenging activity of DPPH was seen on the red side of the red fruit compared to both sides of the green fruit (Figure 3A–C). However, no significant differences were observed among the different temperature treatments (Figure 3A–C). The increase in antioxidant activity of the red fruit on its red side was well correlated to the increase in flavonoids and anthocyanins (compare Figures 2 and 3). Thus, the red fruit accumulated anthocyanins, flavonoids and antioxidant activity mostly at their red side.

Accumulation of ROS was detected by fluorescence microscopy after DCF staining of mango peels stored at 5, 8 or 12 °C for 3 weeks and an additional 7 days of shelf life. The ROS fluorescence intensity was significantly increased (2-fold) in the green peel compared to the low fluorescence and ROS levels detected in the red peel (Figure 3D–F). It was also noted that the relative ROS levels increased with time from harvest to cold storage to shelf life under all temperature regimes (Figure 3D–F), and this increase was negatively correlated to the reduction of antioxidant activity during storage and fruit ripening (Figure 3A–C).

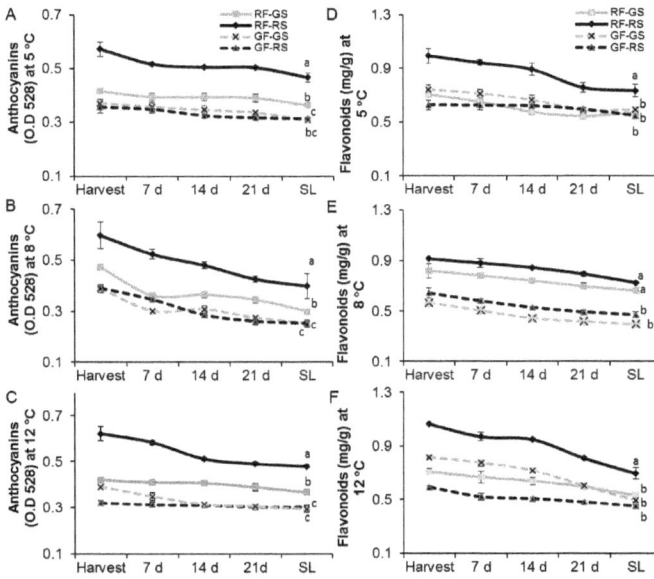

Figure 2. Quantification of total anthocyanins and flavonoids by chemical methods in the peel of red mango fruit (RF) or green fruit (GF) on their red side (RS) or green side (GS). (**A**)Total anthocyanin content during cold storage at 5 °C. (**B**) Total anthocyanin content during cold storage at 8 °C. (**C**) Total anthocyanin content during cold storage at 12 °C. (**D**) Total flavonoid content after cold storage at 5 °C. (**E**) Total flavonoid content after cold storage at 8 °C. (**F**) Total flavonoid content after cold storage at 12 °C. SL, shelf life. Average and SE are presented. Different letters indicate significant difference ($p < 0.05$).

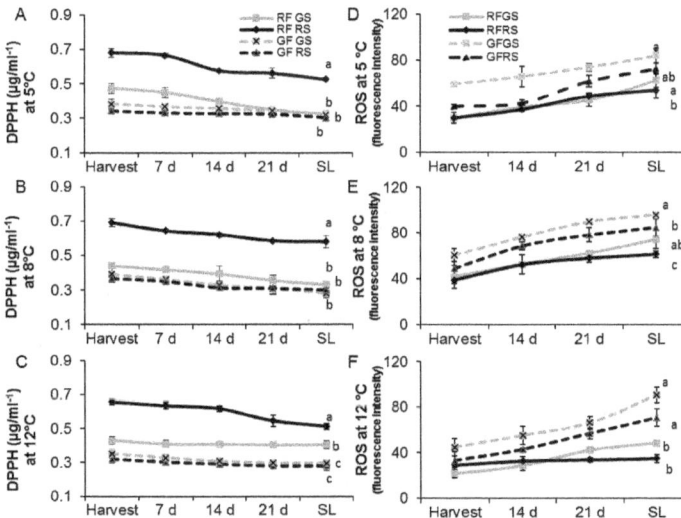

Figure 3. Antioxidant activity and ROS in the peel of red mango fruit (RF) or green fruit (GF) on their red side (RS) or green side (GS). (**A–C**) Antioxidant activity at different storage temperatures (5, 8 and 12 °C, respectively). (**D–F**) Reactive oxygen species (ROS), as quantified by relative fluorescence intensity, in red and green peel of 'Shelly' mango stored at 5, 8 or 12 °C, respectively. SL, shelf life. Average and SE are presented. Different letters indicate significant difference ($p < 0.05$).

2.3. Chilling Tolerance of Red Mango Fruit

Chilling tolerance of red mango fruit was assessed on the basis of severity of black spots and pitting (scale of 0–10) in red and green mango fruit after 3 weeks of cold storage at 5, 8 or 12 °C and an additional 7 days of shelf life at 20 °C. As expected, fruit stored at the suboptimal temperature of 5 °C had more chilling injuries, expressed as black spots and pitting, than fruit stored at higher temperatures. However, red fruit had less black spots and pitting damage than green fruit after cold storage and after additional shelf life (Figure 4A,B), whereas more red spots were observed on red vs. green fruit (data not shown).

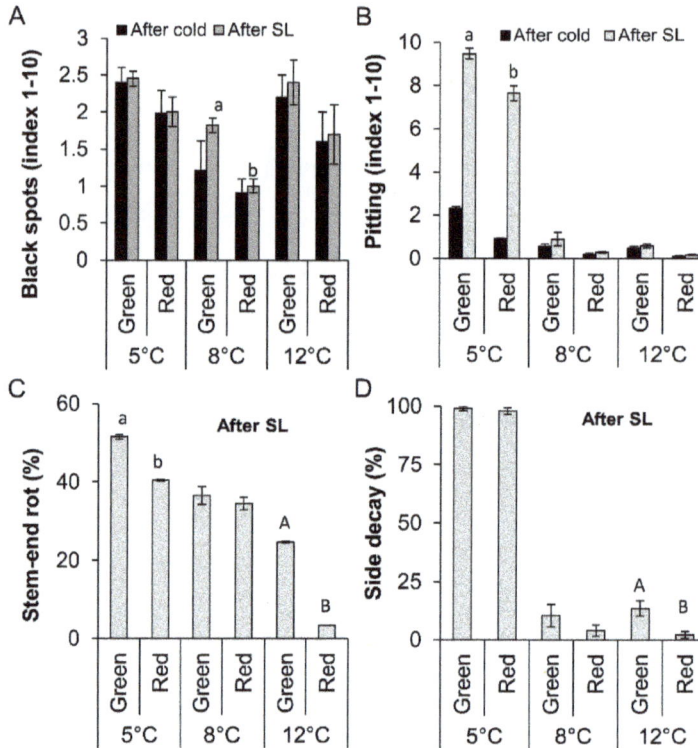

Figure 4. Evaluation of chilling injuries of red and green mango fruit after 3 weeks of cold storage (5 °C, 8 °C or 12 °C) and a further 7 days of shelf life (SL) (20 °C). (**A**) Black spot severity (index 1–10). (**B**) Pitting severity (index 1–10). (**C**) Incidence of stem-end rot after SL in red and green fruit. (**D**) Incidence of side decay after SL in red and green fruit. Average and SE are presented. Different letters indicate significant difference ($p < 0.05$).

Natural stem decay and side decay were evaluated after cold storage at the different temperatures and additional shelf life. After shelf life, when the fruit ripened, a correlation was found in which fruit stored at lower temperature developed more stem-end rot than those stored at higher temperature (Figure 4C,D). Furthermore, red fruit had less decay incidence than green fruit (Figure 4C). Likely, due to the severe pitting induced upon cold storage at 5 °C, a higher incidence of side decay was found in fruit stored at 5 °C than in those stored at higher temperatures (Figure 4B,D).

2.4. Lipid Peroxidation during Cold Storage

Lipid peroxidation is known to occur with chilling injuries and can be detected by bioluminescence [6]. During cold storage of red and green mango fruit, the luminescence of the whole fruit was measured with an *in-vivo* imaging system (IVIS). The fruit stored at lower temperatures (5 or 8 °C) showed more luminescence than those stored at the optimal temperature of 12 °C (Figure 5). Interestingly, green fruit had considerably more chilling injury symptoms (Figure 4A,B) and more luminescence than the red fruit stored at suboptimal temperatures (Figure 5). Indeed, red fruit stored at 8 °C showed very low luminescence, which was correlated with low levels of lipid peroxidation and chilling injury.

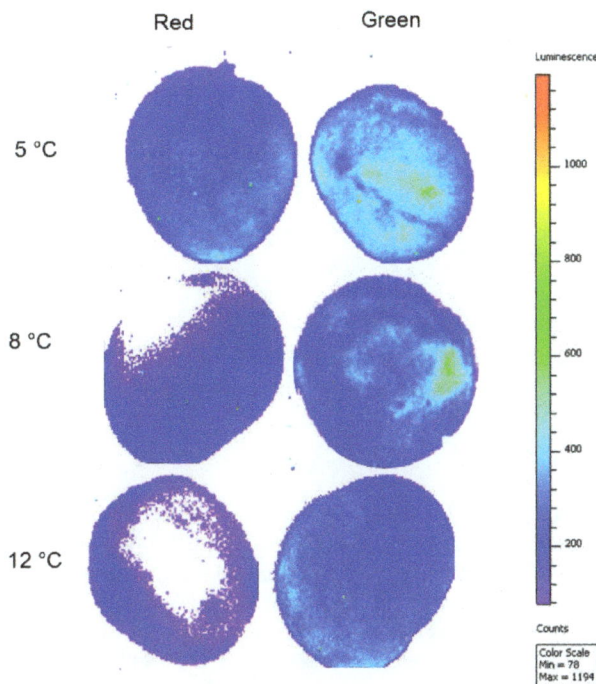

Figure 5. Mango fruit luminescence. Evaluation of luminescence and lipid peroxidation in red and green mango fruit after 3 weeks of cold storage at 5 °C, 8 °C or 12 °C by *in-vivo* imaging system (IVIS). Luminescence of lipid peroxidation of red fruit (left panels) and green fruit (right panels) after 3 weeks storage at 5 °C (top panel), 8 °C (middle panel) and 12 °C (bottom panel).

2.5. GC–MS Analysis of Mango Volatiles

Volatile compounds were identified in the peel of red and green 'Shelly' mango fruit after 3 weeks of cold storage; a total of 28 putative volatile compounds were identified and quantified using calibration graph and internal standard (Table S1). Among them, 11 volatiles were significantly altered during cold storage of red fruit compared to green fruit (Figure 6). Four volatile compounds (limonene, (E)-β-damascenone, 1,7-di-epi-α-cedrene, *n*-heptanal) were upregulated in green fruit peel, and found in a concentration of 0.3–1 µg/gFW, and were not detected in the peel of red fruit. While, seven volatile compounds ((Z)-β-ocimene, eugenol, β-bourbonene, β-elemene, methyleugenol, epi-cubebol, cadina-1,4-diene) were elevated in the red fruit peel and found in a concentration of 0.9–21 µg/gFW and were not detected in the peel of the green fruit (Figure 6). Interestingly, compounds as eugenol and methyleugenol that were increased in red fruit peel are products of the phenylpropanoid pathway, while other compounds are not. Thus, a significant change in volatiles profile was observed between the peels of red and green mango fruits.

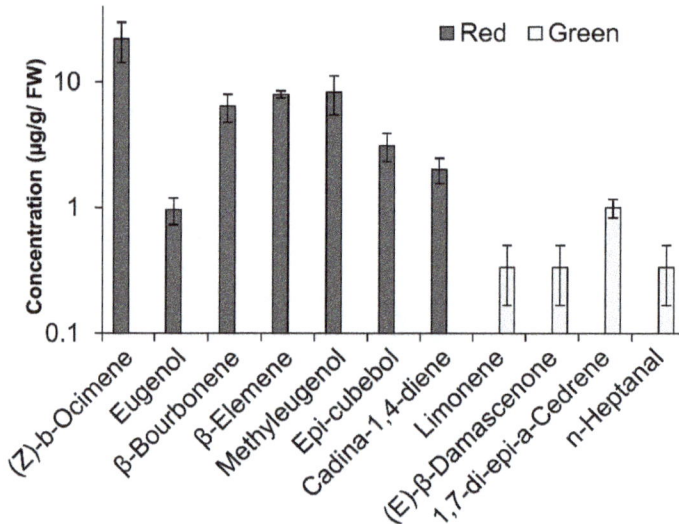

Figure 6. GC-MS analysis of volatile compounds of red and green peels of 'Shelly' mango after 3 weeks of cold storage at 12 °C. Graphical representation of the concentration of unique volatiles of red and green mango peel.

2.6. Storage Elongation

To increase mango fruit storage, red and green fruits were stored at 10 °C (instead of the optimal 12 °C) for 4 weeks, 5 weeks or 6 weeks (instead of the usual 3 weeks). Then the fruit were transferred to 1 week of shelf life at 20 °C. Both red and green fruit showed more or less similar ripening parameters during storage, including: (i) decreasing firmness; (ii) increasing total soluble solids (TSS); and (iii) decreasing citric acid content (Figure S2). Supporting the similar ripening of red and green mango fruit, ethylene and respiration (CO_2) were similar after 3 weeks of cold storage at 5, 8 or 12 °C (Figure S3). Interestingly, while the ripening parameters were relatively similar, the red fruit had higher TSS and lower acid concentration than the green fruit after extended cold storage and additional shelf life (Figure S2). These increased sugars and decreased acids are correlated to better mango fruit quality and taste. A similar increase in TSS and reduction in acids was recorded in red vs. green mango fruit stored at 12 °C (Figure S4).

As expected, red fruits showed fewer chilling injuries, expressed as black spots and pitting, after 5 and 6 weeks of cold storage (Figure 7A,B). Moreover, red fruit showed a non-significant reduction in natural postharvest stem-end rot after long storage (Figure 7C), whereas a significant reduction was observed in natural side decay of red fruit compared to green fruit after 4 and 6 weeks of storage (Figure 7D). Thus, red fruit had less chilling injuries and postharvest decay and could be stored for longer periods of time, as demonstrated by representative pictures (Figures S5 and S6).

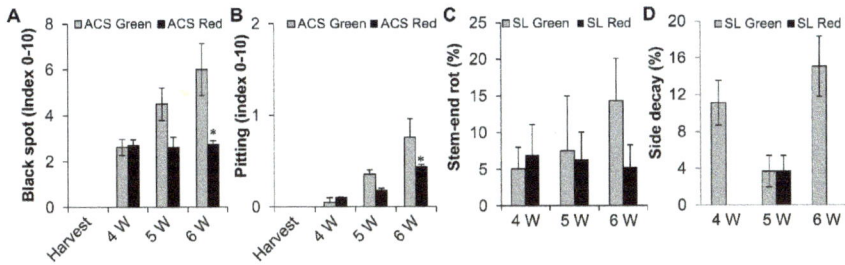

Figure 7. Postharvest chilling and decay in red and green 'Shelly' mango fruit stored at 10 °C for 4–6 weeks (W). (**A**) Black spot (index 0–10) after cold storage (ACS) at 10 °C for 4, 5 or 6 weeks. (**B**) Pitting (index 0–10) after cold storage at 10 °C for 4, 5 or 6 weeks. (**C**) Stem-end rot incidence (percentage) after cold storage at 10 °C for 4, 5 or 6 weeks and further shelf life (SL). (**D**) Side decay incidence (percentage) after cold storage at 10 °C for 4, 5 or 6 weeks and further shelf life.

3. Discussion

In response to light, 'Shelly' mango fruit that grew on the outer part of the tree canopy accumulated anthocyanin and red color in their peel, whereas the non-exposed surfaces of red fruit or fruit that developed inside the canopy were less exposed to sunlight and remained green [12]. Both anthocyanin and flavonols have pleotropic effects and are known to be involved in plant protection against pathogens [16,17].

In this study, 'Shelly' mango fruit were harvested from outside (red fruit) and inside (green fruit) the tree canopy. Analysis of total content of flavonoids and anthocyanin showed that both increased by 2-fold in the red mango fruit on their red side, which was exposed to sunlight (Figure 1). These results suggest that exposure to light increases the phenylpropanoid pathway in mango fruit leading to increased flavonols and anthocyanin. Anthocyanin is induced by a number of environmental factors, including high light, UV radiation, cold temperatures and water stress, and it has been proposed as an important compound in the plant's response to abiotic stress [18–20].

The storage temperature (5, 8 or 12 °C) did not affect the flavonoid or anthocyanin levels in the mango fruit. However, both flavonol and anthocyanin levels gradually decreased with time of storage.

Both flavonols and anthocyanins are known to have antioxidant activity [21]. As expected, red fruit had 2- to 3-time increased antioxidant activity compared to green fruit (Figure 3). Indeed, studies have shown that tomato *Delila* and *Rosea1* mutants accumulate anthocyanins, resulting in reduced ROS accumulation and reduced susceptibility to gray mold [22]. Similarly, in the present study, both sides of the red fruit had reduced ROS levels at harvest relative to green fruit (Figure 3). Exposure of plants to low temperatures can cause an increase in ROS production in various plant tissues [1]. Accordingly, ROS were accumulated during cold storage, and more significantly at suboptimal temperature (5 or 8 °C). However, ROS accumulation in response to cold storage was inhibited on both sides of the red fruit as compared to green fruit (Figure 3). This reduction in ROS was tightly correlated with the high antioxidant activity in the red mango fruit, which had higher levels of anthocyanins and flavonols (Figures 2 and 3).

ROS accumulation during storage at suboptimal temperature activates lipid peroxidation, which is a main characteristic of initial chilling injuries in mango fruit [5]. This lipid peroxidation can be visualized by bioluminescence [5]. As expected, fruit stored at lower temperature had higher luminescence. Furthermore, green fruit, with higher levels of ROS, showed a 2- to 3-timeincrease in luminescence and lipid peroxidation compared to the red mango fruit (Figure 5). The reduced lipid peroxidation in red mango fruit was correlated with reduced chilling injuries in these fruit (Figure 4). Thus, we suggest that the mechanism of increased tolerance of red mango fruit to chilling injury involves higher antioxidant activity and high levels of anthocyanin and flavonols that lead to reduced

ROS in response to cold storage; this, in turn, leads to reduced lipid peroxidation and reduced chilling injuries. Alternatively, exposure of the red fruit to direct sunlight in the orchard might induce fruit resistance to chilling injuries during storage.

After cold storage at optimal temperature (12 °C), 28 putative volatile compounds of mango fruit [23,24] were detected. However, the overall volatile profile of the red fruit was significantly different from that of the green fruit in response to cold storage, showing seven unique compounds in red fruit and four unique compounds in green fruit (Figure 6). As expected, several of the volatiles that were unique to red fruit, such as eugenol and methyleugenol, are products of the phenylpropanoid pathway, which is known for conferring a good flavor, and antioxidant and anti-infective activity [25]. In the present study, the major compounds detected in the peel of red mango cv. 'Shelly' were (Z)-β–ocimene, β-elemene, epi-cubebol, cadina-1,4-diene and β-bourbonene, which are known as aroma volatiles of monoterpenes and sesquiterpenes. Those volatiles were observed in previous studies on different mango cultivars and their volatile concentration were in a similar range. For example, volatiles as cadina-1,4-diene were observed in cv. 'Paulista', Epi-cubebol in cv. 'Espada', β –ocimene in cv. 'Espada', β-elemene in cv. '*Coquinho*' and β-bourbonene in cv. 'Nam Dok Mai' [26–28]. Interestingly, all of those volatiles compounds are known to have antifungal and antibacterial activity [29–31]. Red mango fruit, which were exposed to light and accumulated anthocyanin, were more tolerant to postharvest chilling. To examine longer storage times, mango fruit were stored at 10 °C for 4–6 weeks instead of the common practice of 3 weeks' storage at 12 °C. The red and green fruit, picked from the same orchard on the same day, showed no significant difference in ripening parameters: all of the fruit turned yellow, and showed increased TSS and reduced citric acid as storage and ripening progressed (Figure S2). Interestingly, the red fruit had a higher level of TSS and lower acidity after shelf life. This is probably because the red fruit were exposed to light during their growth and accumulated higher levels of starch. Increased sugar and reduced acidity are correlated to better taste and acceptance [3].

While both red and green mango fruit presented similar ripening progress, fruit with a high percentage of anthocyanin and red color in their peel were more resistant to chilling injuries and presented less black spots and pitting (Figure 7). Our previous study suggested that black spots and pitting in mango fruit are actually discolored lenticels that accumulate dead cells in response to chilling [32]. This programmed cell death was correlated with the increase in ROS. A similar increase in ROS was seen in green mango stored at suboptimal temperatures (Figure 3). Interestingly, as chilling injuries is due to discolored lenticels accumulated in the green mango, more natural opening occurred. This natural opening could be a suitable penetration site for pathogenic fungi [6]. Indeed, more side decay was observed in the green fruit that had more black spots and pitting.

To summarize, this study showed that 'Shelly' mango fruit exposed to sunlight in the orchard activate the phenylpropanoid pathway and accumulate flavonols and anthocyanin. Accumulation of these flavonoids was correlated with increased antioxidant activity and reduced ROS, leading to reduced lipid peroxidation, thereby enhancing the fruit's tolerance to chilling during storage at suboptimal temperature. Following these results, red mango fruit stored at 10 °C had reduced chilling injury and reduced decay and could be stored for longer periods of time than the green fruit. This manuscript findings should result in a future applicative research that will induce the accumulation of anthocyanins in the peel of mango fruit in order to modify the fruit to a more attractive and tolerant fruit to both biotic and abiotic stress.

4. Materials and Methods

4.1. Plant Material and Storage Conditions

Mango (*Mangifera indica* L., cv. Shelly) fruit were harvested in July 2016 and 2017. Fruits were picked from two different canopy positions in the orchard: the red-colored fruit from the exterior position with direct exposure to sunlight, and the green-colored fruit from the interior position in the shaded part of the canopy. Five hours after harvest, the fruit were transported from 'Mor-Hasharon' storage house, Israel to the Agricultural Research Organization, Volcani Center, Israel (1 h transport time). Uniform, unblemished fruit weighing approximately 400 g were selected on the basis of peel color: red fruit had more than 60% of the fruit peel colored in red and green fruit had less than 10% of the fruit peel colored in red. The fruit were washed with tap water and air-dried, and then stored at 5, 8 or 12 °C for 3 weeks in the cold-storage rooms. Each treatment consisted of 6 cardboard boxes with 10–12 fruits in each. After cold storage, the fruit were stored for 7 days at 20 °C to mimic shelf life. In the second experiment, red or green mango fruit were stored at 10 °C for 4 weeks, 5 weeks or 6 weeks and then transferred to 20 °C for 7 days to mimic shelf life. Each treatment consisted of six cardboard boxes with 10–12 fruits in each. The temperature in the cold-storage room was monitored by a DAQ tool—double-strand wire logger/data acquisition control system (T.M.I. Barak Ltd., Ramat Gan, Israel). Each experiment was repeated in the years 2016 and 2017.

4.2. Evaluation of Physiological Parameters of Red and Green Mango Fruit

Ripening and physiological parameters—firmness (Newton), TSS, and acidity (citric acid equivalence)—of red and green mango fruit were measured at harvest, after cold storage and after the additional 7 days of shelf life. Fruit firmness was determined by an electronic penetrometer force gauge LT-Lutron FG-20 KG (Jakarta, Indonesia) with an 11-mm probe at two points on the equatorial line of each fruit (20 measurements per treatment). Percentage of TSS was measured from the juice of fruit pulp using Palette digital-refractometer PR-1 (Model DBX-55, Atago, Tokyo, Japan), 10 fruits per treatment. The acidity was determined as citric acid equivalent mass in 1 mL of pulp juice of red and green mango that was dissolved in 40 ml of double distilled water, using an automatic titrimeter (Model 719 s, Titrino Metrohm Ion Analysis Ltd., Herisau, Switzerland), 10 measurements per treatment.

4.3. Mango Fruit Physiological and Pathological Parameters

Physiological and pathological parameters of the mango fruit were assessed on the basis of severity of black spots, red spots, pitting, stem decay and side decay in red and green mango fruit after cold storage (5, 8 or 12 °C) and a further 7 days of shelf life at 20 °C. Black spot and red spot severity indices were evaluated on a relative scale of 0–10 (1 representing mild black spots and 10 representing severe black spots, 40 evaluations per treatment). Pitting severity index was presented on a relative scale of 0–10 (1 representing mild pitting and 10 representing severe pitting, 40 evaluations per treatment). Stem decay and side decay incidence was evaluated as percentage of decayed red or green mango fruit in each box.

4.4. Analysis of Peel Color of Red and Green Mango Fruit

The mango peel color was measured after harvest, after cold storage and after shelf life using a chromometer model CR-400/410 (Konicka Minolta, Osaka, Japan) at two points on the equatorial line of each fruit, at the red side and at the green side (20 measurements per treatment) and presented as hue angle, where 0 = red, 60 = yellow, 120 = green. Anthocyanin and flavonoid contents were measured in a Multiplex III fluorescence detector (Force A, Orsay, France), consisting of 12 fluorescent signals. The ratios between these signals in different mathematical expressions were correlated to the fluorescence of major chemical groups, e.g., anthocyanin (FER_RG, the ratio of far-red emission excited by red or green light) and flavonoids (FLAV, the logarithm of the red to ultraviolet [UV] excitation ratio of far-red chlorophyll fluorescence) [33]. The methodology was as described in Bahar et al. [33]. Ten fruit were measured from both sides for each treatment after harvest and after cold storage.

4.5. Determination of Total Anthocyanin and Flavonol Contents in Red and Green Mango Fruit

Total anthocyanin content was determined for peel extract of red and green mango fruit stored in the cold (5, 8 or 12 °C) for 3 weeks and an additional 7 days of shelf life by spectrophotometry after organic methanol extraction of mango fruit peel and further measurement of absorption at 528 nm [34].

The total flavonoid content in the red and green mango peel was extracted and measured for fruits stored under different cold-storage conditions (5, 8 or 12 °C) for 3 weeks and an additional 7 days of shelf life using the aluminum chloride colorimetric method [35] and different concentrations of quercetin was used as a standard for quantification. The absorbance was measured at 415 nm using a UV spectrophotometer, and the levels of total flavonol contents were determined in triplicates, respectively. Results were expressed as mg quercetin in 1 g of sample (mg QE/g).

4.6. DPPH Radical-Scavenging Capacity

The DPPH radical-scavenging activity of red and green mango peel extracts was estimated according to the method followed by Cheung et al. [36] with slight modifications. In this assay, antioxidants present in the sample reduced the DPPH radicals, which absorb at 517 nm. Several concentration of ascorbic acid was used as a reference standard for quantification, the reaction was carried out in triplicate. The 'Shelly' mango peel was extracted following different cold-storage conditions (5, 8 or 12 °C for 3 weeks) and an additional 7 days of shelf life.

4.7. Analysis of ROS Production and Confocal Microscopy

Mango fruit peel (200-μm thickness) was taken from red fruit–red side and green side, and green fruit–red side and green side, and incubated with 10 μM of 2,7-dichlorodihydro fluorescein diacetate (H_2DCF-DA) in phosphate buffered saline (PBS) for 15 min in the dark, then washed twice with PBS. The stained peels were observed under fluorescence microscope (Olympus-BX53, Tokyo, Japan) using GFP3 excitation and emission wavelengths. The relative intensity of the fluorescent signal was calculated, using Image J software, as the average intensity from three focal planes in three biological repeats for each sample of red or green sides of the red or green fruit stored at 5, 8 or 12 °C for 3 weeks and an additional 7 days of shelf life.

4.8. Evaluation of Lipid Peroxidation by IVIS

Red and green mango fruit were randomly selected after 3 weeks of cold storage at 5, 8 or 12 °C to detect lipid peroxidation level using a pre-clinical IVIS (PerkinElmer, Waltham, MA, USA). Fruit were kept in the dark for 2 h prior to evaluation. Lipid peroxidation in the fruit was detected and visualized by auto-luminescence of peroxide lipids as described previously [6], using auto-luminescence for 20 min with emission at 640–770 nm. The auto-luminescence was recorded with a highly sensitive charge-coupled-device (CCD) camera. The optical luminescent image data are presented as intensity in terms of radiance ($photons^{-1}$ cm^{-2} steradian). The measurements were repeated three times with different fruits.

4.9. GC-MS Analysis of Mango Volatiles

Peel (1 g) was sampled from seven red and seven green 'Shelly' mango fruit for each replicate, and three biological replicates were collected. Each replicate was collected in a 20-mL amber vial (LaPhaPack, Langerwehe, Germany) prepared in advance with 5 mL of 20% (*w/v*) NaCl (Sigma-Aldrich, St Louis, MO, USA), 0.6 g NaCl, and 25 μL of 10 ppm S-2-octanol (Sigma-Aldrich) added as an internal standard. On the day of analysis, samples were prewarmed for 1 h at 30 °C on an orbital shaker at 250 rpm.

A solid-phase microextraction (HS-SPME) holder (Agilent, Palo Alto, CA, USA) assembled with a fused silica fiber (Supelco, Bellefonte, PA, USA) coated with polydimethylsiloxane (50/30-μM thickness, 1 cm) was used to absorb the volatile compounds. The fiber was desorbed at 250 °C for 3 min in splitless injection mode in an Agilent gas chromatograph series 7890A fitted with an Agilent HP-5MS fused silica capillary column (30 m long × 0.25 mm ID × 0.25 μm film thickness) coupled to a 5975C mass spectrometer detector (Agilent). The temperature conditions were adjusted as follows: 40 °C for 2 min, raised at 10 °C/min to 150 °C, then at 15 °C/min to 220 °C with holding for 5 min; injector temperature was 250 °C. The total run time was 23 min with helium as the carrier gas adjusted to a flow rate of 0.794

mL/min in splitless mode, with ionization energy of 70 eV. Compounds were tentatively identified using the NIST mass spectral library (Version 5) with Chemstation version E.02.00.493. MS description was based on the NIST library, Version 05 (Agilent). Volatile identification was based on the RI and mass spectra as indicated. Specific compounds were also identified by authentic standards. Relative quantification was done according to the retention time of each compound.

4.10. Ethylene and Respiration

Red and green mango fruit cv. Shelly were stored at 5, 8 or 12 °C. Eight red-colored and eight green-colored fruit were enclosed in glass jars (one fruit per jar). The respiration production rates of the mango fruit were measured by closing the bottles for 1 h. Samples were taken using syringes and subsequently analyzed by GC for CO_2 (GC-2014 gas chromatograph, Shimadzu, Tokyo, Japan) and for ethylene (Varian Model 3300 GC, Agilent Technologies, Santa Clara, CA, USA). The respiration rate was measured after harvest, and after 1 week and 3 weeks in cold storage.

4.11. Statistical Analysis

Data were analyzed for significance of differences by Student's t-test or ANOVA with Tukey–Kramer HSD using JMP software (SAS, Cary, NC, USA). $p < 0.05$ was considered statistically significant.

Supplementary Materials: The following are available online at http://www.mdpi.com/1420-3049/23/7/1832/s1. F Figure S1–S6 and Table S1.

Author Contributions: S.P.K. conducted experiments, analysed the data and prepared the manuscript. D.M. conducted the experiments and analysed the data. O.F. conducted the experiments and analysed the data. N.A. coordinated the experiments, data analysis and the manuscript preparation.

Funding: This research was funded by the Chief Scientist of the Israeli Ministry of Agriculture (Grant No. 430-0592-18).

Acknowledgments: This manuscript is contribution no. 822/18 from the Agricultural Research Organization, Volcani Center, Rishon LeZion 7505101, Israel.

Conflicts of Interest: The authors declare no conflict of interest.

References

1. Manthey, J.A.; Perkins-Veazie, P. Influences of harvest date and location on the levels of β-carotene, ascorbic acid, total phenols, the in vitro antioxidant capacity, and phenolic profiles of five commercial varieties of mango (*Mangifera indica* L.). *J. Agric. Food Chem.* **2009**, *57*, 10825–10830. [CrossRef] [PubMed]
2. Ma, X.; Wu, H.; Liu, L.; Yao, Q.; Wang, S.; Zhan, R.; Xing, S.; Zhou, Y. Polyphenolic compounds and antioxidant properties in mango fruits. *Sci. Horticult.* **2011**, *129*, 102–107. [CrossRef]
3. Alkan, N.; Kumar, P. Postharvest storage management of mango fruit. In *Achieving Sustainable Cultivation of Mango*; Galán Saúco, V., Lu, P., Eds.; Burleigh Dodds Science Publishing: Cambridge, UK, 2018.
4. Narayana, C.; Rao, D.; Roy, S.K. Mango production, postharvest physiology and storage. In *Tropical and Subtropical Fruits: Postharvest Physiology, Processing and Packaging*; Wiley: Hoboken, NJ, USA, 2012; pp. 259–276.
5. Sivankalyani, V.; Maoz, I.; Feygenberg, O.; Maurer, D.; Alkan, N. Chilling Stress Upregulates α-Linolenic Acid-Oxidation Pathway and Induces Volatiles of C6 and C9 Aldehydes in Mango Fruit. *J. Agric. Food Chem.* **2017**, *65*, 632–638. [CrossRef] [PubMed]
6. Sivankalyani, V.; Sela, N.; Feygenberg, O.; Zemach, H.; Maurer, D.; Alkan, N. Transcriptome dynamics in mango fruit peel reveals mechanisms of chilling stress. *Front. Plant Sci.* **2016**, *7*, 1579. [CrossRef] [PubMed]
7. Grotewold, E. The genetics and biochemistry of floral pigments. *Annu. Rev. Plant Biol.* **2006**, *57*, 761–780. [CrossRef] [PubMed]
8. Ajila, C.; Naidu, K.; Bhat, S.; Rao, U.P. Bioactive compounds and antioxidant potential of mango peel extract. *Food Chem.* **2007**, *105*, 982–988. [CrossRef]

9. Al-Farsi, M.; Alasalvar, C.; Morris, A.; Baron, M.; Shahidi, F. Comparison of antioxidant activity, anthocyanins, carotenoids, and phenolics of three native fresh and sun-dried date (*Phoenix dactylifera* L.) varieties grown in Oman. *J. Agric. Food Chem.* **2005**, *53*, 7592–7599. [CrossRef] [PubMed]

10. De Pascual-Teresa, S.; Moreno, D.A.; García-Viguera, C. Flavanols and anthocyanins in cardiovascular health: A review of current evidence. *Int. J. Mol. Sci.* **2010**, *11*, 1679–1703. [CrossRef] [PubMed]

11. Spencer, J.P. The impact of fruit flavonoids on memory and cognition. *Br. J. Nutr.* **2010**, *104*, S40–S47. [CrossRef] [PubMed]

12. Sivankalyani, V.; Feygenberg, O.; Diskin, S.; Wright, B.; Alkan, N. Increased anthocyanin and flavonoids in mango fruit peel are associated with cold and pathogen resistance. *Postharvest Biol. Technol.* **2016**, *111*, 132–139. [CrossRef]

13. Kumar, S.P.; Feygenberg, O.; Maurer, D.; Diskins, S.; Saada, D.; Cohen, Y.; Luria, S.; Alkan, N. Anthocyanin in mango fruit peel is associated with cold and pathogen resistance. *ACTA Horticult.* **2018**, in press.

14. Lo Piero, A.R.; Puglisi, I.; Rapisarda, P.; Petrone, G. Anthocyanins accumulation and related gene expression in red orange fruit induced by low temperature storage. *J. Agric. Food Chem.* **2005**, *53*, 9083–9088. [CrossRef] [PubMed]

15. Sanchez-Ballesta, M.T.; Romero, I.; Jiménez, J.B.; Orea, J.M.; González-Ureña, Á.; Escribano, M.I.; Merodio, C. Involvement of the phenylpropanoid pathway in the response of table grapes to low temperature and high CO_2 levels. *Postharvest Biol. Technol.* **2007**, *46*, 29–35. [CrossRef]

16. Hammerschmidt, R. Phytoalexins: What have we learned after 60 years? *Annu. Rev. Phytopathol.* **1999**, *37*, 285–306. [CrossRef] [PubMed]

17. Treutter, D. Significance of flavonoids in plant resistance: A review. *Environ. Chem. Lett.* **2006**, *4*, 147. [CrossRef]

18. Nozzolillo, C.; Isabelle, P.; Andersen, Ø.M.; Abou-Zaid, M. Anthocyanins of jack pine (*Pinus banksiana*) seedlings. *Can. J. Bot.* **2002**, *80*, 796–801. [CrossRef]

19. Oberbaueri, S.F.; Starr, G. The role of anthocyanins for photosynthesis of Alaskan arctic evergreens during snowmelt. *Adv. Bot. Res.* **2002**, *37*, 129–145.

20. Gould, K.S. Nature's Swiss army knife: The diverse protective roles of anthocyanins in leaves. *BioMed Res. Int.* **2004**, *2004*, 14–320. [CrossRef] [PubMed]

21. Skrovankova, S.; Sumczynski, D.; Mlcek, J.; Jurikova, T.; Sochor, J. Bioactive compounds and antioxidant activity in different types of berries. *Int. J. Mol. Sci.* **2015**, *16*, 24673–24706. [CrossRef] [PubMed]

22. Zhang, Y.; Butelli, E.; De Stefano, R.; Schoonbeek, H.-J.; Magusin, A.; Pagliarani, C.; Wellner, N.; Hill, L.; Orzaez, D.; Granell, A. Anthocyanins double the shelf life of tomatoes by delaying overripening and reducing susceptibility to gray mold. *Curr. Biol.* **2013**, *23*, 1094–1100. [CrossRef] [PubMed]

23. Lalel, H.J.; Singh, Z.; Tan, S.C. Aroma volatiles production during fruit ripening of 'Kensington Pride' mango. *Postharvest Biol. Technol.* **2003**, *27*, 323–336. [CrossRef]

24. Pandit, S.S.; Chidley, H.G.; Kulkarni, R.S.; Pujari, K.H.; Giri, A.P.; Gupta, V.S. Cultivar relationships in mango based on fruit volatile profiles. *Food Chem.* **2009**, *114*, 363–372. [CrossRef]

25. Kamatou, G.P.; Vermaak, I.; Viljoen, A.M. Eugenol-from the remote Maluku Islands to the international market place: A review of a remarkable and versatile molecule. *Molecules* **2012**, *17*, 6953–6981. [CrossRef] [PubMed]

26. Andrade, E.H.A.; Maia, J.G.S.; Maria das Graças, B.Z. Aroma volatile constituents of Brazilian varieties of mango fruit. *J. Food Compos. Anal.* **2000**, *13*, 27–33. [CrossRef]

27. Laohaprasit, N.; Kukreja, R.K.; Arunrat, A. Extraction of volatile compounds from 'Nam Dok Mai' and 'Maha Chanok' mangoes. *Int. Food Res. J.* **2012**, *19*, 1445–1448.

28. Gebara, S.S.; de Oliveira Ferreira, W.; Ré-Poppi, N.; Simionatto, E.; Carasek, E. Volatile compounds of leaves and fruits of *Mangifera indica var. coquinho* (*Anacardiaceae*) obtained using solid phase microextraction and hydrodistillation. *Food Chem.* **2011**, *127*, 689–693. [CrossRef] [PubMed]

29. Simic, A.; Sokovic, M.D.; Ristic, M.; Grujic-Jovanovic, S.; Vukojevic, J.; Marin, P.D. The chemical composition of some *Lauraceae* essential oils and their antifungal activities. *Phytother. Res.* **2004**, *18*, 713–717. [CrossRef] [PubMed]

30. Vukovic, N.; Milosevic, T.; Sukdolak, S.; Solujic, S. Antimicrobial activities of essential oil and methanol extract of Teucrium montanum. *Evid.-Based Complement. Altern. Med.* **2007**, *4*, 17–20. [CrossRef] [PubMed]

Molecules **2018**, *23*, 1832

31. Tsao, R.; Zhou, T. Antifungal activity of monoterpenoids against postharvest pathogens *Botrytis cinerea* and *Monilinia fructicola*. *J. Essent. Oil Res.* **2000**, *12*, 113–121. [CrossRef]

32. Sivakumar, D.; Jiang, Y.; Yahia, E.M. Maintaining mango (*Mangifera indica* L.) fruit quality during the export chain. *Food Res. Int.* **2011**, *44*, 1254–1263. [CrossRef]

33. Bahar, A.; Kaplunov, T.; Zutahy, Y.; Daus, A.; Lurie, S.; Lichter, A. Auto-fluorescence for analysis of ripening in Thompson Seedless and colour in Crimson Seedless table grapes. *Aust. J. Grape Wine Res.* **2012**, *18*, 353–359. [CrossRef]

34. Xie, D.-Y.; Sharma, S.B.; Paiva, N.L.; Ferreira, D.; Dixon, R.A. Role of anthocyanidin reductase, encoded by BANYULS in plant flavonoid biosynthesis. *Science* **2003**, *299*, 396–399. [CrossRef] [PubMed]

35. Chang, C.-C.; Yang, M.-H.; Wen, H.-M.; Chern, J.-C. Estimation of total flavonoid content in propolis by two complementary colorimetric methods. *J. Food Drug Anal.* **2002**, *10*, 178–182.

36. Cheung, L.; Cheung, P.C.; Ooi, V.E. Antioxidant activity and total phenolics of edible mushroom extracts. *Food Chem.* **2003**, *81*, 249–255. [CrossRef]

Sample Availability: Samples of the compounds are available from the authors.

![molecules logo] ![MDPI logo]

Article

Anthocyanin Composition and Content in Rye Plants with Different Grain Color

Pavel A. Zykin [1], Elena A. Andreeva [1,2,*], Anna N. Lykholay [1,2], Natalia V. Tsvetkova [1,2] and Anatoly V. Voylokov [2]

[1] Faculty of Biology, St. Petersburg State University, Universiteskaya nab.7/9, St. Petersburg 199034, Russia; pavel.zykin@gmail.com (P.A.Z.); lankira@mail.ru (A.N.L.); ntsvetkova@mail.ru (N.V.T.)

[2] Laboratory of Plant Genetics and Biotechnology, Vavilov Institute of General Genetics Russian Academy of Sciences, St. Petersburg branch, Universiteskaya nab. 7/9, St. Petersburg 199034, Russia; av_voylokov@mail.ru

* Correspondence: elena.alex.andreeva@gmail.com; Tel.: +7-921-747-9981

Received: 29 March 2018; Accepted: 17 April 2018; Published: 19 April 2018

Abstract: The color of grain in cereals is determined mainly by anthocyanin pigments. A large level of genetic diversity for anthocyanin content and composition in the grain of different species was observed. In rye, recessive mutations in six genes (vi1...vi6) lead to the absence of anthocyanins in all parts of the plant. Moreover, dominant genes of anthocyanin synthesis in aleurone (gene C) and pericarp (gene Vs) also affect the color of the grain. Reverse phase high-performance liquid chromatography and mass spectrometry were used to study anthocyanins in 24 rye samples. A lack of anthocyanins in the lines with yellow and brown grain was determined. Delphinidin rutinoside and cyanidin rutinoside were found in the green-seeded lines. Six samples with violet grains significantly varied in terms of anthocyanin composition and content. However, the main aglycone was cyanidin or peonidin in all of them. Monosaccharide glucose and disaccharide rutinose served as the glycoside units. Violet-seeded accession forms differ in the ratio of the main anthocyanins and the range of their acylated derivatives. The acyl groups were presented mainly by radicals of malonic and sinapic acids. For the colored forms, a profile of the revealed anthocyanins with the indication of their contents was given. The obtained results are discussed in connection to similar data in rice, barley, and wheat, which will provide a perspective for future investigations.

Keywords: rye inbred lines; genes of grain color; anthocyanin identification; comparative genetics; HPLC-MS analysis

1. Introduction

Due to the peculiarities of their chemical structure, anthocyanins have certain biological activity, which manifests itself in the composition of plant food. In cereals, the edible part is the grain (caryopsis). The grain is a single-seeded fruit, the tissues of which accumulate specific metabolites, including anthocyanins. Anthocyanins and the structurally similar proanthocyanidins are one of the end products of the branched chain of flavonoid biosynthesis [1]. Their accumulation in the caryopsis of cereals occurs in the cells of the fruit (pericarp) and seed (testa) coats as well as in the aleurone layer of the endosperm. The synthesis of anthocyanins and their intra- and intercellular movement is controlled by structural and regulatory genes [1]. The mutational variability of these genes leads to a change in the qualitative and quantitative composition of anthocyanins in grain, which is largely influenced by environmental conditions. During the course of domestication and further breeding, the composition of metabolites, including anthocyanins, significantly changed in cereal grains [2]. This has led to most modern varieties of wheat, barley, maize, and rice being represented by genotypes that are not capable of synthesizing anthocyanins in the grain. Currently, taking into account the aim to improve health through grain nutrients, work has been conducted to recreate the genotypes that can accumulate anthocyanins and their precursors in the grain of cereals [3]. The source of colorful genes for cereal improvement include local varieties, genetic collections, and wild relatives. For the accumulation of anthocyanins, genetic engineering approaches have also been developed [1]. The composition and concentration of anthocyanins in small cereals largely reflect the color of the grain. Wheat and barley produce yellow (white), blue (green), red (brown), and purple (violet, black) grains. The cultural and weedy rye revealed the same range of variation in coloration of grain. This implies a similar genetic control and approximately the same composition of anthocyanins in all the cereals with variously colored grain. The first data on the composition of anthocyanins in small cereals were obtained in rye. The analysis of anthocyanin pigments in green [4] and purple (violet) [5] grain was carried out using paper chromatography. This established that green grain contains delphinidin 3-*O*-rutinoside as the main pigment. In the purple grain, the composition of anthocyanins was richer. Cyanidin 3-*O*-glucoside and peonidin 3-*O*-glucoside in addition to trace amounts of cyanidin 3-*O*-rutinoside and peonidin 3-*O*-rutinoside were found. In addition, acylated forms were found for all four glycosides, but the composition of acyl residues was not established. Similar results on other cereals were subsequently obtained [6]. The improvement of separation and identification techniques has significantly expanded the understanding of anthocyanin composition in related cereals [7–9], although this has not occurred in rye. The purpose of this paper is to fill this gap and to discuss new data on anthocyanin composition in related species of wheat, barley and rice.

2. Results

The high-performance liquid chromatography-mass spectrometry (HPLC-MS) analysis of 24 rye samples with different grain colors showed an absence of a detectable level of anthocyanins in ten accessions with yellow grain and in three accessions with brown grain. A uniform pattern was established for the five lines of independent origin, which is described as green-seeded (Table 1, Figure 1a). Delphinidin rutinoside was found in all five lines, cyanidin rutinoside was detected in four of them and delphinidin 3-*O*-glucoside was found in one line (L301). A specific feature of rye as a cross-pollinated species is inbred depression, which is manifested as a sharp decrease for all quantitative traits of plants during inbred propagation. The amount of anthocyanins is not an exception. All studied inbred lines with green and violet (GC-14/1, GC-14/2) seed colors have low anthocyanin content, which does not exceed 1 mg/kg (Table 1).

Table 1. Identification and tentative quantification of anthocyanin composition in rye grain.

Anthocyanin [1]	M/Z Molecular Ion	M/Z Fragment Ions	RT, min	Content mg/kg of Dry Matter [3]
Green grain L8, 0.18 mg/kg				
Delphinidin Rutinoside	611.16	303.05, 465.11	17.8	0.15
Cyanidin Rutinoside	595.17	287.05, 449.11	20.3	0.03
Green grain L87, 0.07 mg/kg				
Delphinidin Rutinoside	611.16	303.05, 465.11	17.9	0.07
Green grain L301, 0.88 mg/kg				
Delphinidin Rutinoside	611.16	303.05, 465.11	17.9	0.88
Cyanidin Rutinoside	595.17	287.05, 449.11	20.4	<LOQ [2]
Delphinidin 3-*O*-Glucoside	465.10	303.05	15.9	<LOQ
Green grain RMu1 green, >0.0 mg/kg				
Cyanidin Rutinoside	595.17	287.05, 449.11	20.4	<LOQ
Delphinidin Rutinoside	611.16	303.05, 465.11	17.8	<LOQ
Green grain RMu5, >0.0 mg/kg				
Delphinidin Rutinoside	611.16	303.05, 465.11	17.9	<LOQ
Cyanidin Rutinoside	595.17	287.05, 449.11	20.5	<LOQ

Table 1. *Cont.*

Anthocyanin [1]	M/Z Molecular Ion	M/Z Fragment Ions	RT, min	Content mg/kg of Dry Matter [3]
Violet grain S10, 32.6 mg/kg				
Cyanidin (malonyl)hexoside)	535.11	287.06	25.0	7.39
Cyanidin 3-*O*-Glucoside	449.11	287.06	18.7	6.47
Peonidin (malonyl)hexoside	549.12	301.07	29.3	6.03
Peonidin 3-*O*-Glucoside	463.12	301.07	22.9	5.23
Cyanidin (dimalonyl)hexoside	621.11	287.06	29.1	3.57
Peonidin (dimalonyl)hexoside	635.13	301.07	33.3	2.04
Cyanidin Rutinoside	595.17	287.05, 449.11	20.6	0.72
Peonidin Rutinoside	609.18	301.07, 463.12	24.6	0.72
Cyanidin (sinapoyl)hexoside	655.17	287.05	36.6	0.43
Cyanidin (malonyl)hexoside	535.11	287.05	25.1	<LOQ
Violet grain RMu12, 403.75 mg/kg				
Peonidin Rutinoside	609.18	301.07, 463.12	25.0	115.37
Cyanidin Rutinoside	595.17	287.05, 449.11	20.5	90.98
Cyanidin (succinyl)hexoside	549.12	287.06	30.1	55.27
Cyanidin 3-*O*-Glucoside	449.11	287.06	18.2	40.10
Cyanidin (malonyl)hexoside	535.11	287.06	25.5	35.87
Peonidin 3-*O*-Glucoside	463.12	301.07	23.1	34.03
Unidentified Peonidin	635.13	301.07	34.6	11.79
Cyanidin (dimalonyl)hexoside	621.11	287.06	29.6	9.08
Peonidin (malonyl)hexoside	549.12	301.07	27.9	7.85
Cyanidin (sinapoyl)hexoside	655.17	287.06	37.7	3.41
Unidentified Cyanidin	625.18	287.05	38.3	<LOQ
Violet grain RMu13, 281.46 mg/kg				
Cyanidin Rutinoside	595.17	287.05, 449.11	20.6	101.36
Peonidin Rutinoside	609.18	301.07, 463.12	24.6	52.92
Cyanidin (malonyl)hexoside	535.11	287.06	25.0	37.48
Cyanidin 3-*O*-Glucoside	449.11	287.06	18.7	27.07
Peonidin (malonyl)hexoside	549.12	301.07	29.1	26.13
Peonidin 3-*O*-Glucoside	463.12	301.07	23.0	21.45
Cyanidin (dimalonyl)hexoside	621.11	287.06	28.9	9.11
Peonidin (dimalonyl)hexoside	635.13	301.07	33.1	5.94
Pelargonidin Rutinoside	579.17	271.06, 443.10	22.8	<LOQ
Cyanidin (sinapoyl)hexoside	655.17	287.06	38.3	<LOQ
Violet grain GC-14, 1.63 mg/kg				
Peonidin 3-*O*-Glucoside	463.12	301.07	22.8	1.33
Peonidin (malonyl)hexoside	549.12	301.07	29.3	0.30
Malvidin 3-*O*-Galactoside	496.74	331.08	22.9	<LOQ
Peonidin Rutinoside	609.18	301.07, 463.12	24.6	<LOQ
Violet grain GC-14/1, 0.11 mg/kg				
Peonidin 3-*O*-Glucoside	463.12	301.07	23.4	0.11
Violet grain GC-14/2, 0.2 mg/kg				
Cyanidin Rutinoside	595.17	287.05, 449.11	20.6	0.10
Peonidin 3-*O*-Glucoside	463.12	301.07	23.2	0.10
Peonidin Rutinoside	609.19	301.07, 463.12	25.0	<LOQ
Cyanidin 3-*O*-Glucoside	449.11	287.05	17.8	<LOQ

[1] Semi-bold font denotes the major anthocyanins in sample; [2] LOQ—values are lower than the limit of quantification (SN < 10), but higher than the limit of detection (SN > 3); [3] Content calculated as cyanidin equivalent.

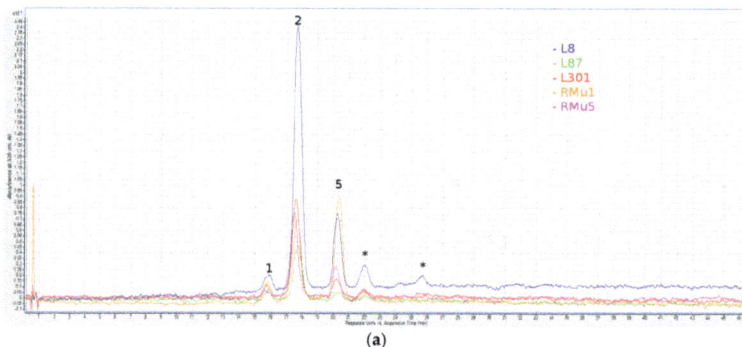

(a)

Figure 1. *Cont.*

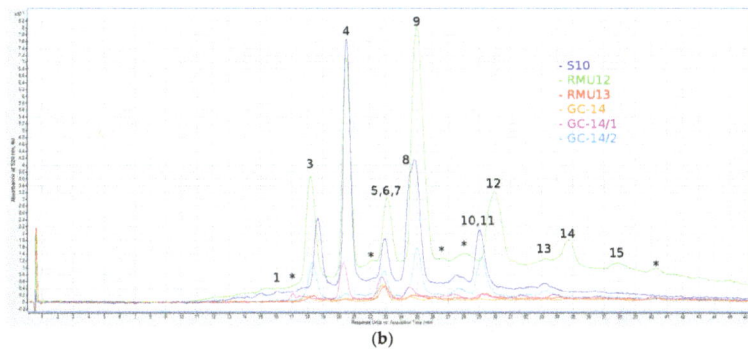

(b)

Figure 1. Liquid chromatography with spectrophotometric absorption detection in visible spectrum (LC-VIS) chromatograms at 520 ± 4 nm of green grain (**a**) and violet grain (**b**) seeds extracts. 1—Delphinidin 3-*O*-Glucoside; 2—Delphinidin Rutinoside; 3—Cyanidin 3-*O*-Glucoside; 4—Cyanidin Rutinoside; 5—Pelargonidin Rutinoside; 6—Malvidin 3-*O*-Galactoside; 7—Peonidin 3-*O*-Glucoside; 8—Peonidin Rutinoside; 9—Cyanidin (malonyl)hexoside; 10—Peonidin (malonyl)hexoside; 11—Cyanidin (dimalonyl)hexoside; 12—Cyanidin (succinyl)hexoside; 13—Peonidin (dimalonyl)hexoside; 14—Unidentified Peonidin; 15—Cyanidin (sinapoyl)hexoside; and *—peaks for which no anthocyanin aglycone was identified.

Through a decrease in the total content of anthocyanins, inbred depression can have a non-specific effect on the composition of detected anthocyanins, especially on the detection of minor components. Between the three related samples (GC-14, GC-14/1 and GC-14/2), differences in the composition of the minor anthocyanins were found (Table 1, Figure 1b). Peonidin 3-*O*-glucoside was found in all three forms. In GC-14/1, only this anthocyanin was found, while additional peonidin derivatives were also found in the original sample of GC-14, which were namely peonidin 3-*O*-glucoside and peonidin (malonyl)hexoside. GC-14 was distinguished by the presence of malvidin 3-*O*-galactoside, which was not found in other violet forms. Differences, such as the aglycone (malvidin) and glycosyl residue (galactoside), represent the genetic heterogeneity of GC-14, which has been reproduced from the moment of its isolation by cross-pollination of plants within the plot. GC-14/2 is distinguished from both related forms by the presence of two cyanidin derivatives: cyanidin rutinoside (0.10 mg/kg); and cyanidin 3-*O*-glucoside detected at the limit of sensitivity. The inbred lines RMu12 and RMu13 are similar in the expression of intense anthocyanin coloration of all parts of plants, including grains. The similarity of line genotypes was indicated by the very close profile of anthocyanins found in these forms (Table 1, Figure 1). In both lines, the rutinosides of cyanidin and peonidin dominated, while lower concentrations of glucosides of both anthocyanidins and their derivatives with one or two residues of malonic acid were found. The composition of anthocyanins in the grain of these lines was distinguished by the presence of a significant concentration of another anthocyanin with an aliphatic acyl group in RMu12, which was namely cyanidin (succinyl)hexoside. Furthermore, another distinguishing characteristic is the presence of pelargonidin rutinoside as a minor component in RMu13. In all three lines with a high concentration of anthocyanins (S10, the RMu12 and RMu13) anthocyanin-containing acyl groups of an aromatic sinapic acid was found namely, cyanidin (sinapoyl)hexoside. Line S10 differs by the high content of glycosides attached to cyanidin and peonidin cores, added by one or two residues of malonic acid. Moreover, the predominant glycoside residue in line S10 is monosaccharide glucose, while disaccharide rutinose is more common for the violet-seeded lines RMu12 and RMu13.

To provide an alternative quantitative approach, we measured the concentration of anthocyanins in the anthocyanin-rich forms of S10, RMu13 and RMu12 by the spectrophotometric pH–differential method. The results and data used for calculations are summarized in Table 2.

Table 2. Quantification of anthocyanins by spectrophotometric pH—differential method.

Accession	Major Anthocyanin	$\lambda_{\text{vis-max}}$ (nm) [1]	ε	MW	Content mg/kg of Dry Matter
S10	Cyanidin (malonyl)glucoside	528	32,360	535.11	25.33
RMu13	Cyanidin Rutinoside	510	7000	595.17	328.2
RMu12	Peonidin Rutinoside	512	14,100	609.18	451.14

[1] Maximum absorption in the visible region (400–700 nm).

The variations in data between the two methods used could be explained by the use of the spectrophotometric method only for the major anthocyanin, despite the use of complex mixtures; and the use of cyanidin ($\lambda_{\text{vis-max}}$ = 516) equivalents of anthocyanins for HPLC-MS due to the lack of other quantitative standards. Nevertheless, the difference is small and consistent, allowing us to conclude that the line RM12 is exceptionally high in anthocyanin content. The total anthocyanin content calculated using both methods is on average 15 times more in the RMu12 than in the S10, which is significantly higher than the other three violet-seeded accessions that only contain 0.11–1.63 mg/kg.

3. Discussion

The synthesis of anthocyanin pigments in grain of cereals is localized in the maternal tissues—fruit (pericarp) and seed (testa) coats as well as in the hybrid tissue—aleurone layer of the endosperm. Each of these tissues is characterized by a certain composition of anthocyanins and related compounds. The qualitative and quantitative composition of pigments is controlled by structural and regulatory genes (transcription factors). Along with similarity in the genetic control of anthocyanins in all cereals, the species specificity and significant intraspecific variability due to non-lethality of mutations in genes of anthocyanin biosynthesis were established [10]. Dominant alleles of regulatory genes coloring aleurone and pericarp in bread wheat were introgressed in its genome on the basis of distant hybridization [11]. The native genes are only homeological loci (R1–R3) of the red coloration of the testa, which controls three MYB-type transcription factors [12]. In barley, the variability for grain color exists due to the presence of many spontaneous and induced mutants [13]. In rice, this variability is due to the diversity of grain color in local varieties. Open-pollinated rye varieties are heterozygous and heterogeneous populations, due to the action of a rigid system of gametophytic incompatibility. They are also heterogeneous in terms of the color of the grains. In most varieties, there are green, yellow and rarely brown grains, between which there are grains of intermediate colors. The purple (violet) grain is typical for some samples of weedy rye [14]. The use of self-compatible (self-fertile) mutants allows differentiation of populations for inbred lines and to fix mutant alleles of genes in them, including those for grain color [14]. No anthocyanins were detected in brown-seeded and yellow-seeded lines, including those without anthocyanin. This corresponds to the absence or very low content of anthocyanins in white, red and brown rice grain; yellow grain in barley; and yellow and red grain in bread wheat [6]. The absence or low content of anthocyanins in all layers of the grain explains their yellow color in all four species. The red color of caryopsis in rice and wheat is associated with the color of testa. The color of testa in rice and wheat grain, as well as the brown color of the pericarp in certain genotypes of rice, is explained by the synthesis of proanthocyanidins—oligomeric and polymeric compounds related to leucoanthocyanidins, which are colorless precursors of colored anthocyanidins. In barley, proanthocyanidins (procyanidin B3, prodelphynidin B3) and their precursor (+)-catechin were detected well before other small cereals [13]. In the grains of barley, they do not have any color. Oligomers and polymers of proanthocyanidins and their precursors flavan-3-ol units are detected by chemical agents (solution of vanillin in hydrochloric acid). The red color of the grain in wheat is explained by the accumulation in the testa of closely related flavonoid compounds, which are converted into a red-brown insoluble pigment as the kernels mature [15]. It was established that the accumulation in the testa of proanthocyanidins and their precursors in barley and wheat control the orthologous MYB-type transcription factors Hvmyb10 and Tamyb10, respectively [12]. In rye, the variability in the color of testa was not revealed as the grain of the three main colors have brown testa.

We have shown that the testa in rye reacts with reagents for proanthocyanidins and their precursors (unpublished). In the presence of vanillin, the brown color of the testa changes to red and when processed with the 4-(Dimethylamino)-cinnamaldehyde (DMACA) solution, the testa becomes blue. However, the nature of the pigment in rye grain with a superficial brown color, which is characteristic of the samples studied by us, has not been established. A previous study [16] showed that the brown color of grain in the line GK-37 arises from the interaction of two genes, with the brown grain producing plants with vi1 vi1, VsVs genotype. A similar interaction of the anthocyaninless mutation (rd) and the dominant allele (Rc–) of the purple grain coloring gene was established in rice [17]. The complementary Rd and Rc rice genes control the structure of dihydroflavonol-4-reductase and MYB-type transcription factor, respectively. As a result of the action of dihydroflavonol-4-reductase leucoanthocyanidins, the precursors of both anthocyanins and proanthocyanidins are formed. It is assumed that the formation of proanthocyanidins in brown pericarp is a consequence of incomplete loss of enzyme activity. The resulting intermediates accumulate in the genotypes of rd rd, Rc– and create a brown pigment. The function of genes vi1 and Vs in rye is unknown.

The green color of the grain in rye and the blue in bread, wheat, and barley are associated with the synthesis of anthocyanins in the aleurone layer of the endosperm. Data on the presence of anthocyanins in aleurone of rice caryopsis are absent in the available literature. Genes of blue aleurone in bread wheat are transferred to its genome, which is composed of whole chromosomes of the fourth homological group or their fragments from related wild species [11]. In barley, a combination of dominant genes (alleles) controlling three types of transcription factors [18] is responsible for the synthesis of anthocyanins in aleurone. The presence of anthocyanins in green grain in rye is associated with a dominant gene C. For the manifestation of the green coloring of rye grain, the presence of dominant alleles in all the loci vi1–vi6 [14,16] is obligatory. The synthesis of anthocyanins in pericarp of barley, wheat, and rice is controlled by corresponding transcription factors [19–21]. In rye, the gene Vs acting in a similar manner at the morphological level was identified and mapped [16]. The composition of anthocyanins in colored aleurone of barley, wheat, and rye is related to the major aglycone, which is delphinidin. Our data confirmed the previously obtained data [4] for green-seeded rye. Furthemore, detection of cyanidin rutinoside in addition to delphinidine rutinoside in green rye grain was done for the first time. Cyanidin derivatives typically dominate in the colored pericarp of barley, wheat, and rice [8,9].

In blue and purple barley, the derivatives of all six major anthocyanidins were identified, which are accumulated in different concentrations in aleurone or pericarp mainly in the form of monoglucosides. Two of them, which are namely the cyanidin 3-*O*-glucoside and peonidin 3-*O*-glucoside, are dominant over the others in the purple barley pericarp [6,7,22]. Only these two anthocyanins with total anthocyanin content of 52.1–1683.6 µg/kg were detected in rice, which has a grain color that varies from light purple to black [23]. Similar results were obtained by other researchers [24,25]. However, the review of 25 works reported 18 different anthocyanins identified in purple rice. Among them, only four could be quantified, which are two major anthocyanins, namely cyanidin-3-*O*-glucoside (51–84%) and peonidin-3-*O*-glucoside (6–16%); and two minor anthocyanins, which are cyanidin-rutinoside (3–5%) and cyanidin-3-*O*-galactoside (1–2%). The glycosides of other four main anthocyanidins (pelargonidin, delphinidin, petunidin and malvidin) besides cyanidin and peonidin were also discovered in some rice genotypes. These anthocyanins can prevail in some rare genotypes. Thus, in the cultivar Chinakuromai, petunidin 3-*O*-glucoside composed almost half of the anthocyanin content. It is interesting that all anthocyanins identified in purple (black) rice are comprised of mono- and disaccharides of six major anthocyanidins, although the acyl derivatives were not found [8].

For the first time, anthocyanins in purple grain were studied in rye [5]. Cyanidin 3-*O*-glucoside and peonidin 3-*O*-glucoside were found to be the main anthocyanins, while there were trace amounts of cyanidin rutinoside, peonidin rutinoside and acylated forms of these glycosides in five inbred lines with purple (violet) grain. The authors highlighted the variation in the ratio of aglycones and acylated derivatives in the studied lines of rye. Our data confirmed that the main aglycones in purple rye are cyanidin and peonidin, which is consistent with other studied cereals with purple grains. For the first time, acyl residues were identified in anthocyanins and their quantitative ratio was established in five rye forms with violet (purple) grains. Significant differences in the qualitative and quantitative composition of anthocyanins in violet-seeded lines were found. Self-fertility of the studied accessions allows us to select a constant sub-line with a high proportion of acylated anthocyanins and high total anthocyanin concentration. It can be used for segregation analysis of anthocyanins in hybrid progenies, with the objective of identifying the genes responsible for inter-line differences. The determination of molecular function of genes which were discovered at the morphological level is of great interest. The available data on candidate anthocyanin genes identified in other cereals can be effectively used through partial sequencing of the rye genome [26]. On the basis of such investigations, a new method will appear for constructing gene markers to obtain the different combinations of anthocyanin genes. For example, it is possible to produce double mutants for vi1–vi6 mutations using markers; to combine each of them with dominant C and Vs genes; and to study anthocyanins and their precursors in the constructed genotypes. Such new genotypes may be useful to resolve different problems, in particular to clarify the chemical nature of brown pigment in the rye grain and to describe the pleiotropic effect of regulatory genes on the metabolism in different kernel compartments. This will allow us to produce breeding material for production of rye varieties with health-promoting effects.

4. Materials and Methods

4.1. Plant Material

The composition of anthocyanins was analyzed in 24 forms and inbred lines of rye, which differed in terms of grain color and the presence of anthocyanins on other parts of the plant (Table 3). Every lacking anthocyanin form carries one of the six recessive mutation (vi1, vi2, vi3, vi4, vi5 or vi6) leading to the absence of anthocyanin coloration of coleoptile, nodes of the stem, glumes and awns of ears and grain. Segregation analysis revealed non-allelic nature of these mutations. The mutant genes have been designated by Latin name viridis (vi) [14] in accordance to the absence of anthocyanin on mutant seedlings, which are green for this reason.

Table 3. Plant material used in the work.

Characteristic	Accession [1]
Lacking anthocyanin	Esto (vi1), GC-22yellow (vi2), GC-23 (vi3), GC-149 (vi4), GC-151 (vi5), GC-150 (vi6)
Yellow-seeded	GC-12, L2, L5, L7 (c)
Brown-seeded	RMu1 brown (vi1, Vs?), GC-22 brown (vi2, Vs?), GC-37(vi1, Vs)
Green-seeded	L8, L87, L301, RMu5, RMu1 green (C)
Violet-seeded	GC-14, GC-14/1, GC-14/2, RMu12, RMu13, S10 (Vs)

[1] Identified or assumed genes are shown in brackets [16].

The variety Esto (vi1) used in the work is a population consisting of self-incompatible genotypes. The accession of genetic collection GC-22 (vi2) was originally created by cross-pollination of plants without anthocyanin, which was selected in two inbred progenies. The form of GC-23 (vi3) was produced under self-pollination of the cloned plant of Vyatka variety. The forms of GC-149 (vi4), GC-151 (vi5) and GC-150 (vi6) were obtained on the basis of hybrids of three different plants of Vyatka variety with self-fertile lines. For many years, forms without anthocyanin were propagated by cross-pollination and their homozygosity was checked by the absence of anthocyanin coloration. Furthermore, for each accession, inbred lines were produced using single seed descent from the original plant. In the course of this work, from the accession GC-22, two sublines were produced that differed in color of kernels: line GC-22 yellow with yellow grains and line GC-22 brown with a light-brown grains. The yellow-seeded forms with anthocyanin color of coleoptiles, nodes and glumes were presented by the accession GC-12 and three highly-inbred lines of independent origin (L2, L5, L7). The inbred lines L2, L5 and L7 were the accessions from Peterhof genetic collection of rye. They originated from plants of three different rye varieties and reproduced by selfing for a long time. The uniformity of the yellow coloring of grain in the hybrids between these lines allows us to identify these lines as the recessive homozygote cc [14]. The dominant allele of this gene C leads to the green color of grain due to the presence of anthocyanins in aleurone. The green-seeded rye with this gene was represented by three highly-inbred lines (L87, L301, L8), line RMu5 and line RMu1 green, which was isolated from the sample of variety Esto. From the same sample, the inbred line RMu1 brown was isolated, which had a constant brown color of the grain. In the test crosses, it was shown that this line is homozygous for allele vi1, which causes the absence of anthocyanin color of plants in the variety Esto. Brown grain is a characteristic of line GC-37, which is derived from a hybrid between violet-seeded (gene Vs) and rye lacking anthocyanin (gene vi1). The genotype of this line, which is vi1vi1, VsVs=, causes the brown color of grain and the absence of anthocyanin color of the normally colored vegetative parts of the plant. Of the six violet-seeded samples (gene Vs), three are related. This is a sample of a weedy rye GC-14 and the two related inbred lines (GC-14/1 and GC-14/2), which were obtained from one initial plant of this population. RMu12 and RMu13 lines in addition to the purple color of the grain are characterized by highly colored red stems. The constant manifestation of these traits was observed in plants of three inbred generations. A grain sample of the spring purple rye S10, obtained from J. Sybenga (Agricultural University, Wageningen, The Netherlands), was heterogeneous and included a green and purple grain. Within two generations of inbred propagation, constant lines were obtained with green and purple grains. All forms with the designation RMu and line L301 have been kindly provided by P. Wehling (JKI, Groß Lüsewitz, Germany) and line L87 by A. Börner (IPK, Gatersleben, Germany). The other accessions and lines are original and are reproduced in the Peterhof genetic collection [14].

4.2. Sample Preparation and Chemicals

A total of 24 rye accessions with different grain colors harvested in 2016 and 2017 were selected for the analysis. For each accession, about 100 grains were weighted, vacuum-dried in a centrifugal vacuum concentrator (Labconco Centrivap) at room temperature until no further change in mass was observed. On average, drying to total dryness took 36 h. Dry mass was recorded for further reference. Grains were ground with household grain mill (Zepter MixSy) for 5 min. Standards of cyanidin chloride and malvidin 3-*O*-galactoside were from Sigma-Aldrich (St. Louis, MO, USA), HPLC-MS grade acetonitrile was purchased from Panreac (Barcelona, Spain), while HPLC-grade formic acid and other chemicals were from Vecton (St.-Petersburg, Russia).

4.3. Extraction of Anthocyanins

The anthocyanins were extracted according to a previous study [27], with some modifications. We chose 70% ethanol acidified by 1% (*w/v*) citric acid as an extraction solvent to avoid esterification of the free carboxyl group of acylated anthocyanins and to prevent their deacylation. Furthermore, all subsequent steps up to HPLC-MS were carried out at a low temperature (<30 °C) for the same reason.

The extraction was done with three portions of extraction solvent for 24 h at +4 °C in the dark. The supernatant was combined and dried to 3 mL on vacuum concentrator (Labconco Centrivap) at room temperature.

4.4. Anthocyanins Purification

The crude extract was purified by selective removal of non-anthocyanin compounds in stages (Figure 2).

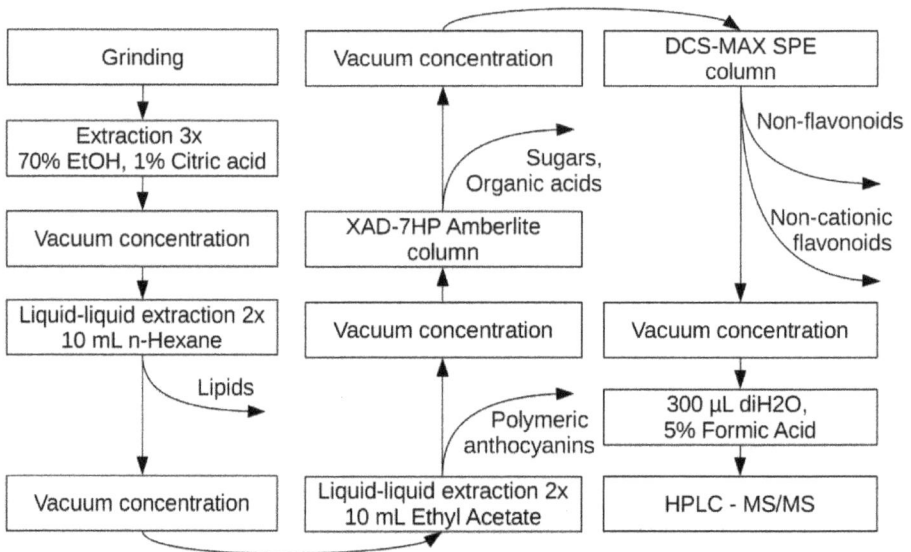

Figure 2. Stages of monomeric anthocyanins purification by selective removal of non-anthocyanin compounds.

Lipids were removed by liquid/liquid extraction with Hexane as follows, which was repeated twice. A total of 10 mL of hexane was added to 3 mL of extract, which was vortexed on BioSan V-32, Multi-Vortex for 15 min, centrifuged at $7000\times g$ at +4 °C for 10 min with the water phase maintained. The solution was evaporated to dryness in the centrifugal vacuum concentrator and resuspended in 3 mL of deionized water. Other non-polar compounds, including the polymeric anthocyanidins (proanthocyanidins) were removed by liquid/liquid extraction with ethyl acetate twice as follows: 10 mL of water-impregnated ethyl acetate was added to 3 mL of extract, which was vortexed on BioSan V-32, Multi-Vortex for 15 min, centrifuged at $7000\times g$ at +4 °C for 10 min with maintenance of the water phase. The solution was evaporated to dryness in the centrifugal vacuum concentrator and resuspended in 1.5 mL of deionized water with 0.1% of formic acid. To remove sugars and organic acids, the solution was loaded on self-made single-use columns loaded with 1.5 mL of Amberlite HAD-7HP resin, which was equilibrated with 5 mL of 0.1% of formic acid on deionized water. Columns were rinsed with 10 mL deionized water and eluted with 3 mL of methanol acidified with 0.1% formic acid. The solution was evaporated to dryness in centrifugal vacuum concentrator and resuspended in 1.5 mL of deionized water with 0.1% of formic acid. To separate positively charged compounds, such as anthocyanins from other flavonoids, samples were loaded on a DSC-MCAX SPE cartridge, which was pre-conditioned with 1.5 mL of methanol and equilibrated with 5 mL of water with 0.1% formic acid. The cartridge was rinsed with 3 mL water with 0.1% formic acid solution. The flavanol glycosides were eluted with 3 mL methanol, while the anthocyanins were eluted with 3 mL of a 50:50 solution of 5 mM bicarbonate-ammonium buffer at a pH of 6.0 with added methanol. The choice of buffer was dictated by its compatibility with ESI-MS. The solution was evaporated to dryness in centrifugal vacuum concentrator and resuspended with 300 µL of deionized water with 5% of formic acid in chromatography micro-vials.

4.5. Identification and Semi-Quantification of Anthocyanins by HPLC-ESI-MS

An Agilent 6538 quadrupole-TOF mass spectrometer (Agilent Technologies, Palo Alto, CA, USA) with Agilent 1200 series high-performance liquid chromatography was used. Chromatography was conducted on Agilent Zorbax SB-C18 column (1.8 µm; 1 × 150 mm) because this type of column could withstand a low pH of 1. We used gradient HPLC, solvent A of 5% formic acid in water (v/v) and solvent B of acetonitrile. The gradient elution program was used as follows: 0–3 min, 0% B; 3–5 min, 0–3% B; 5–55 min, 3–30% B; 55–60 min, 100% B. The column temperature was 55 °C, flow rate was 180 μL/min and injection volume was 5 μL. For the identification of anthocyanins, electrospray ionization (ESI) was operated in the positive mode with a mass range of 100–1500 (m/z), drying gas temperature of 350 °C, flow rate of 7.0 L/min, nebulizer pressure of 30 psig, capillary voltage of 3500 V, fragmentary voltage of 175 V and skimmer voltage of 65 V. The device was operated in automated tandem mode with an isolation window of 1.3 a.m.u., collision energy of 30 V, molecular ions selection in the range of 260–1500 Da, one linear and 3 MS/MS spectra were acquired every 2.1 s. Simultaneous spectrophotometric detection was conducted with Agilent 1200 Series G1365D MWD at 520 nm with a slit width of 8 nm, speed of 2.5 Hz and flow cell volume of 0.5 μL.

To prevent carry-over between probes, each probe was followed by blank injection (solvent "A"), which showed no detectable peaks.

Due to the absence of standards for all identified anthocyanins, the blueberry extract containing 20 well-characterized peaks was used as a retention time and fragmentation pattern standard (Supplementary Material Table S1: The database of anthocyanine standards and theoretical values used for the identification in this study).

For semi-quantitative analysis, quantification of the sample's anthocyanin content was conducted using the external standard calibration curve, which was obtained by triplicate analysis of cyanine and malvidin-3-galactoside standards sequentially diluted in 5% formic acid at the concentration from 0.5 to 0.0001 mg/mL. The amount of anthocyanins in the samples was determined by the liquid chromatography-electrospray ionization mass spectrometry (LC-ESI-MS) method, the extracted ion peak areas of each anthocyanin were expressed as equivalents of cyanidin chloride. The limit of detection (LOD) and limit of quantification (LOQ) were set as concentrations that gave signal-to-noise ratio (SNR) of 3 and 10, respectively. Results with SNR that were lower than LOD were excluded from the table, while results with an SNR between LOD and LOQ, being non-quantitative, were marked as "<LOQ".

The positive identification of anthocyanin was reported when: (i) molecular ion m/z and aglycon fragment were ±0.05 Da of that of standards, or in absence of standards, ±0.05 Da of that of theoretical mono-isotopic mass; (ii) the retention time was ±1 min of that for the standard.

The total amount of anthocyanins in high concentration samples was also determined by spectrophotometric pH–differential method according to a study of Giusti and Wrolstad [28]. Briefly, 20 μL of sample was mixed with 80 μL of 0.025 M potassium chloride buffer (pH of 1.0) and 20 μL of sample was mixed with 80 μL of 0.4 M sodium acetate buffer (pH of 4.5). The absorbance of the mixture was measured at two wavelengths: at the maximum absorbance of major anthocyanin, as determined by ESI-MS ($\lambda_{\text{vis-max}}$); and at 700 nm using a UV–Vis spectrophotometer BioRad xMarkII.

The corrected absorbance was calculated as:

$$A = \left(A_{\lambda\text{vis}-\text{max}}^{\text{pH1.0}} - A_{700} \right) - \left(A_{\lambda\text{vis}-\text{max}}^{\text{pH4.5}} - A_{700} \right). \tag{1}$$

The concentration was calculated as:

$$\text{Conc}_{\text{mG}/L} = \frac{A \times \text{MW} \times \text{DF} \times 1000}{\varepsilon \times 1} \tag{2}$$

where A is corrected absorbance; MW is molecular weight of the major anthocyanin; DF is the dilution factor; and ε is the molar absorbance of the major anthocyanin.

5. Conclusions

The improvement of the isolation procedures, the use of modern analytical methods and the involvement of diverse genetic material has allowed for the gathering of new data on the composition of anthocyanins in the colored grain of cereals. In pericarp and aleurone in wheat and barley as well as in pericarp of rice, anthocyanins derived from all six major anthocyanidins were identified: pelargonidin, cyanidin, peonidin, delphinidin, petunidin, and malvidin. In green and violet rye kernels the all of the anthocyanidins except for petunidin were identified. Anthocyanins differ in position, number, and type of glycosyl and acyl residues that are associated with an anthocyanid core. A common rule is the predominance of delphinidin as the major aglycone in blue (green in rye) grain and cyanidin derivatives in the purple (violet in rye) grain [9]. The so-called black color of grain may be the result of a very high concentration of anthocyanins in pericarp or its presence in pericarp and aleurone simultaneously. An additional rule is the accumulation of proanthocyanidins in the testa of all four discussed species. In certain genotypes of rice, the accumulation of proanthocyanidins is also established in the pericarp. Obviously, modification of the anthocyanin structure requires the presence of corresponding regulatory and structural genes in the plant genome. Mutational variability of these genes and influence of the environment lead to a wide variety of composition and concentration of anthocyanins in cereal grain

Supplementary Materials: Supplementary materials are available online.

Acknowledgments: This study was partially supported by RFBR (grant No 16-04-00411) and state budget (project "Genetics and breeding of rye on the base of natural hereditary diversity"—AAAA-A16-116111610177-3). Part of the research was done with the use of "Center for Molecular and Cell Technologies", Research Park, St. Petersburg State University.

Author Contributions: Pavel A. Zykin and Anatoly V. Voylokov designed and coordinated the study; Pavel A. Zykin and Elena A. Andreeva performed the chromatographic analysis and analyzed the data; Anna N. Lykholai performed MS analysis and analyzed its data; Natalia V. Tsvetkova produced and contributed plant material; Pavel A. Zykin and Anatoly V. Voylokov wrote different parts of the paper; all authors approved the final version of the manuscript.

Conflicts of Interest: The authors declare no conflict of interest.

References

1. Winkel-Shirley, B. Flavonoid biosynthesis. A colorful model for genetics, biochemistry, cell biology, and biotechnology. *Plant Physiol.* **2001**, *126*, 485–493. [CrossRef] [PubMed]
2. Sang, T.; Li, J. Molecular Genetic Basis of the Domestication Syndrome in Cereals. In *Cereal Genomics II*; Gupta, P.K., Varshney, R.K., Eds.; Springer: Dordrecht, The Netherlands, 2013; pp. 319–340, ISBN 978-94-007-6400-2.
3. Casas, M.I.; Duarte, S.; Doseff, A.I.; Grotewold, E. Flavone-rich maize: An opportunity to improve the nutritional value of an important commodity crop. *Front. Plant Sci.* **2014**, *5*, 440. [CrossRef] [PubMed]
4. Dedio, W.; Kaltsikes, P.J.; Larter, E.N. The anthocyanins of *Secale cereale*. *Phytochemistry* **1969**, *8*, 2351–2352. [CrossRef]
5. Dedio, W.; Hill, R.D.; Evans, L.E. Anthocyanins in the pericarp and coleoptiles of purple-seeded rye. *Can. J. Plant Sci.* **1972**, *52*, 981–983. [CrossRef]
6. Abdel-Aal, E.-S.M.; Young, J.C.; Rabalski, I. Anthocyanin Composition in Black, Blue, Pink, Purple, and Red Cereal Grains. *J. Agric. Food Chem.* **2006**, *54*, 4696–4704. [CrossRef] [PubMed]
7. Kim, M.-J.; Hyun, J.-N.; Kim, J.-A.; Park, J.-C.; Kim, M.-Y.; Kim, J.-G.; Lee, S.-J.; Chun, S.-C.; Chung, I.-M. Relationship between phenolic compounds, anthocyanins content and antioxidant activity in colored barley germplasm. *J. Agric. Food Chem.* **2007**, *55*, 4802–4809. [CrossRef] [PubMed]
8. Goufo, P.; Trindade, H. Rice antioxidants: Phenolic acids, flavonoids, anthocyanins, proanthocyanidins, tocopherols, tocotrienols, γ-oryzanol, and phytic acid. *Food Sci. Nutr.* **2014**, *2*, 75–104. [CrossRef] [PubMed]
9. Abdel-Aal, E.-S.M.; Hucl, P.; Shipp, J.; Rabalski, I. Compositional Differences in Anthocyanins from Blue- and Purple-Grained Spring Wheat Grown in Four Environments in Central Saskatchewan. *Cereal Chem. J.* **2016**, *93*, 32–38. [CrossRef]
10. Khlestkina, E.K. Genes determining the coloration of different organs in wheat. *Russ. J. Genet. Appl. Res.* **2013**, *3*, 54–65. [CrossRef]

11.   Havrlentová, M.; Pšenáková, I.; Žofajová, A.; Rückschloss, L.; Kraic, J. Anthocyanins in Wheat Seed—A Mini Review. *Nova Biotechnol. Chim.* **2014**, *13*, 1–12. [CrossRef]

12.   Himi, E.; Noda, K. Red grain colour gene (R) of wheat is a Myb-type transcription factor. *Euphytica* **2005**, *143*, 239–242. [CrossRef]

13.   Jende-Strid, B. Genetic Control of Flavonoid Biosynthesis in Barley. *Hereditas* **1993**, *119*, 187–204. [CrossRef]

14.   Smirnov, V.G.; Sosnikhina, S.P. *Genetika rzhi (Genetics of Rye)*; Izd-vo Leningradskogo Universiteta: Leningrad, USSR, 1984; 263p.

15.   Kohyama, N.; Chono, M.; Nakagawa, H.; Matsuo, Y.; Ono, H.; Matsunaka, H. Flavonoid compounds related to seed coat color of wheat. *Biosci. Biotechnol. Biochem.* **2017**, *81*, 2112–2118. [CrossRef] [PubMed]

16.   Voylokov, A.V.; Lykholay, A.N.; Smirnov, V.G. Genetic control of anthocyanin coloration in rye. *Russ. J. Genet. Appl. Res.* **2015**, *5*, 262–267. [CrossRef]

17.   Furukawa, T.; Maekawa, M.; Oki, T.; Suda, I.; Iida, S.; Shimada, H.; Takamure, I.; Kadowaki, K. The Rc and Rd genes are involved in proanthocyanidin synthesis in rice pericarp. *Plant J.* **2007**, *49*, 91–102. [CrossRef] [PubMed]

18.   Strygina, K.V.; Börner, A.; Khlestkina, E.K. Identification and characterization of regulatory network components for anthocyanin synthesis in barley aleurone. *BMC Plant Biol.* **2017**, *17*, 184. [CrossRef] [PubMed]

19.   Shoeva, O.Y.; Mock, H.-P.; Kukoeva, T.V.; Börner, A.; Khlestkina, E.K. Regulation of the Flavonoid Biosynthesis Pathway Genes in Purple and Black Grains of *Hordeum vulgare*. *PLoS ONE* **2016**, *11*, e0163782. [CrossRef] [PubMed]

20.   Shoeva, O.; Gordeeva, E.; Khlestkina, E. The Regulation of Anthocyanin Synthesis in the Wheat Pericarp. *Molecules* **2014**, *19*, 20266–20279. [CrossRef] [PubMed]

21.   Rahman, M.M.; Lee, K.E.; Lee, E.S.; Matin, M.N.; Lee, D.S.; Yun, J.S.; Kim, J.B.; Kang, S.G. The genetic constitutions of complementary genes *Pp* and *Pb* determine the purple color variation in pericarps with cyanidin-3-*O*-glucoside depositions in black rice. *J. Plant Biol.* **2013**, *56*, 24–31. [CrossRef]

22.   Zhang, X.-W.; Jiang, Q.-T.; Wei, Y.-M.; Liu, C. Inheritance analysis and mapping of quantitative trait loci (QTL) controlling individual anthocyanin compounds in purple barley (*Hordeum vulgare* L.) grains. *PLOS ONE* **2017**, *12*, e0183704. [CrossRef] [PubMed]

23.   Lee, J.H. Identification and quantification of anthocyanins from the grains of black rice (*Oryza sativa* L.) varieties. *Food Sci. Biotechnol.* **2010**, *19*, 391–397. [CrossRef]

24.   Kim, M.-K.; Kim, H.; Koh, K.; Kim, H.-S.; Lee, Y.S.; Kim, Y.H. Identification and quantification of anthocyanin pigments in colored rice. *Nutr. Res. Pract.* **2008**, *2*, 46–49. [CrossRef] [PubMed]

25.   Park, Y.S.; Kim, S.J.; Chang, H.I. Isolation of anthocyanin from black rice (Heugjinjubyeo) and screening of its antioxidant activities. *Korean J. Microbiol. Biotechnol.* **2008**, *36*, 55–60.

26.   Bauer, E.; Schmutzer, T.; Barilar, I.; Mascher, M.; Gundlach, H.; Martis, M.M.; Twardziok, S.O.; Hackauf, B.; Gordillo, A.; Wilde, P.; et al. Towards a whole-genome sequence for rye (*Secale cereale* L.). *Plant J.* **2017**, *89*, 853–869. [CrossRef] [PubMed]

27.   Lees, D.H.; Francis, F.J. Standardization of pigment analyses in cranberries. *Hortscience* **1972**, *7*, 83–84.

28.   Giusti, M.M.; Wrolstad, R.E. Characterization and measurement of anthocyanins by UV-visible spectroscopy. *Curr. Protoc. Food Anal. Chem.* **2001**. [CrossRef]

Sample Availability: Samples of the seeds are available from the authors.

![molecules logo]

![MDPI logo]

Article

Roselle Anthocyanins: Antioxidant Properties and Stability to Heat and pH

Hai-Yao Wu, Kai-Min Yang and Po-Yuan Chiang *

Department of Food Science and Biotechnology, National Chung Hsing University, 145 Xingda Rd., South Dist., Taichung City 40227, Taiwan; j490560511@gmail.com (H.-Y.W.); a9241128@gmail.com (K.-M.Y.)
* Correspondence: pychiang@nchu.edu.tw; Tel.: +886-4-2285-1665

Received: 1 May 2018; Accepted: 31 May 2018; Published: 5 June 2018

Abstract: Roselle is rich in anthocyanins and is traditionally used to prepare a bright red beverage by decoction. However, heat treatment and different pH environments are often encountered during food processing, and these factors are often detrimental to anthocyanins. Therefore, it is very important to understand the influence of pH and heat treatment on anthocyanins for the application of roselle. This study determined the antioxidant properties of roselle extract, explored changes in the color and anthocyanin content in different pH environments, and evaluated the thermal stability of roselle anthocyanins using kinetic equations. The results showed that the roselle extract is rich in anthocyanins and has good antioxidant capacity (DPPH IC50 = 4.06 mg/mL, ABTS IC50 = 3.7 mg/mL). The anthocyanins themselves exhibited a certain degree of heat resistance and good color stability in an acidic environment. In contrast, they degraded very quickly and exhibited significant changes in color in a low-acid environment. The activation energy (Ea) ranges of the anthocyanins in the acidic and low-acid environments were quite different at 55.8–95.7 and 31.4–74.9 kJ/mol, respectively. Thus, it can be concluded that roselle anthocyanins are susceptible to heat treatment in a low-acid environment, affecting their quality and appearance; however, they can serve as a good source of functional ingredients and color in an acidic environment.

Keywords: roselle; anthocyanins; heat; pH

---

1. Introduction

Roselle (*Hibiscus sabdariffa* L.) is recognized as a tropical shrub which belongs to the family *Malvaceae*. Roselle can be found in tropical and sub-tropical regions such as India, Indonesia, and Malaysia, among others [1]. The main planting area for the shrub in Taiwan is in the eastern region of the island, which is why roselle is one of the most important crops in Taitung, Taiwan. The roselle calyx is brightly colored and rich in nutrients such as anthocyanins, organic acids, pectin, etc. [2]. The calyces of roselle have been widely used in medicines and foods such as syrup, refreshing drinks, wines, jams, and natural food colorants [2,3]. The leaves or calyces are traditionally prepared in beverage as they are rich in anthocyanins, which have antioxidant properties and are useful in diuretic and sedative treatments [4]. Anthocyanidin is one of the natural water-soluble pigments and is among the derivatives of the flavonoid compounds found in phenolic compounds [2]. The chemistry of anthocyanins has been reviewed extensively. So far, studies have reported that the most common anthocyanins, which include pelargonidin, peonidin, cyanidin, malvidin, petunidin, and delphinidin, are widely present in fruits and vegetables such as grapes, raspberries, roselle, and purple cabbage, among others [5]. Besides their vibrant colors, anthocyanins also have anti-oxidant and bioactive properties linked to certain health benefits; for example, they have properties linked to anti-diabetic, anti-inflammatory, and anti-cancer effects [6–8]. More and more people recognize that food not only satisfies hunger but also provides a variety of health benefits [9]. Health foods include a variety of foods that not only provide direct health benefits by reducing the risk of chronic disease in relatively short periods of time, but also foods enriched with vitamins and polyphenols, as well as foods that improve the well-being of consumers [10]. Functional beverages are widely consumed worldwide and are a fast-growing segment of the health foods category. Since consumers have shown an increasing awareness of different health issues, these compounds have increasingly become the focus not only of scientific research but also of marketing considerations. Many studies have pointed out that phytochemical-rich foods, such as the catechins in green tea, the anthocyanins in grape juice, and the flavonoids in citrus fruit juices, among others, have good health effects [11–13]. In short, the anti-inflammatory, anti-oxidant, and cancer-preventing properties of such phytochemicals are among their many proposed health-promoting properties [6,11,13]. Anthocyanin-rich plant materials are increasingly being labeled as functional materials in recent years as they contain high amounts of secondary plant metabolites and traditionally constitute a high proportion in the human diet. Meanwhile, the anthocyanin contents of beverages not only influence the potential health effects of those beverages but also their sensory qualities [5].

Thermal processing and pH adjustment are among the commonly used unit operations in the food industry. For example, thermal processing is the most widely used preservation method in industrial beverage production [14]. However, thermal treatment can cause organoleptic and nutritional loss, as well as changes in the levels of ascorbic acid, phenolic compounds, and carotenoids, thereby leading to decreased antioxidant capacity and other effects on bioactivities.

In nature, anthocyanins are extremely unstable and susceptible to degradation by external factors such as pH, temperature, light, oxygen, enzymes, metal ions, and other factors. In addition to affecting food products directly, anthocyanin degradation may also result in the production of aldehyde substances with benzene rings that can affect human health [15]. While thermal processing and pH adjustment are among the commonly used unit operations in the food industry. For example, thermal processing is the most widely used preservation method in industrial beverage production [14]. However, thermal treatment can cause organoleptic and nutritional loss, thereby leading to decreased antioxidant capacity and other effects on bioactivities. Color fading and off-flavor formation limits the shelf life of commercial products containing anthocyanins, in addition to restricting the utilization of anthocyanins for certain applications. Therefore, the aim of this study was to determine the antioxidant properties of the roselle extract, in addition to exploring changes in the color and anthocyanin content of the extract in different pH environments. Finally, the thermal stability of roselle anthocyanins was evaluated using kinetic equations.

2. Results and Discussion

2.1. Total Anthocyanins and Antioxidant Capacity of Roselle Extract

Recognition of the potential health effects of antioxidants more than twenty years ago stimulated what could be called the "Antioxidant Bandwagon" of research seeking to determine which natural materials contain the highest levels of the most active antioxidants, the addition of antioxidants to beverages and other forms of foods, prophylactic and therapeutic medical applications, and hyper marketing [16]. Polyphenols are among the well-known antioxidants. They transfer an electron to free radicals, which thus become stable as their electrons are paired. This prevents damage to cells and tissue caused by oxidant stress. Consequently, a diet that is rich in polyphenols could potentially modulate certain secondary physiological effects of oxidant stress, prevent obesity, or optimize the treatment of diabetes [8]. In the present study, dehydrated roselle calyces were found to have a total polyphenol concentration (TPC) of about 683.13 mg gallic acid equivalent (GAE)/100 g. There is a positive correlation between the total content of phenolic compounds and the antioxidant activity of an extract, and over 95% of the antioxidant capacity of extracts is due to their phenolic components [17,18].

Using the pH-differential method, we were able to determine that the total anthocyanin content (TAC) of dehydrated roselle calyces was 361.99 mg CGE/100 g. In other studies, the TAC of different varieties of black rice has been reported to range between 4.1 and 256.5 mg/100 g [19]. It can thus be seen that roselle is a good source of anthocyanins. In addition, many researchers have pointed out that roselle and its extract possess functional properties that can be used to develop new products with additional nutritious characteristics that may provide health benefits to people [1,2]. At the same time, numerous in vitro antioxidant assays have been developed to measure radical scavenging activity, for example, 1,1-diphenyl-2-picrylhydrazyl (DPPH) radical scavenging activity, 2,2′-azino-bis(3-ethylbenzothiazoline)-6-sulfonic acid (ABTS) radical scavenging activity, and ferric ion reducing antioxidant power (FRAP), such that various kinds of oxygen radical absorbance capacity assay (ORAC) are now household words in scientific and health food publications and on internet websites [16]. The abundant pigments in roselle are responsible for its red color and are the main source of its antioxidant capacity. In this study, the antioxidant capacities of roselle extract were determined by measuring its DPPH radical scavenging activity, ABTS radical scavenging activity, and FRAP.

Figure 1a–c represent the DPPH radical scavenging activity, the ABTS radical scavenging activity and ferric ion reduction antioxidant power, respectively. And their results collectively indicate the antioxidant capacity of the sample. As shown in Figure 1a, the DPPH radical scavenging activity was 20–60% when the sample concentration was 1–5 mg/mL. When the sample concentration reached 7.5 mg/mL, the DPPH radical scavenging activity was above 80%. The calculated IC50 values of the roselle extract and Trolox (6-hydroxy-2,5,7,8-tetramethylchroman-2-carboxylic acid) in the DPPH assays were 4.06 and 0.05 mg/mL, respectively. The IC50 value, defined as the concentration of the extract required for 50% scavenging of radicals under the experimental condition employed, is a parameter widely used to measure free radical scavenging activity. A smaller IC50 value corresponds to a higher antioxidant activity. The IC50 values of the obtained bilberry, blackberry, strawberry, and raspberry pomace extracts were 4.0, 1.7, 3.8, and 4.0 mg/100 mL, respectively [20].

From the ABTS assays, it could be seen that the ABTS radical scavenging activity was 20–80% when the sample concentration was 1–7.5 mg/mL. The calculated IC50 values of the roselle extract and Trolox in the ABTS assays were 3.7 and 0.07 mg/mL, respectively. DPPH is hydrophobic so its reactions must be run in organic solvents. The previous literature has reported that DPPH reactions are mostly attributable to hydrogen atom transfer. In contrast to DPPH, ABTS$^+$ is water-soluble, so it can reflect antioxidant capacity in non-organic solvent environments [16]. The ABTS$^+$ assay can be used to assess whether antioxidants are hydrogen atom transfer-dominant or single electron transfer-dominant in their reactions, and it can be used to compare changes in the same antioxidant during processing or storage. For example, ABTS$^+$ has been used to monitor changes in tocopherol activity after heat exposure in frying oils and in the extrusion of packaging films and to determine the loss of antioxidant activity in strawberries dried with different methods [21,22].

Anthocyanins have been demonstrated to have high antioxidant capacity. In particular, the level of hydroxylation on the 3′ and 4′ positions of the B-ring structure is a fundamental determinant of their radical scavenging activity [18]. In the FRAP assays, the roselle extract exhibited good ferric ion reducing antioxidant power (33.98 mg Trolox equivalents (TE)/100 g sample). Salvador Fernández-Arroyo et al. [23] measured the antioxidant capacity of H. sabdariffa aqueous extract, and the experimental data showed that the values obtained in the FRAP assay for H. sabdariffa aqueous extract doubled those obtained for olive leaf extract. The high FRAP value of the H. sabdariffa aqueous extract could be explained through the reported efficacy of chlorogenic acid and its derivatives, which are the main compounds in H. sabdariffa aqueous extract, as reductants [24].

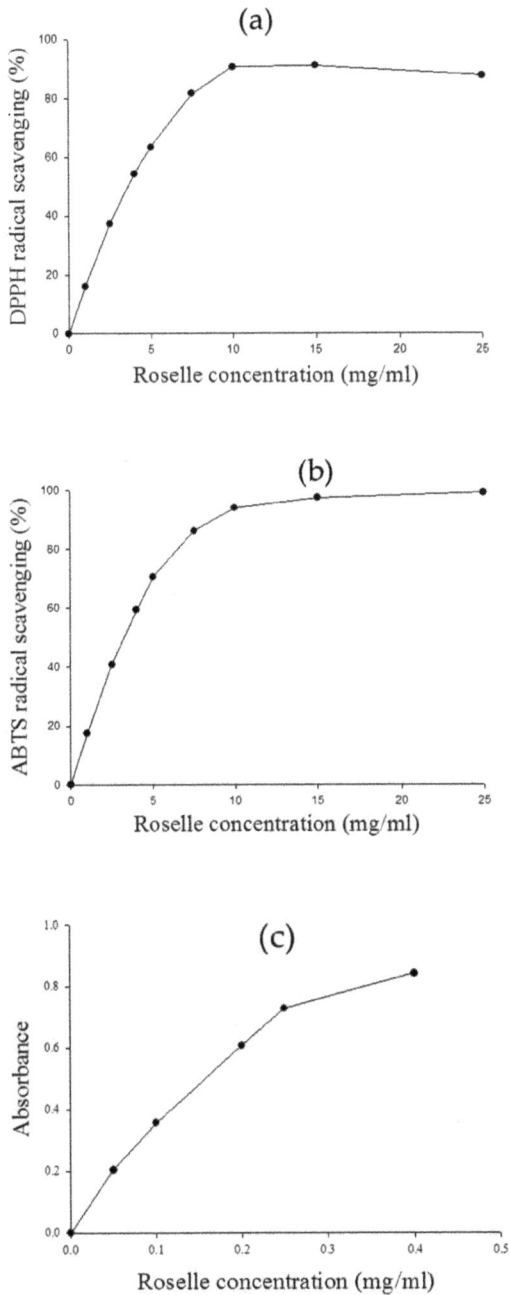

Figure 1. The antioxidant capacity results for the roselle extract of (**a**) DPPH radical scavenging; (**b**) ABTS radical scavenging; (**c**) FRAP.

2.2. Color Stability

Anthocyanins are one of the most commonly utilized water-soluble natural colorants because they exhibit vibrant colors that range from red to blue. The color of anthocyanins is strongly dependent on the pH of the surrounding aqueous phase [5]. Relatedly, the fate of anthocyanins during the production of beverages is determined by countless factors, and all of these factors need to be taken into consideration to optimize beverage production processes. Table 1 shows the color changes of roselle extract in the pH range of 1–7 before heating. It can be seen that the color of the extract varied in the different pH buffers, gradually shifting from dark red to light red at pH values of 1–4. When the solution had a pH of 5, the extract presented as nearly colorless, whereas its color changed to blue at a pH of 7. The intensity of the acidity of a food is expressed by its pH value. The pH of a food is one of several important factors that determine the survival and growth of microorganisms during processing, storage, and distribution.

Table 1. The value of lightness (L), redness (a), and yellowness (b) of the extracts in different pH buffer solutions.

	pH 1	pH 2	pH 3	pH 4	pH 5	pH 6	pH 7
L	54.34 ± 0.03	58.32 ± 0.04	57.72 ± 0.03	69.36 ± 0.04	78.83 ± 0.07	75.25 ± 0.09	46.43 ± 0.12
a	57.85 ± 0.14	57.72 ± 0.06	50.21 ± 0.03	35.88 ± 0.04	16.52 ± 0.05	11.14 ± 0.09	19.26 ± 0.06
b	26.70 ± 0.03	24.80 ± 0.09	20.18 ± 0.08	12.61 ± 0.02	10.56 ± 0.01	12.46 ± 0.10	6.44 ± 0.17

Data presented are in mean ± SD ($n = 3$).

Due to the different composition of food raw materials, the pH value of the beverage will different, so the heat sterilization conditions are not the same. Low-acid foods (pH greater than 4.6 and less than 7.0) need a higher sterilization temperature than acidic foods (pH of 4.6 or below). The purpose of this study was to investigate whether the anthocyanin in roselle is suitable as a nutrient additive and pigment source in beverages. Therefore, we divided the pH range discussed in this study into an acidic environment and a low-acid environment.

The UV-visible spectra of roselle extract in different pH buffers were recorded, as shown in Figure 2. When the extract was placed in the acidic environment, the maximum absorption peak was obtained at 520 nm, and the absorbance gradually decreased with increases in the pH value. As the pH increased, the maximum absorption peak showed a slight shift to the right, which was accompanied by a decrease in the maximum absorbance. The bright red color of the roselle extract under certain conditions is due to the main anthocyanins present in the calyces of the plant: delphinidin 3-*O*-sambubioside and cyanidin 3-*O*-sambubioside [25]. In this study, the a value at the initial time was decreased as the pH was increased from 1–6 (Table 1). As the heating time was increased, the a value (redness) of the extract in the acidic environment was decreased, while the b value (yellowness) was increased over the same period of time (data not shown). The main reason for these changes was the occurrence of anthocyanin cleavage and the Maillard reaction [26]. In contrast, the changes in the a value and b value of the extract in the low-acid environment lacked any regularity, possibly because the anthocyanin degraded rapidly and produced numerous degradation products. When the extract was heated in a different pH buffer, the color became unpleasant. Furthermore, as the temperature was increased, the more obvious the above situation was.

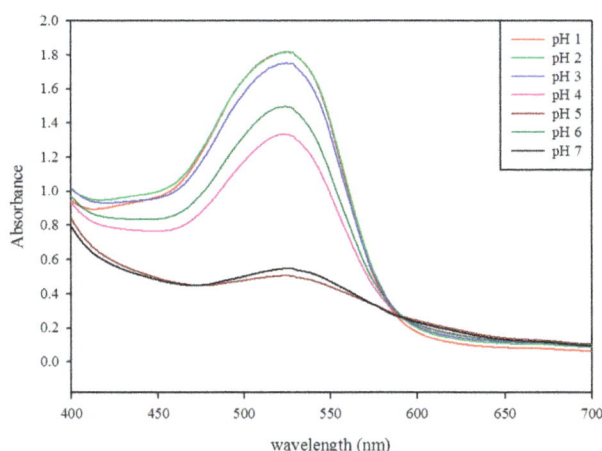

Figure 2. UV-vis spectra of extracts in different buffer solutions ranging from pH 1 to 7.

The value of the total color difference (ΔE) can express the difference in color of an extract from before heating to after heating in different environments. The larger the value, the more obvious the change in color from before heating to after heating. The results of ΔE of the extract with different heating times and temperatures in different buffer environments are shown in Table 2. It can be seen that the color of the extract changed by varying degrees under heating in different buffer environments. It was pointed out by Kim et al. [27] that human eyes can distinguish a color difference when $\Delta E > 12$. Thus, we can ascertain that when heated at a lower temperature (70 °C), the color of the extract in the acidic environment remains relatively stable. We can draw this conclusion because the color change after 2 h of heating was very small. In contrast, when the extract was placed in an environment with a pH of 7, the color changed obviously even when low temperature heating was applied. The reason for this change is mainly due to the degradation of anthocyanins during heating. When the extract was heated at 70, 80, and 90 °C for 30 minutes at a pH of 7, the residual rates were 29.10, 21.48, and 17.24%, respectively. When the extract was heated at 70, 80, and 90 °C for 2 h at a pH of 2, meanwhile, the residual rates were 87.69, 77.79, and 60.21%, respectively.

Table 2. Effects of temperatures on the total color difference (ΔE) of the extracts in different pH buffer solutions.

Temperature (°C)	Time (min)	pH 1	pH 2	pH 3	pH 4	pH 5	pH 6	pH 7
70	30	1.78	0.88	2.83	4.11	3.46	6.57	30.03
	60	0.65	1.33	3.81	4.56	3.57	7.42	34.79
	90	0.68	1.21	4.25	5.35	4.49	8.54	36.67
	120	0.69	1.97	5.31	5.73	5.01	9.32	36.47
80	30	1.29	1.00	3.14	2.67	4.83	7.48	24.56
	60	1.49	2.02	5.17	3.70	6.14	9.86	27.05
	90	14.46	3.32	6.71	5.53	7.02	11.75	32.84
	120	17.08	4.50	8.92	6.66	7.54	13.20	33.51
90	30	3.51	2.19	12.09	10.07	8.10	10.26	22.76
	60	6.45	4.96	18.02	12.13	9.10	12.45	22.05
	90	10.27	7.37	22.62	14.87	10.17	12.26	22.36
	120	13.93	13.41	26.03	15.41	11.76	16.47	25.96

The current interest in natural antioxidants from plant sources has become substantial, particularly with respect to bioactive antioxidants such as polyphenols and flavonoids. Roselle anthocyanins have not only good antioxidant properties, but also bright red color, which is suitable as a source of natural pigments in food. However, the low-acid foods need higher temperature and longer time to sterilize, which will cause the degradation of anthocyanins. Meanwhile, the cleavage of an anthocyanin leads to colorless compounds, and polymerization is accompanied by browning [28]. This is what we need to pay attention to.

2.3. Thermal Kinetic Degradation

The thermodynamic parameters allowed a deeper understanding of the thermal degradation kinetics of anthocyanins to minimize undesired degradation and to optimize quality of foods. Figure 3 shows the changes in anthocyanin content when the extract was heat-treated in different pH environments. The anthocyanin degradation increased with increasing temperature and time, and the trend of degradation was more obvious with increasing time. After being heated at 70 °C for 2 h in a pH 7 environment, there was almost no anthocyanin content left in the solution, while under heating at 70 °C for 2 h in the acidic environment, the anthocyanin residues still exceeded 80%. Increasing the temperature of the heat treatment is more detrimental to the anthocyanin, but heating at a higher temperature (90 °C) for 2 h in the acidic environment still allows for more than 50% of the anthocyanin content to remain. In this study, the thermal degradation of roselle anthocyanins in buffers with pH values of 1–7 buffer followed first-order reaction kinetics ($R^2 > 0.9$). These results were similar to those of previous research indicating that the degradation of anthocyanin follows a first-order model [29].

The first-order reaction rate constants (k), half-life values of anthocyanins ($t_{1/2}$), and Ea values for a defined temperature range are shown in Table 3. It is clear from Table 3 that as the temperature and pH were increased, the k values increased. The greater the k value, the faster the reaction rate and the faster the degradation of anthocyanins. As expected, the degradation was dependent on temperature and pH, being faster at high temperatures and in low-acid environments. Table 3 shows that the $t_{1/2}$ values for anthocyanin degradation were 1155.2, 385.1, and 182.4 min in a pH of 1 at 70, 80, and 90 °C, respectively, while the $t_{1/2}$ values for anthocyanin degradation were 13.5, 19.4, and 24.8 min in a pH of 7 at 70, 80, and 90 °C, respectively. As the temperature and pH were increased, the $t_{1/2}$ values decreased in a manner consistent with faster reactions accompanied by higher k values. Aurelio et al. [30] previously reported that $t_{1/2}$ values of anthocyanins in a roselle infusion were 11.5, 7.22, and 3.21 h at 70, 80, and 90 °C, respectively. These results are similar to those for the extract in this study at a pH of 2, while the original pH of the extract was about 2.2.

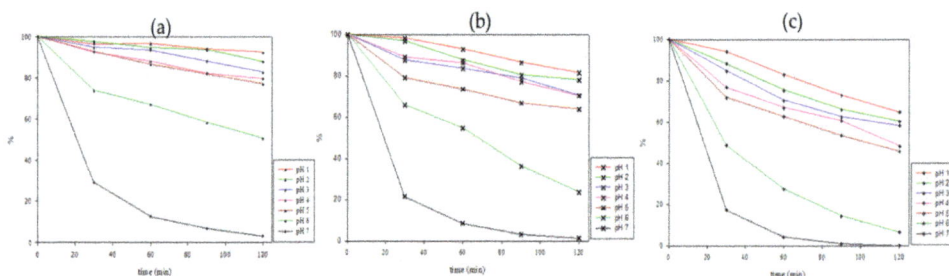

Figure 3. The anthocyanins residual rate (%) during heating at (**a**) 70 °C; (**b**) 80 °C; and (**c**) 90 °C in different pH buffer solutions.

Table 3. Effects of temperatures on the k, $t_{1/2}$, and Ea values of anthocyanin degradation in different pH buffer solutions.

pH	Temperatures (°C)	k (min^{-1})	$t_{1/2}$ (min)	Ea (kJ/mol)	Arrhenius Equation (R^2)
1	70	6.0×10^{-4}	1155.2		
	80	1.8×10^{-3}	385.1	95.7	y = −11509x + 26.18 (0.991)
	90	3.8×10^{-3}	182.4		
2	70	1.0×10^{-3}	693.1		
	80	2.3×10^{-3}	301.4	75.6	y = −9090.3x + 19.62 (0.996)
	90	4.3×10^{-3}	161.2		
3	70	1.4×10^{-3}	495.1		
	80	2.7×10^{-3}	256.7	61.6	y = −7411x + 15.05 (0.998)
	90	4.6×10^{-3}	150.7		
4	70	1.9×10^{-3}	364.8		
	80	2.8×10^{-3}	247.6	55.8	y = −6709.4x + 13.24 (0.969)
	90	5.6×10^{-3}	123.8		
5	70	2.2×10^{-3}	315.1		
	80	3.6×10^{-3}	192.5	53.6	y = −6445.4x + 12.66 (0.998)
	90	6.2×10^{-3}	111.8		
6	70	5.3×10^{-3}	130.8		
	80	1.2×10^{-2}	59.2	74.9	y = −9000.3x + 21.00 (0.999)
	90	2.3×10^{-2}	30.8		
7	70	2.8×10^{-2}	24.8		
	80	3.6×10^{-2}	19.4	31.4	y = −3773.7x + 7.41 (0.984)
	90	5.1×10^{-2}	13.5		

In the study by Aramwit et al. [31] on the stability of mulberry (*Morus alba*) anthocyanins at 40, 50, and 70 °C, the authors found that the anthocyanin content was significantly decreased after exposure to heating at 70 °C, and they thus advised that mulberry fruit extracts be processed at a temperature lower than 70 °C, a view that supports our own findings. Moreover, Wang and Xu [32] reported that the $t_{1/2}$ values for anthocyanin degradation in blackberry juice were 16.7, 8.8, and 4.7 h at 60, 70, and 80 °C, respectively. Meanwhile, Cemeroğlu et al. [33] reported that the $t_{1/2}$ values for anthocyanin degradation in sour cherry concentrate were 24.0, 10.9, and 4.4 h at 60, 70, and 80 °C, respectively. Comparing these results indicates that anthocyanins from different sources were susceptible to high temperatures. Results from previous research show that when the heating temperature is kept below 80 °C, the rate of anthocyanin degradation is decreased [12].

In general, Ea is used to describe the energy required to reach the active state of a reaction [34]. In this study, the Ea range of the roselle extract in the acidic environment was 55.8–95.7 kJ/mol, while its activation energy range in the low-acid environment was 31.4–74.9 kJ/mol. In the acidic conditions, the activation energy decreased with increasing pH, and the activation energy was lowest at a pH of 4 (Ea = 55.8 kJ / mol), which is similar to the activation energy of grape skin anthocyanins at pH 3.7 (Ea = 51.0 kJ/mol) in other studies [35]. The high Ea values revealed that the anthocyanin degradation was slower at the same temperature. These results proved that roselle anthocyanins are more stable in acidic environments than in low-acid environments. Therefore, roselle is suitable for addition to acidic beverages, in which it not only has good functionality but also a pleasant color.

3. Materials and Methods

3.1. Materials and Reagents

Dehydrated roselle (*H. sabdariffa* L.) calyces were purchased from the Taitung County Farmers' Association, Taiwan. ABTS, DPPH, Trolox, and gallic acid were purchased from Sigma Chemical Co.

(St. Louis, MO, USA). All other analytical grade chemicals and ethanol were purchased from Echo Chemical Co., Ltd. (Miaoli, Taiwan).

3.2. Roselle Extract Preparation

Dehydrated roselle calyces were ground into powder with milling, then mixed with 30% (*v/v*) ethanol in a 1:20(*w/v*) ratio, which was followed by extraction at 75 °C for 20 min in a water bath. After that, the ethanol extract was filtered and stored in a dark bottle at 4 °C until use.

3.3. Antioxidant Assays

3.3.1. Total Anthocyanin Concentration (TAC)

Monomeric TAC was determined using the pH differential method [20]. Briefly, two solutions were prepared at pH 1.0 (0.025 M KCl) and pH 4.5 (0.4 M CH_3COONa). The extract samples were diluted accordingly to be within a measurable absorbance range. The two samples diluted to pH 1.0 and pH 4.5 were then shaken and equilibrated in the dark for 15 minutes. Next, the absorbance values of the samples were measured by spectrophotometer (Hitachi, Tokyo, Japan) at 700 nm (A_{700}), and the maximum absorption wavelength ($A_{\lambda \text{ vis-max}}$) of each was used to calculate their respective TAC values. Each TAC value was expressed as mg cyanidin-3-glucoside (C3G) equivalents per liter of concentrate according to the following equations:

$$\text{Absorbance (A)} = (A_{\lambda \text{ vis-max}} - A_{700})\, \text{pH } 1.0 - (A_{\lambda \text{ vis-max}} - A_{700})\, \text{pH } 4.5$$

$$\text{TAC (mg/L)} = (A \times MW \times DF \times 1000)/(\varepsilon \times l)$$

MW: the molecular weight, calculated as cyanidin-3-glucoside (449.2); DF: the dilution factor; l: the cuvette radius, 1 cm; ε: the molar absorptivity, calculated as cyanidin-3-glucoside (26,900).

3.3.2. Total Polyphenol Concentration (TPC)

The total phenolic content values were determined colorimetrically using the Folin–Ciocalteu method [20]. One mL of sample (diluted 1:20 (*v:v*) with 30% ethanol) was mixed with 1 mL of Folin–Ciocalteu reagent at room temperature. After waiting for 3 min, 0.1 mL of sodium carbonate (10% *w/v*) was added to adjust to the optimum pH for the reaction. The mixture was vortexed and incubated at room temperature for 1 h, and then the absorbance was measured by spectrophotometer at 735 nm. Gallic acid was used as a standard, and the total phenolic content was expressed in mg gallic acid equivalents (GAE) per liter of concentrate. A mixture of water and reagents was used as a blank. All analyses were done in triplicate (*n* = 3).

3.3.3. DPPH Method

The DPPH assay was conducted according to the method of previous research [20] with some modifications. A working solution of DPPH was prepared daily, and the concentration of 0.1 mM of diluted sample was mixed with 1 mL of DPPH. The mixture was then vortexed and left to stand for 10 min in a dark place at room temperature. The absorbance was then measured spectrophotometrically at 517 nm. Trolox was used as a reference standard. The percent of reduction of DPPH was calculated by the formulation:

$$\%\text{DPPH reduction} = (1 - As/Ac) \times 100$$

where Ac = absorbance of a control, As = absorbance of sample.

3.3.4. ABTS Method

The ABTS radical scavenging activity was assayed as previously reported [22]. ABTS radical cation (ABTS $^+$) stock solution was prepared by mixing ABTS (7 mmol/L) with potassium persulfate

(2.45 mmol/L). The mixture was then allowed to stand in the dark for 12–16 h at room temperature prior to use. The stock solution was subsequently diluted with deionized water until the absorbance was 0.7 ± 0.02 at 734 nm. The diluted sample (0.1 mL) was mixed with 2 mL of ABTS$^+$ working solution. The mixture was vortexed and left to stand for 10 min in a dark place at room temperature, and then the absorbance was measured spectrophotometrically at 734 nm. ABTS radical scavenging activity was calculated in the same way as DPPH radical scavenging activity.

3.3.5. FRAP Method

The ferric reducing ability of the extracts was determined using the method of previous research [23]. To perform the assay, 0.2 mL of FRAP reagent and 1 mL of the diluted sample were mixed. The absorbance was then measured at 593 nm by spectrophotometer after it was left to stand for 4 min in a dark place at room temperature. The FRAP reagent working solution was used as the blank. The ferric reducing ability was expressed as mg 100 g^{-1} TE.

3.4. Effect of Temperature and pH on the Roselle Extract

3.4.1. Extract in Different pH Buffers

Samples of roselle extract and citrate–phosphate buffer solutions with different pH (1–7) were mixed in a 1:4 ratio, and then put in a water bath (70, 80, or 90 °C) with shaking (90 rpm). The samples were removed from the water bath at 30 min intervals and cooled rapidly to determine the anthocyanin content and color. Three replicates per treatment group were performed.

3.4.2. Wavelength Scanning

Samples of roselle extract and buffers with different pH (1–7) were mixed in an appropriate ratio, then measured with a spectrophotometer (Hitachi U-2800-A, Hitachi, Tokyo, Japan) for wavelength scanning (400–700 nm) after 15 min.

3.4.3. Color Measurement

The samples were placed into a color meter (Color meter, NE-4000, Nippon Denshku Industries Co., Osaka, Japan) with Hunter LAB coordinates (L, a, b) to determine their lightness (L), redness (a), and yellowness (b). The instrument (transmittance, C illuminant, 2° observer) was calibrated with black and white (X = 92.81, Y = 94.83, Z = 111.71) reference tiles and then the color difference (ΔE) of each sample was calculated.

$$\Delta E = [(L - L_0)^2 + (a - a_0)^2 + (b - b_0)^2]^{1/2}$$

where L_0, a_0, b_0 is the value of L, a, b at time zero.

3.5. Kinetics of Degradation

The degradation reaction of anthocyanins follows the first-order reaction [29], and the rate of the degradation reaction is related to the anthocyanin content in the heating process. The equation can be expressed as follows:

$$\ln(C_t/C_0) = -k \times t$$

$$t_{1/2} = -\ln 0.5 \times k^{-1}$$

C_0: initial anthocyanin content. C_t: heat t time anthocyanin content; k: reaction constant. t: heat time (min). $t_{1/2}$: the half-life.

Activation energy calculation: The Arrhenius equation can be expressed as the relation between the reaction rate constant and the temperature.

$$k = A \times e^{-Ea/RT}$$

$$\ln k = \ln A - (Ea/RT)$$

A: proportional constant of the reaction. Ea: the activation energy (kJ/mol); R: the gas universal constant (8.314 J/mol/k). T: the temperature (K).

The *Ea* value was calculated from the slope of the straight lines using a linear regression procedure of the SigmaPlot (SigmaPlot 10.0 Windows version, SPSS Inc., Chicago, IL, USA).

3.6. Statistical Analysis

The experimental data were subjected to a two-way analysis of variance, and the means were compared by Duncan's multiple range test at a $p < 0.05$ significance level using SAS (SAS Inst., Inc., Cary, NC, USA). The parameters of the kinetic models and the Arrhenius equation were estimated by either the linear regression procedure or non-linear regression iterative procedure of SigmaPlot (SigmaPlot 10.0 Windows version, SPSS Inc.). All analyses were done in triplicate ($n = 3$).

4. Conclusions

In this study, we determined the total anthocyanin content, antioxidant capacity, and changes in the color of roselle extract during the roselle extract degradation process. The anthocyanin degradation followed first-order reaction kinetics. A higher degradation rate was observed in low-acid environments at higher temperatures. However, anthocyanin content was not only affected by temperature but also by pH. Roselle anthocyanins have a certain degree of heat resistance in acidic environments and thus have potential for application in acidic beverages. For example, it can be used as a nutritional supplement and pigment in drinks. Studies such as this one are important for providing insights into the application of roselle in industrial production.

Author Contributions: H.-Y.W. carried out all the experiments; H.-Y.W., K.-M.Y., and P.-Y.C. designed all the experiments and analyzed the data; H.-Y.W. and P.-Y.C. wrote the manuscript.

Acknowledgments: The authors wish to acknowledge the help of Wei-Tzu Shih and Chun-Hsiang Hsu in color difference data analysis. We would like to thank two reviewers and the editor for their comments and suggestions.

Conflicts of Interest: The authors declare no conflict of interest.

References

1. Dhar, P.; Kar, C.S.; Ojha, D.; Pandey, S.K.; Mitra, J. Chemistry, phytotechnology, pharmacology and nutraceutical functions of kenaf (*Hibiscus cannabinus* L.) and roselle (*Hibiscus sabdariffa* L.) seed oil: An overview. *Ind. Crops Prod.* **2015**, *77*, 323–332. [CrossRef]
2. Da-Costa-Rocha, I.; Bonnlaender, B.; Sievers, H.; Pischel, I.; Heinrich, M. *Hibiscus sabdariffa* L.—A phytochemical and pharmacological review. *Food Chem.* **2014**, *165*, 424–443. [CrossRef] [PubMed]
3. Ifie, I.; Abrankó, L.; Villa-Rodriguez, J.A.; Papp, N.; Ho, P.; Williamson, G.; Marshall, L.J. The effect of ageing temperature on the physicochemical properties, phytochemical profile and α-glucosidase inhibition of Hibiscus sabdariffa (roselle) wine. *Food Chem.* **2017**. [CrossRef]
4. Carvajal-Zarrabal, O.; Waliszewski, S.M.; Barradas-Dermitz, D.M.; Orta-Flores, Z.; Hayward-Jones, P.M.; Nolasco-Hipolito, C.; Angulo-Guerrero, O.; Sanchez-Ricano, R.; Infanzon, R.M.; Trujillo, P.R. The consumption of Hibiscus sabdariffa dried calyx ethanolic extract reduced lipid profile in rats. *Plant Foods Hum. Nutr.* **2005**, *60*, 153–159. [CrossRef] [PubMed]
5. Clifford, M.N. Anthocyanins-nature, occurrence and dietary burden. *J. Sci. Food Agric.* **2000**, *80*, 1063–1072. [CrossRef]

6. Mozaffari-Khosravi, H.; Jalali-Khanabadi, B.A.; Afkhami-Ardekani, M.; Fatehi, F. Effects of sour tea (Hibiscus sabdariffa) on lipid profile and lipoproteins in patients with type II diabetes. *J. Altern. Complem. Med.* **2009**, *15*, 899–903. [CrossRef] [PubMed]

7. Yang, M.Y.; Peng, C.H.; Chan, K.C.; Yang, Y.S.; Huang, C.N.; Wang, C.J. The hypolipidemic effect of Hibiscus sabdariffa polyphenols via inhibiting lipogenesis and promoting hepatic lipid clearance. *J. Agric. Food Chem.* **2010**, *58*, 850–859. [CrossRef] [PubMed]

8. Tsuda, T.; Horio, F.; Uchida, K.; Aoki, H.; Osawa, T. Dietary cyanidin 3-*O*-beta-D-glucoside-rich purple corn color prevents obesity and ameliorates hyperglycemia in mice. *J. Nutr.* **2003**, *133*, 2125–2130. [CrossRef] [PubMed]

9. Siro, I.; Kapolna, E.; Kapolna, B.; Lugasi, A. Functional food. product development, marketing and consumer acceptance-a review. *Appetite* **2008**, *51*, 456–467. [CrossRef] [PubMed]

10. Bagchi, D.; Nair, S. *Developing New Functional Food and Nutraceutical Products*; Bagchi, D., Nair, S., Eds.; Academic Press: Cambridge, MA, USA, 2016.

11. Bhardwaj, P.; Khanna, D. Green tea catechins: Defensive role in cardiovascular disorders. *Chin. J. Nat. Med.* **2013**, *11*, 345–353. [CrossRef]

12. Marchese, D. Citrus consumers trend in Europe. new tastes sensation: The blood orange juice case. In *Citrus Processing Short Course Proceedings*; University of Florida: Gainesville, FL, USA, 1995; pp. 19–39.

13. Santhakumar, A.B.; Kundur, A.R.; Fanning, K.; Netzel, M.; Stanley, R.; Singh, I. Consumption of anthocyanin-rich queen garnet plum juice reduces platelet activation related thrombogenesis in healthy volunteers. *J. Funct. Foods* **2015**, *12*, 11–22. [CrossRef]

14. Andres, V.; Villanueva, M.-J.; Mateos-Aparicio, I.; Tenorio, M.-D. Colour, bioactive compounds and antioxidant capacity of mixed beverages based on fruit juices with milk or soya. *J. Food Nutr. Res.* **2014**, *53*, 71–80.

15. Peng, B.; Li, H.; Deng, Z. Degradation of anthocyanins in foods during heating process and its mechanism. *J. Food Safe Qual.* **2016**, *7*, 3851–3858.

16. Schaich, K.M.; Tian, X.; Xie, J. Hurdles and pitfalls in measuring antioxidant efficacy: A critical evaluation of ABTS, DPPH, and ORAC assays. *J. Funct. Foods* **2015**, *14*, 111–125, reprint in *J. Funct. Foods* **2015**, *18*, 782–796. [CrossRef]

17. Turkmen, N.; Velioglu, Y.S.; Sari, F.; Polat, G. Effect of extraction conditions on measured total polyphenol contents and antioxidant and antibacterial activities of black tea. *Molecules* **2007**, *12*, 484–496. [CrossRef] [PubMed]

18. Wang, S.; Chu, Z.; Ren, M.; Jia, R.; Zhao, C.; Fei, D.; Su, H.; Fan, X.; Zhang, X.; Li, Y. Identification of anthocyanin composition and functional analysis of an anthocyanin activator in solanum nigrum fruits. *Molecules* **2017**, *22*, 876. [CrossRef] [PubMed]

19. Goufo, P.; Trindade, H. Rice antioxidants: Phenolic acids, flavonoids, anthocyanins, proanthocyanidins, tocopherols, tocotrienols, γ-oryzanol, and phytic acid. *Food Sci. Nutr.* **2014**, *2*, 75–104. [CrossRef] [PubMed]

20. Vulić, J.J.; Tumbas, V.T.; Savatović, S.M.; Đilas, S.M.; Ćetković, G.S.; Čanadanović-Brunet, J.M. Polyphenolic content and antioxidant activity of the four berry fruits pomace extracts. *Acta Period. Technol.* **2011**, *42*, 271–279. [CrossRef]

21. Lang, J.C. Tocopherol stability in controlled release packaging films. Master's Thesis, The State University of New Jersey, New Brunswick, NJ, USA, January 2009.

22. Wojdyło, A.; Figiel, A.; Oszmiański, J. Effect of drying methods with the application of vacuum microwaves on the bioactive compounds, color, and antioxidant activity of strawberry fruits. *J. Sci. Food Agric.* **2009**, *57*, 1337–1343. [CrossRef] [PubMed]

23. Fernández-Arroyo, S.; Rodríguez-Medina, I.C.; Beltrán-Debón, R.; Pasini, F.; Joven, J.; Micol, V.; Segura-Carretero, A.; Fernández-Gutiérrez, A. Quantification of the polyphenolic fraction and in vitro antioxidant and in vivo anti-hyperlipemic activities of Hibiscus sabdariffa aqueous extract. *Food Res. Int.* **2011**, *44*, 1490–1495. [CrossRef]

24. De Leonardis, A.; Pizzella, L.; Macciola, V. Evaluation of chlorogenic acid and its metabolites as potential antioxidants for fish oils. *Eur. J. Lipid Sci. Technol.* **2008**, *110*, 941–948. [CrossRef]

25. Tonfack, L.B.; Bernadac, A.; Youmbi, E.; Mbouapouognigni, V.P.; Ngueguim, M.; Akoa, A. Impact of organic and inorganic fertilizers on tomato vigor, yield and fruit composition under tropical andosol soil conditions. *Fruits* **2009**, *64*, 167–177. [CrossRef]

26. Kara, Ş.; Erçelebi, E.A. Thermal degradation kinetics of anthocyanins and visual colour of Urmu mulberry (*Morus nigra* L.). *J. Food Eng.* **2013**, *116*, 541–547. [CrossRef]

27. Kim, S.; Park, J.B.; Hwang, I.K. Quality attributes of various varieties of Korean red pepper powders (*Capsicum annuum* L.) and color stability during sunlight exposure. *J. Food Sci.* **2002**, *67*, 2957–2961. [CrossRef]

28. He, J.; Giusti, M.M. Anthocyanins: Natural colorants with health-promoting properties. *Annu. Rev. Food Sci. Technol.* **2010**, *1*, 163–187. [CrossRef] [PubMed]

29. Fernández-López, J.A.; Angosto, J.M.; Giménez, P.J.; León, G. Thermal stability of selected natural red extracts used as food colorants. *Plant Foods Hum. Nutr.* **2013**, *68*, 11–17. [CrossRef] [PubMed]

30. Aurelio, D.L.; Edgardo, R.G.; Navarro-Galindo, S. Thermal kinetic degradation of anthocyanins in a roselle (*Hibiscus sabdariffa* L. cv. 'Criollo') infusion. *Int. J. Food Sci. Technol.* **2008**, *43*, 322–325. [CrossRef]

31. Aramwit, P.; Bang, N.; Srichana, T. The properties and stability of anthocyanins in mulberry fruits. *Food Res. Int.* **2010**, *43*, 1093–1097. [CrossRef]

32. Wang, W.-D.; Xu, S.-Y. Degradation kinetics of anthocyanins in blackberry juice and concentrate. *J. Food Eng.* **2007**, *82*, 271–275. [CrossRef]

33. Cemeroglu, B.; Velioglu, S.; Isik, S. Degradation kinetics of anthocyanins in sour cherry juice and concentrate. *J. Food Sci.* **1994**, *59*, 1216–1218. [CrossRef]

34. Qiu, G.; Wang, D.; Song, X.; Deng, Y.; Zhao, Y. Degradation kinetics and antioxidant capacity of anthocyanins in air-impingement jet dried purple potato slices. *Food Res. Int.* **2018**, *105*, 121–128. [CrossRef] [PubMed]

35. Lavelli, V.; Sri Harsha, P.S.C.; Spigno, G. Modelling the stability of maltodextrin-encapsulated grape skin phenolics used as a new ingredient in apple puree. *Food Chem.* **2016**, *209*, 323–331. [CrossRef] [PubMed]

Sample Availability: not available.

molecules

MDPI

Article

The Microencapsulation of Maqui (*Aristotelia chilensis* (Mol.) Stuntz) Juice by Spray-Drying and Freeze-Drying Produces Powders with Similar Anthocyanin Stability and Bioaccessibility

Carolina Fredes [1], Camila Becerra [1], Javier Parada [2] and Paz Robert [1,*]

[1] Departamento de Ciencia de los Alimentos y Tecnología Química,
Facultad de Ciencias Químicas y Farmacéuticas, Universidad de Chile,
Sergio Livingstone Pohlhammer 1007, Independencia 8380492, Chile;
carolina.fredes@postqyf.uchile.cl (C.F.); camilaj.bc@gmail.com (C.B.)
[2] Institute of Food Science and Technology, Faculty of Agricultural Sciences, Austral University of Chile,
Valdivia 5090000, Chile; javier.parada@uach.cl
* Correspondence: proberts@uchile.cl; Tel.: +56-2-29781666

Academic Editor: Gregory T. Sigurdson
Received: 31 March 2018; Accepted: 16 May 2018; Published: 20 May 2018

Abstract: The microencapsulation of maqui juice by spray-drying and freeze-drying was studied as a strategy to protect anthocyanins in new food formulations in order to improve the anthocyanin retention before consumption and the bioaccessibility. It is well known that the encapsulation method affects both the shape and size of powders, being assumed that undefined forms of freeze-drying powders might affect their stability due to the high permeability to oxygen. The objective of this study was to compare the microencapsulation of maqui juice by spray-drying and freeze-drying, evaluating the stability of specific anthocyanins in yogurt and after in vitro digestion. Results indicated that most relevant differences between spray-drying and freeze-drying powders were the morphology and particle size that affect their solubility (70.4–59.5%) when they were reconstituted in water. Nevertheless these differences did not affect the stability of anthocyanins as other research have proposed. Both encapsulation methods generated powders with a high stability of 3-*O*-monoglycosylated anthocyanins in yogurt (half-life values of 75–69 days for delphinidin-3-sambubioside). Furthermore, no significant differences in the bioaccessibility of anthocyanins between maqui juice powders (44.1–43.8%) were found. In conclusion, the microencapsulation of maqui juice by freeze-drying is as effective as spray-drying to produce new value-added food formulations with stable anthocyanins.

Keywords: encapsulation methods; in vitro digestion; maltodextrin; soy protein isolate; yogurt

1. Introduction

Maqui (*Aristotelia chilensis* (Mol.) Stuntz, Elaeocarpaceae) is a native berry from Chile, recognized worldwide as a rich source of health-promoting anthocyanins with antioxidant, anti-inflammatory, anti-diabetic, and hypoglycaemic activities [1]. New maqui fruit-based products appear frequently on the international market correlating with an increase in Chilean exports of maqui fruit (frozen, juice, dehydrated, canned, and other fruit preparations) worth US $308,562 in 2017, with a mean value of US $4.3–6.5/kg [2]. While the potential health properties of maqui fruit are well known, the stability of anthocyanins in maqui products currently on the market has been scarcely studied. For this reason, it is necessary to study microencapsulation methods to develop new value-added maqui fruit foods with stable anthocyanins that convey health benefits to consumers.

The anthocyanin content of maqui is significantly higher than in other berry-type fruits [3]; where eight anthocyanins (70% delphinidin and 30% cyanidin glucosides) have been identified in the anthocyanin profile of the maqui fruit [4]. However, anthocyanins in general are susceptible to degradation when they are exposed to environmental conditions such as heat, oxygen, and changes in pH, among others [5]. Therefore, encapsulation technology can be used as a strategy to protect maqui anthocyanins in value-added products such as health food ingredients.

Spray-drying (SD) is the most common technique used to encapsulate anthocyanins [5,6]. It is useful for encapsulating heat-sensitive materials because of its short drying times (5–30 s) [7]. Moreover, spray dryers are commonly available in the food and pharmaceutical industries. However, during the last 10 years, freeze-drying (FD) has become more widespread in the food industry [6]. Freeze-drying technique is based on the removal of water from a frozen product by sublimation and has been used as an alternative method to encapsulate anthocyanins [8]. Although at least two previous studies have compared SD and FD as methods to encapsulate anthocyanins from black glutinous rice bran [8] and blueberries [9], a comparison of the stability of encapsulated maqui anthocyanins in food matrices and the gastrointestinal (GI) tract using these two methods has not been evaluated.

Anthocyanin encapsulation has been studied as protection method against environmental conditions during storage, in order to extend shelf life. However, to develop more effective health food ingredients, the anthocyanin stability in food matrices and simulated GI tracts should be assessed [6]. In this context, the selection of the encapsulation method plays an important role in the powder properties that would influence the anthocyanin retention before consumption and the bioaccessibility (BA). For example, Laokuldilok and Kanha [8] indicated that the morphology of FD powders might affect their stability during storage due to the high permeability to oxygen. Maltodextrins (MD) of different dextrose equivalents are biopolymers widely used as encapsulating agent (EA) in the encapsulation of anthocyanins, because MD have low viscosity at high solid content, bland flavor and colourless solutions [6,7]. MD is water soluble and when it is added to liquid foods (as yogurt) the active compounds are released and they are exposed to food conditions. In this study, maqui-juice microparticles were designed with maltodextrin (MD) blend with soy protein isolate (SPI), since SPI is partially soluble, which would lead to a decrease in the solubility of the maqui-juice microparticles. Thus, the objective of this study was to evaluate two different encapsulation methods, spray-drying and freeze-drying, regarding stability of each specific anthocyanin in yogurt and their bioaccessibility in vitro.

2. Results and Discussion

2.1. The Encapsulation of Anthocyanins from Maqui Juice (MJ) by SD and FD

In order to find the optimal conditions for encapsulation by SD, a response surface methodology (RSM) experiment was carried out. The experimental data are shown in Table 1. Equations were fitted to predict yield, recovery, and encapsulation efficiency (EE) of total anthocyanins (TA) and specific anthocyanins. Table 2 shows the predictive models following the general equation:

$$y_n = \beta_{n0} + \beta_{n1}x_1 + \beta_{n2}x_2 + \beta_{n3}x_1^2 + \beta_{n4}x_1x_2 + \beta_{n5}x_2^2 \qquad (1)$$

where β_n are the regression coefficients for each response variable y_n, x_1 is the inlet air temperature, and x_2 the MJ/maltodextrin (MD)-soy protein isolate (SPI) ratio.

Table 1. Orthogonal central composite design showing factor's level combinations and results for encapsulation of MJ anthocyanins by SD.

Run	Inlet Air Temperature (°C)	MJ/EA Ratio	Yield (%)	Recovery (%)	TA	EE (%)					
						Del-3-sa-5-glu	Del-3,5-diglu	Cy-3-sa-5-glu+Cy-3,5-diglu	Del-3-sa	Del-3-glu	Cy-3-glu
1	120	1:2	76.3	88.1	77.7	84.0	79.4	81.1	79.9	77.1	73.4
2	180	1:2	74.6	93.3	78.8	83.1	79.0	80.6	79.0	77.1	74.2
3	120	1:6	69.1	90.3	91.9	93.5	92.7	91.7	90.3	89.3	88.8
4	180	1:6	72.1	93.3	89.4	90.9	88.8	88.1	87.2	84.6	84.6
5	114	1:4	73.7	91.4	90.7	92.6	91.3	90.6	87.8	87.7	87.4
6	186	1:4	73.9	96.1	90.0	90.4	88.9	89.1	87.8	86.4	86.1
7	150	1:1.6	74.8	100.0	74.7	76.6	71.1	71.5	71.5	66.5	65.3
8	150	1:6.4	70.5	82.6	91.6	92.2	90.9	90.8	88.9	87.7	87.4
9	150	1:4	71.0	91.8	89.7	91.0	88.8	88.8	86.9	83.8	84.7
10	150	1:4	69.4	90.1	87.3	89.5	87.6	86.6	85.1	82.2	83.2
11	150	1:4	71.5	94.0	89.3	90.6	89.3	88.6	86.6	84.8	86.1
12	150	1:4	69.5	90.1	90.1	90.9	89.7	89.6	86.0	85.7	85.0

TA = total anthocyanins; del-3-sa-5-glu = delphinidin-3-sambubioside-5-glucoside; del-3,5-diglu = delphinidin-3,5-diglucoside; cy-3-sa-5-glu+cy-3,5-diglu = cyanidin-3-sambubioside-5-glucoside+cyanidin-3,5-diglucoside; del-3-sa 5 = delphinidin-3-sambubioside; del-3-glu = delphinidin-3-glucoside; cy-3-sa = cyanidin-3-sambubioside; cy-3-glu = cyanidin-3-glucoside.

Table 2. Regression coefficients of fitted equations for each response variables.

	y_n	β_{n0}	β_{n1}	β_{n2}	β_{n3}	β_{n4}	β_{n5}	R^2	R^2 (Adjusted for d.f.)
	Yield (%)	136.3140	-0.6917	-6.1670	0.0021	0.0196	0.2692	0.902	0.820
	Recovery (%)	100.6240	-0.1919	1.5889	0.0010	-0.0092	-0.1968	0.429	0.00
	TA	57.4578	-0.0461	14.7790	0.0003	-0.0150	-1.1579	0.979	0.961
EE%	del-3-sa-5-glu	90.3751	-0.3418	11.3291	0.0011	-0.0071	-0.9569	0.963	0.933
	del-3,5-diglu	78.7725	-0.3441	15.8078	0.0012	-0.0146	-1.2780	0.972	0.949
	cy-3-sa-5-glu+cy-3,5-diglu	88.4993	-0.4300	14.0942	0.0015	-0.0129	-1.1456	0.930	0.872
	del-3-glu	94.0431	-0.4748	11.6651	0.0016	-0.0092	-0.9285	0.951	0.909
	cy-3-sa	102.4870	-0.7099	15.2545	0.0025	-0.0196	-1.1305	0.927	0.866
	cy-3-glu	70.4168	-0.3790	18.4930	0.0015	-0.0208	-1.4470	0.971	0.947

TA = total anthocyanins; del-3-sa-5-glu = delphinidin-3-sambubioside-5-glucoside; del-3,5-diglu = delphinidin-3,5-diglucoside; cy-3-sa-5-glu+cy-3,5-diglu = cyanidin-3-sambubioside-5-glucoside+cyanidin-3,5-diglucoside; del-3-sa 5 = delphinidin-3-sambubioside; del-3-glu = delphinidin-3-glucoside; cy-3-sa = cyanidin-3-sambubioside; cy-3-glu = cyanidin-3-glucoside.

The coefficients of determination (R^2) and R^2-adjusted for the predictive models were higher than 0.90 and 0.82, respectively, for each response variable except for recovery (0.42 and 0.0, respectively), suggesting that the predictive models seemed to reasonably represent the observed values, with the exception of recovery. Therefore, 8 of the 9 responses could be sufficiently explained by the models, while for recovery, results suggest that its optimal condition is outside the experimental zone. Anthocyanin recovery reveals the effect of the drying process on anthocyanin content. In this study, high recovery results ($\geq$82.6%) indicate that the temperatures used in SD did not have a great effect on the degradation of anthocyanins.

Multiple response optimization methodology was performed in order to obtain optimal conditions that maximize all variables. These optimal conditions were an inlet air temperature of 186 °C and a MJ/MD-SPI ratio of 1:5. The corresponding predictive values were a yield of 73.4%, a recovery of 93.5%, and an EE of TA of 90.9%. The predictive yield was relatively low but in line with previous reports [6], whereas recovery (93.5%) and anthocyanin EE (>87%) were higher than expected from previous research [10]. In order to compare the properties of SD powders obtained under optimal conditions with FD powders, a MJ/MD-SPI ratio of 1:5 was used to formulate the FD powders.

2.2. Characterization of the MJ Powders Obtained by SD and FD

The anthocyanin profile of MJ was similar than those previously described [4,11–15], showing eight anthocyanins (del-3-sa-5-glu, del-3,5-diglu, cy-3,5-diglu+cy-3-sa-5-glu (coeluted in peak 3–4), del-3-sa, del-3-glu, cy-3-sa and cy-3-glu). Additionally, cy-3-sa was found under the limit of quantification (0.16 µg/mL). The anthocyanin profiles for MJ powders (SD and FD) were similar to MJ where 3,5-*O*-diglycosylated anthocyanins ($\approx$70%) predominated over 3-*O*-glycosylated anthocyanins ($\approx$30%) in both MJ and the MJ powders (Table 3). In concordance with the high anthocyanin content of MJ (17.5 mg/g) as a raw material, both MJ powders showed high anthocyanin content (SD, 3.3 mg/g and FD, 3.3 mg/g) in comparison to microparticles in powder from other raw materials such as Andes berry (0.1 mg/g) [16] and bayberry (0.6 mg/g) [17].

Table 3. Anthocyanins content in MJ and MJ powders and the BA of MJ anthocyanins after an in vitro digestion model.

	MJ		SD		FD	
	mg/g	BA %	mg/g	BA %	mg/g	BA %
Del-3-sa-5-glu	7.4 ± 0.02	35.1 ± 3.35 [b,A]	1.3 ± 0.04	43.2 ± 0.61 [c,AB]	1.4 ± 0.07	40.8 ± 2.00 [cd,B]
Del-3,5-diglu	3.3 ± 0.01	36.4 ± 1.12 [b,A]	0.6 ± 0.02	34.6 ± 1.16 [b,A]	0.6 ± 0.03	36.2 ± 1.9 [c,A]
Cy-3-sa-5-glu + cy-3,5 diglu	3.4 ± 0.01	48.9 ± 1.58 [c,A]	0.7 ± 0.01	73.5 ± 3.04 [d,B]	0.6 ± 0.03	76.5 ± 2.39 [e,B]
Del-3-sa	1.5 ± 0.02	21.9 ± 2.83 [a,A]	0.3 ± 0.01	24.8 ± 0.85 [a,AB]	0.3 ± 0.01	28.5 ± 2.56 [b,B]
Del-3-glu	2.4 ± 0.01	21.2 ± 0.50 [a,A]	0.4 ± 0.01	20.8 ± 0.41 [a,A]	0.4 ± 0.02	22.6 ± 0.47 [a,B]
Cy-3-glu	1.2 ± 0.02	24.0 ± 2.29 [a,A]	0.2 ± 0.01	43.8 ± 2.93 [c,B]	0.2 ± 0.01	44.8 ± 1.00 [d,B]
3-*O*-monoglycosylated anthocyanins	5.0 ± 0.01	23.3 ± 3.29 [a,A]	0.9 ± 0.02	27.2 ± 0.57 [a,AB]	0.9 ± 0.04	30.8 ± 2.09 [a,B]
3,5-*O*-diglycosylated anthocyanins	14.1 ± 0.04	39.5 ± 3.18 [b,A]	2.6 ± 0.06	50.1 ± 2.07 [b,B]	2.6 ± 0.12	48.4 ± 0.90 [b,B]
TA HPLC	17.5 ± 0.04	35.2 ± 3.21 [A]	3.3 ± 0.20	44.1 ± 1.68 [B]	3.3 ± 0.41	43.8 ± 0.13 [B]

Mean values (*n* = 3) and standard deviation that are followed by different upper case letters in the same row indicate significant differences ($p \leq 0.05$) between encapsulation methods and by different lower case letters in the same column indicate significant differences ($p \leq 0.05$) among anthocyanins. MJ, maqui juice; SD, powders obtained under spray-drying optimal conditions; FD, powders obtained by freeze-drying grinded product; TA, total anthocyanins; BA, bioaccessibility. del-3-sa-5-glu, delphinidin-3-sambubioside-5-glucoside; del-3,5-diglu, delphinidin-3,5-diglucoside; cy-3-sa-5-glu, cyanidin-3-sambubioside-5-glucoside; cy-3,5-diglu, cyanidin-3,5-diglucoside; del-3-sa, delphinidin-3-sambubioside; del-3-glu, delphinidin-3-glucoside; cy-3-glu, cyanidin-3-glucoside.

The EE has been used to determine if the anthocyanins were encapsulated. EE represents anthocyanin-biopolymer interaction due to electrostatic interactions, hydrogen bonding, hydrophobic interactions, or Van der Waals forces [7]. However, encapsulation by spray-drying is considered an immobilization technology rather than a true encapsulation technology because some active compounds remain exposed on the microparticle surface [6]. Thus, the EE is a way to gauge successful anthocyanin encapsulation which should result in a powder that has minimum surface anthocyanin content on the microparticles and maximum retention of active compounds [5]. According to our results, and in line with the predictive value for SD, the EE of TA was high for SD (92.5%) and FD (>93.0%) powders (Table 4) where the encapsulation method did not significantly affect the EE. Similar results showing high EE of TA were described by Santana et al. [18] for jussara microparticles (88.3–99.3%) using a mixture of arabic gum/modified starch/SPI and SD. Furthermore, our results showed that the mixture of polysaccharides and proteins improved the EE of TA, as compared to using either MD (58.5%) or SPI (86.6%) alone, in the microencapsulation of pomegranate juice using SD [10]. In line with the EE of TA results, the EE of specific anthocyanins did not show significant differences between SD and FD methods, although the EE of 3,5-*O*-diglycosylated anthocyanins was significantly higher than EE of 3-*O*-monoglycosylated anthocyanins (Table 4). This suggests a better interaction between the EA (MD-SPI) and 3,5-*O*-diglycosylated anthocyanins by hydrogen bonding, which could be attributed to the greatest number of hydroxyl groups for 3,5-*O*-diglycosylated anthocyanins.

Table 4. MJ powders characteristics and parameters of powders reconstitution.

MJ Powder	SD	FDc	Fdm
EE TA	92.5 ± 0.02 [a]	93.0 ± 0.01 [a]	93.9 ± 0.01 [a]
EE 3-*O*-monoglycosylated anthocyanins	89.5 ± 2.2 [a,A]	91.4 ± 1.3 [ab,A]	92.3 ± 1.4 [b,A]
EE 3,5-*O*-diglycosylated anthocyanins	93.1 ± 1.4 [a,B]	94.6 ± 0.6 [ab,B]	95.4 ± 0.9 [b,B]
Recovery (%)	99.8 ± 0.01 [b]	91.9 ± 0.01 [a]	91.8 ± 0.02 [a]
Yield (%)	64.1 ± 0.01 [a]	94.6 ± 0.01 [b]	94.6 ± 0.01 [b]
a_w	0.3 ± 0.01 [b]	0.1 ± 0.01 [a]	0.1 ± 0.01 [a]
Moisture content (%)	6.4 ± 0.02 [b]	3.2 ± 0.01 [a]	3.2 ± 0.02 [a]
Hygroscopicity (%)	39.4 ± 0.03 [a]	48.0 ± 0.02 [b]	53.1 ± 0.01 [b]
Bulk density (g/mL)	0.4 ± 0.01 [b]	0.3 ± 0.01 [a]	0.3 ± 0.01 [a]
Solubility (%)	70.4 ± 0.01 [b]	59.9 ± 0.02 [a]	59.1 ± 0.02 [a]
Dispersibility (%)	99.1 ± 0.01 [a]	100.0 ± 0.01 [a]	97.7 ± 0.01 [a]
pH	5.3 ± 0.02 [b]	5.2 ± 0.01 [a]	5.2 ± 0.02 [a]

Mean values ($n = 3$) and standard deviation that are followed by different lower case letters in the same row indicate significant differences ($p \leq 0.05$) between encapsulation methods and by different upper case letters in the same column indicate significant differences ($p \leq 0.05$) among anthocyanins. EE, encapsulation efficiency; TA, total anthocyanins; SD, powders obtained under spray-drying optimal conditions; FDc, powders obtained by freeze-drying grinded with coffee grinder; FDm, powders obtained by freeze-drying grinded with porcelain mortar and pestle.

The recovery of anthocyanins in the SD powders (99.8%) was in line with the predictive value and significantly higher than the FD powders (91.8–91.9%) (Table 4). However, these values were higher than those reported by Laokuldilok and Kanha [8] for black glutinous rice bran microparticles using SD (47.7%) and FD (71.9%).

The SD yield (64.1%) was 10% lower than the predictive value and significantly lower than FD (94.6%). Similarly, Laokuldilok and Kanha [8] reported differences in the yield between SD (>64%) and FD (85%). This can be explained by the high viscosity of the feed solution when SPI is used as an EA, which causes more solids to stick to the wall of the dryer chamber, resulting in less powder at the end of the SD process.

The physical characteristics (Table 4) showed that the water activity (a_w) of SD powders was significantly higher than FD powders, but all of the values were low (<0.3), therefore avoiding potential microbial food spoilage [16]. Additionally, the moisture content results were in the range acceptable for food powders (3–10%) but they were also significantly higher in SD powders than FD powders. In concordance with a high moisture content, SD powders had significantly lower hygroscopicity and higher bulk density than FD powders. Moreover, it is important to notice that no significant differences between FD powders obtained by grinded with coffee grinder (FDc) and porcelain mortar (FDm) for any of these physical parameters were found (Table 4), reflecting that the grinding process did not affect a_w, moisture content, hygroscopicity, nor the bulk density of the two types of FD powders.

The ability of the powders to reconstitute in water is significant for the development of food ingredients. As expected, including SPI in the MJ powders decreased solubility in comparison to using only MD [18]. The solubility of SD powders (70.4%) was significantly higher than FD powders (59.9–59.1%), which may be attributed to differences in the morphology of powders and the corresponding contact surface with water. In line with physical parameters, no significant difference in solubility between FDc and FDm were found. Furthermore, the solubility results may also be attributed to the high total sugar content of MJ.

The dispersibility of the MJ powders was over 97.7% and no significant differences among MJ powders were found. The high dispersibility indicates a low tendency to form lumps when MJ powders are added to water, facilitating their reconstitution. Furthermore, it is important to notice that the addition of the MJ powders did not modify the pH of the water.

Scanning electron microscopic photographs showed that SD powders (Figure 1A) have spherical shapes and particles with indented surfaces, obviating their agglomerating tendency. For FDc (Figure 1B) and FDm (Figure 1C) powders, SEM photographs showed undefined forms with more intended surfaces than SD powders, in agreement with the description by Fang and Bhandari [19] who suggested that the encapsulation method as well as the grinding process directly affects both the shape and size of powders.

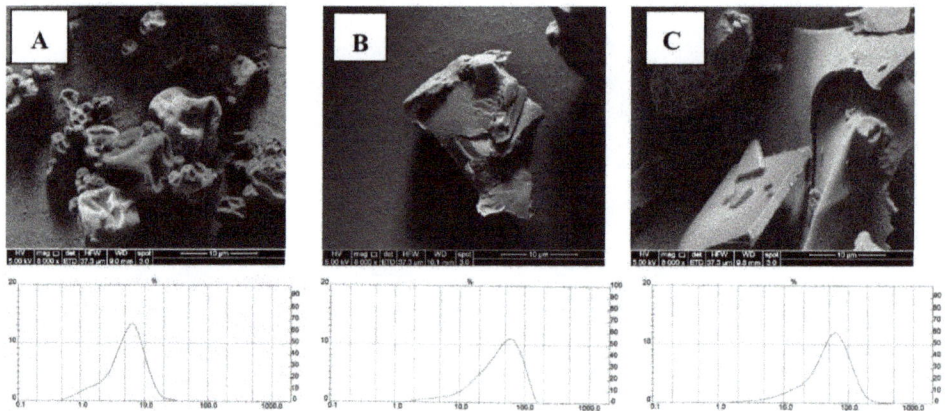

Figure 1. Scanning electron microscopic (SEM) photographs for SD (**A**), FDc (**B**) and FDm (**C**) powders and the distribution of particles diameter using Mastersizer.

The particle size measurement assumes powders have a spherical shape [20]. The particle size of SD powders was 6.4 μm, and 90% of the total sample had a diameter below 10.9 μm (Figure 1A) in line with the particle size of SD powders [16]. The particle size from FD powders (49.4 and 64.0 μm for FDc and FDm, respectively) was almost 10 times higher than the SD powder particles, and 90% of the total sample had a diameter below 93.9 and 117.1 μm, respectively. Therefore, in terms of the characteristics of MJ powders, the most relevant differences between SD and FD powders were the morphology and particle size that affect their solubility in water. However, the differences in particle size between FD powders did not have any impact on their solubility. MD is a water soluble biopolymer but SPI is partially soluble in water. However, MD blend with SPI represents a new system with different properties with respect to each single biopolymer [21], which explains why the solubility of maqui-juice microparticles with MD blend with SPI is lower than MD.

2.3. Stability of MJ Powders in Yogurt

Previous studies [10,22] on the evaluation of anthocyanin stability in yogurt has been focused on the retention of TA by spectrometry where the stability of anthocyanins comparing encapsulated with non-encapsulated anthocyanins has been scarcely studied [10]. The evolution of TA retention (Figure 2) was similar among MJ powders and MJ in agreement with results reported by Robert et al. [10] for anthocyanins from pomegranate juice and pomegranate microparticles. Furthermore, our results showed high retention of TA (SD, 82%, FDc, 88%, and FDm, 81%) after 35 days in line with a study carried out with a freeze-dried mulberry fruit juice [23]. Contrary Coisson et al. [22] and Robert et al. [10] reported that anthocyanins from acai and pomegranate juices, disappeared after 2 and 7 days of storage at 4 °C, respectively. This may be explained by the presence of different types of anthocyanins in fruits justifying the study of the stability of specific anthocyanins in yogurt.

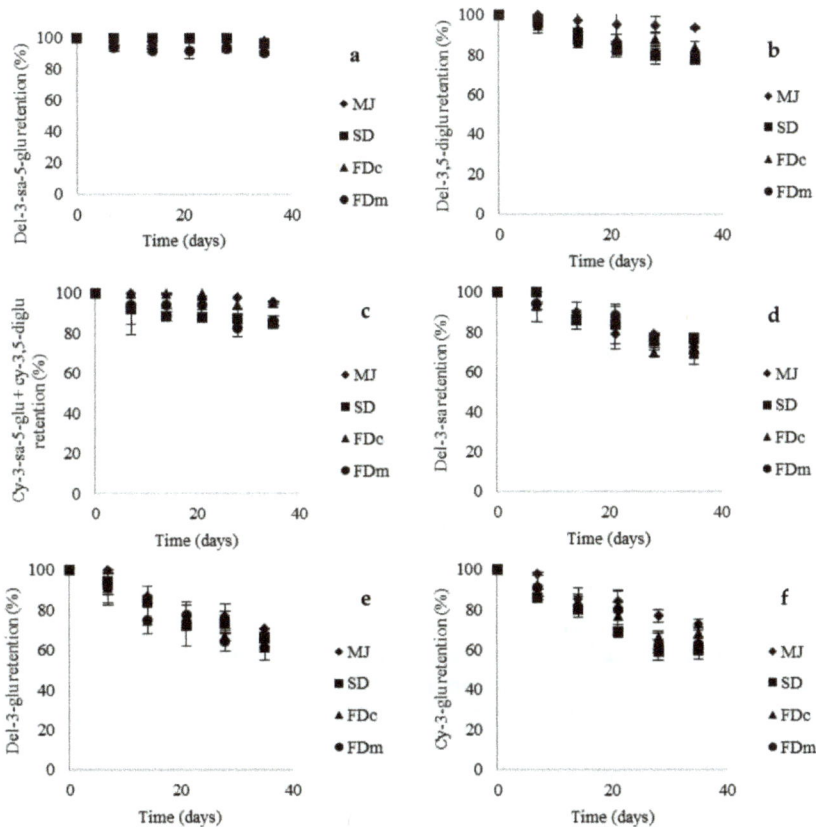

Figure 2. Evolution of the anthocyanins retention of MJ and MJ powders (SD, FDc and FDm) added to yogurt during storage at 5 °C. (**a**) delphinidin-3-sambubioside-5-glucoside retention; (**b**) delphinidin-3,5-diglucoside retention; (**c**) cyanidin-3-sambubioside-5-glucoside + cyanidin-3,5-diglucoside retention; (**d**) delphinidin-3-sambubioside retention; (**e**) delphinidin-3-glucoside retention; (**f**) cyanidin-3-glucoside retention.

Ribeiro et al. [24] indicated that in aqueous solutions, anthocyanins undergo structural rearrangements in response to changes in pH. Anthocyanins are most stable in acidic solutions (pH 1–3), whereas at a pH above 4, anthocyanins undergo chemical degradations to produce phenolic acids. However, as shown in Figure 2, the anthocyanins did not react uniformly; del-3-sa-5-glu showed the highest retention (90–99%) whereas cy-3-glu showed the lowest retention (61–69%) during storage at 5 °C after 35 days. Because the addition of MJ and MJ microparticles into yogurt did not modify the pH of the food matrix (4.3), anthocyanin degradation may be mainly attributed to the pH of the food matrix. Structural features of the anthocyanins influenced their stability in yogurt as reflected in their half-life values (Table 5). In this context, the k_{obs} and the corresponding half-life values were calculated only for 3-*O*-monoglycosylated anthocyanins that reached a degradation of over 30%. The half-life of del-3-sa was significantly higher than del-3-glu and cy-3-glu, suggesting that the extra xylose as a sugar component of del-3-sa favors its stability in yogurt. On the other hand, the encapsulation method did not affect the stability of 3-*O*-monoglycosylated anthocyanins as reflected in the half-life values. This suggests that the differences in particle size and morphology between SD and FD powders in this study did not affect the stability of anthocyanins as other research [8,19] have proposed. The results also showed that MJ powders protected anthocyanins during the shelf-life of yogurt, preserving their stability, suggesting that yogurt might be a good carrier for functionalization.

Table 5. Half-life ($t\frac{1}{2}$) values (days) of anthocyanins and anthocyanins retention (time day 35, t_{35}) from MJ and MJ powders in yogurt during storage at 5 °C.

	MJ (Control)		SD		FDc		FDm	
	$t_{1/2}$ (day)	Retention t_{35} (%)	$t_{1/2}$ (day)	Retention t_{35} (%)	$t_{1/2}$ (day)	Retention t_{35} (%)	$t_{1/2}$ (day)	Retention t_{35} (%)
Del-3-sa-5-glu	nd	100 ± 0.02	nd	97 ± 0.01	nd	99 ± 0.01	nd	90 ± 0.02
Del-3,5-diglu	nd	94 ± 0.02	nd	78 ± 0.02	nd	84 ± 0.03	nd	79 ± 0.03
Cy-3-sa-5-glu+ cy-3,5-diglu	nd	96 ± 0.03	nd	85 ± 0.01	nd	95 ± 0.01	nd	86 ± 0.01
Del-3-sa	75 ± 8 [b,A]	73 ± 0.03	75 ± 11 [b,A]	77 ± 0.01	67 ± 11 [b,A]	69 ± 0.02	70 ± 4 [b,A]	70 ± 0.02
Del-3-glu	65 ± 9 [a,B]	71 ± 0.02	59 ± 5 [ab,AB]	66 ± 0.02	48 ± 8 [a,A]	62 ± 0.01	46 ± 4 [a,A]	62 ± 0.01
Cy-3-glu	68 ± 7 [a,C]	73 ± 0.01	44 ± 2 [a,A]	61 ± 0.02	59 ± 9 [a,BC]	69 ± 0.02	50 ± 7 [a,B]	63 ± 0.02

Mean values (n = 3) and standard deviation that are followed by different upper case letters in the same row indicate significant differences ($p \leq 0.05$) between encapsulation methods and by different lower case letters in the same column indicate significant differences ($p \leq 0.05$) among anthocyanins. del-3-sa-5-glu, delphinidin-3-sambubioside-5-glucoside; del-3,5-diglu, delphinidin-3,5-diglucoside; cy-3-sa-5-glu, cyanidin-3-sambubioside-5-glucoside; cy-3,5-diglu, cyanidin-3,5-diglucoside; del-3-sa, delphinidin-3-sambubioside; del-3-glu, delphinidin-3-glucoside; cy-3-glu cyanidin-3-glucoside; SD, powders obtained under spray-drying optimal conditions; FDc, powders obtained by freeze-drying grinded with coffee grinder; FDm, powders obtained by freeze-drying grinded with porcelain mortar and pestle; nd, not determined.

2.4. Bioaccessibility of MJ Powders after an In Vitro Digestion

The BA has been defined as the amount of compound (anthocyanins) that was released from the food matrix after digestion [25]. In this study, BA was calculated in order to quantify the inhibition of anthocyanin degradation, comparing encapsulated (MJ powders) with non-encapsulated MJ anthocyanins. BA of TA in MJ powders (44.1% for SD and 43.8% for FD powders) was significantly higher than in MJ (35.2%) (Table 3). Nevertheless, no significant differences in the BA of TA between SD and FD powders were found, suggesting that their physical differences would not have an effect on the stability of anthocyanins.

Lucas-González et al. [26] demonstrated that the stability of maqui anthocyanins from a lyophilized fruit was greatly affected in the intestinal phase during in vitro GI digestion. Therefore, in our study, the expected low BA of TA in the MJ can be attributed to the high degradation of anthocyanins in GI conditions. It is well known that anthocyanins are unstable at high pH, and the shift from the acidic pH (pH 2) of the stomach to the almost neutral pH in the duodenum (pH 6) may be responsible for their specific hydrolysis and/or degradation [25–28]. Mouth conditions were considered in this study, although Mosele et al. [28] indicated that anthocyanins suffer negligible modifications due to the short exposure time in the mouth, resulting in a marginal effect of the α-amylase.

When anthocyanins are encapsulated, they are more protected in the GI tract, as was shown in our study. Oidtmann et al. [29] indicated that in bilberry microparticles encapsulated with polysaccharides (pectin amide) or proteins (whey protein isolate), anthocyanins were not degraded under simulated gastric conditions despite the fact that they had a competitive release from the microparticles. The same study indicated that in intestinal conditions, anthocyanins were also released from the microparticles (after 30 min), with degradation occurring after their release [29]. From another point of view, Flores et al. [27] indicated that the enzymatic and pH conditions are highly extractive in the stomach, which may affect the integrity of microparticles, favoring the release of anthocyanins. Nevertheless, the mixture of MD and SPI as EA demonstrated a protective effect, resulting in the enhancement of BA of TA in both SD and FD powders. Thus, our results suggest that this high BA is due to lower exposure times to gut conditions because of the protective effect of MD and SPI.

In concordance with the results of TA, BA of specific anthocyanins from MJ powders was significantly higher than MJ (Table 3). Moreover, there were significant differences in the BA among anthocyanins, where cy-3-sa-5-glu and cy-3,5-diglu had the highest BA (78.5% for SD and 76.5% for FD) and del-3-glu had the lowest BA (20.8% for SD and 22.6% for FD), demonstrating differences in their stability after their release from MJ powders. In agreement with our results, Lila et al. [25] suggest that the fact that cy-3-sa-5-glu plus cy-3,5-diglu demonstrated the highest BA is because two cyanidins combined with two linked sugars generate a more stable anthocyanin structure.

Despite the fact that changes in pH are involved in the degradation of anthocyanins in the GI tract and significant differences between the stability of 3-*O*-monoglycosylated and 3,5-*O*-diglycosylated anthocyanins as in yogurt were found (Table 3), the anthocyanins did not show the same behavior after the in vitro digestion. This can be contributed to the more complex reactions that anthocyanins experience during the digestion process, where the action of different enzymes in combination with different temperatures and time of exposure are all involved.

In this study, the results showed that the BA of both MJ microparticles (44.1–43.8%) and MJ (35.2%) were higher than that described by Lila et al. [25] for a semi-purified maqui extract (4%). These differences may be explained by the additional protector effect that MJ, as a food matrix, gives to anthocyanins. McDougall et al. [30] suggested that polyphenols generate linkages with the food matrix during digestion that may protect more labile anthocyanins against degradation.

3. Materials and Methods

3.1. Raw Material and Encapsulating Agents

Organic concentrated (65 °Brix) MJ (Patagonol™—LE, Bayas del Sur, Purranque, Chile) was microencapsulated using a mixture of MD (Maltodextrin 1520 Prinal S.A., Santiago, Chile) and SPI (Protein HS, min. 88%, Prinal S.A.). The MJ had the following characteristics: moisture content (41.0 ± 0.03%), total soluble solids (660 ± 1 g/kg), pH (3.8 ± 0.03), titratable acidity (58 ± 1 g/kg), and total sugars (610 ± 8 g/kg). Commercial natural yogurt (Soprole®, Santiago, Chile) was purchased in a local market in Santiago, Chile.

3.2. Preparation of MJ Powders by SD

Powders were prepared using MD and SPI (2:1) in 100 g solutions as follows: 2.5 g of MJ was mixed with 2.6–10.73 g of MD, 1.3–5.3 g of SPI and distilled water. The MD-SPI mixture in distilled water was stirred for 12 h and then the MJ was added. Each preparation was homogenized at 11,000 rpm for 5 min using a Polytron PT 2100 (Kinematica A.G, Luzern, Switzerland). The resulting solutions were fed into a mini spray-dryer B290 (Buchi, Flawil, Switzerland). The spray-dryer was operated at an inlet air temperature ranging from 120 to 180 ± 1 °C. Air flow, rate of feeding, and atomization pressure were 600 L/h, 1 mL/min, and 0.14 MPa, respectively. The resulting powders were stored absent of light at −80 °C until analyzed.

The experiment was performed using an orthogonal central composite experimental design with 12 runs (4 factorial points, 4 axial points and 4 central points). Independent variables were the inlet air temperature (120–180 °C) (x_1) and the MJ/MD-SPI ratio (1:2–1:6) (x_2). Dependent variables were the encapsulation efficiency (EE) of anthocyanins, the recovery, and the yield. A response surface methodology (RSM) was applied using the desirability function where 1 represented the maximization of each variable. The data were fitted to a second-order regression model. All experiments were conducted randomly to avoid systematic bias. The linear, quadratic, and interaction effects of the independent variables on the response variables at a confidence level of 95% were considered (Statgraphics Centurion XV, Version 15.1.02, StatPoint, Inc., Warrenton, VA, USA).

3.3. Preparation of MJ Powders by FD

Solutions (100 g) of MJ with MD and SPI (2:1) were prepared using the optimal MJ/MD-SPI ratio obtained for SD powders. Each solution was frozen (−80 °C) for 48 h and then dried in a LABCONCO freeze drier (LABCONCO Corporation, Kansas City, MO, USA) under a pressure below 0.004 mBar, and a temperature of −53 ± 1.7 °C for 48–72 h. The dried product was grinded by two different procedures, a coffee grinder (MKM6003, Bosch, Stuttgart, Germany) (FDc) and a porcelain mortar and pestle (3.5-inch) (FDm). The resulting powders were stored absent of light at −80 °C until analyzed.

3.4. Characterization of the MJ Powders Obtained by SD and FD

3.4.1. Total Anthocyanin Determination in MJ Powders

The coating material structure of the microparticles was completely destructed to determine the total anthocyanins. The MJ powders (200 mg) were dispersed in 2 mL of methanol: acetic acid: water (50:1:49 $v/v/v$). This dispersion was vortexed for 1 min, ultrasonicated for 20 min, centrifuged at 11,000 rpm for 8 min, and then the supernatant was filtered (0.22 μm PTFE membrane filters, VWR International, Atlanta, GA, USA) and injected into the HPLC instrument.

3.4.2. Surface Anthocyanin Determination in MJ Powders

Surface anthocyanins are determined using solvents where anthocyanins are soluble and the EA are insoluble. The MJ powders (400 mg) were dispersed in 2 mL of methanol: acetic acid (99:1 v/v). The dispersion was softly vortexed at room temperature for 1 min and then filtered (0.22 μm PTFE membrane filters, VWR International) and injected into the HPLC instrument.

The yield, recovery, and EE of anthocyanins were calculated according to the following equations:

$$\text{Yield} = \frac{\text{powder after spray drying (g)}}{\text{solids in the feed solution (g)}} \times 100 \tag{2}$$

$$\text{Recovery} = \frac{\text{anthocyanins in the powder (mg/g)}}{\text{anthocyanins in the feed solution (mg/g)}} \times 100 \tag{3}$$

$$\text{EE anthocyanins} = \frac{\text{total anthocyanin}_i \text{ in the powder} - \text{surface anthocyanin}_i \text{ in the powder}}{\text{total anthocyanin}_i \text{ in the powder}} \times 100 \tag{4}$$

where i corresponds to each anthocyanin.

3.4.3. Determination of Anthocyanins by HPLC

The LC-MS-IT-TOF identification of the anthocyanins was reported in our previous study [11]. Anthocyanin quantification was conducted on a HPLC system equipped with a DAD (Flexar, Perkin Elmer, Buckinghamshire, UK) using a C18 column (5 μm × 4.6 mm i.d. × 25 cm, Symmetry, Waters, Ireland), with 5% formic acid in H_2O (A) and 100% methanol (B) as mobile phases. The flow rate was constant at 1 mL/min. Solvent gradient was 10%, 15%, 20%, 25%, 30%, 60%, 10%, and 10% of solvent B at 0, 5, 15, 20, 25, 45, 47, and 60 min, respectively [4]. The absorption was measured at a wavelength of

520 nm. Specific anthocyanins were quantified using a cy-3-glu (Sigma-Aldrich, St. Louis, MO, USA) calibration curve (0.16–24.8 μg/mL; R^2 = 0.9998). Quantification of total anthocyanins (TA) content corresponds to the sum of all of the anthocyanins peaks.

3.4.4. Physical Analysis of MJ Powders and Reconstitution

The moisture content and the a_w of MJ powders were measured according to AOAC method 925.40 [31]. The hygroscopicity was determined by the method described by Cai and Corke [32]. Powder bulk density was determined by loading the powder into a graduated cylinder to 10 mL mark and weighing it [33]. The powder weight and volume were then used to calculate the bulk density, expressed as mass/volume. The solubility and dispersibility as parameters of powder reconstitution were measured following Laokuldilok and Kanha [8]. MJ powders (1 g) were dissolved in 10 mL of distilled water and continuously stirred for 30 min. The suspension was then transferred to a Falcon tube and centrifuged at 6000 rpm for 20 min. The supernatant was dried at 105 °C for 24 h. The dry weight of the soluble solid was measured and used to calculate the solubility as a percentage. For dispersibility, MJ powders (1 g) were added to 10 mL of distilled water and then stirred vigorously by a magnetic stirrer for 25 s. The reconstituted powder was passed through a 150 mm sieve. An aliquot (1 mL) of the sieved solution was dried at 105 °C for 4 h. After that, total solids as a percentage was used to calculate dispersibility.

3.4.5. Powder Morphology

Scanning electron microscopy was used to study the outer structures of the MJ powders. The samples were coated with a 10 nm film of gold by a Sputter Coater 108auto with a thickness Controller MTM-20 (Cressington Scientific Instruments, Watford, UK) and analyzed with a High Resolution Scanning Electron Microscope (HR-SEM) with a secondary electron detector (SED) (model INSPECT-F50, FEI, Thermo Fisher Scientific, Hillsboro, OR, USA) operated at 5.00 KV.

3.4.6. Particle Size

The size of the MJ powders (SD, FDc and FDm) was measured using a MastersizerX (Malvern Instruments, Worcestershire, UK). The samples were dispersed in isopropanol and sonicated for 20 s in enough powder to reach the optimal shade in the equipment. The mean particle size was expressed as the volume mean diameter ($D_{4,3}$).

3.5. Stability of Anthocyanins from MJ Powders in Yogurt

MJ (SD, FDc and FDm) powders (0.25 g) were added to yogurt (25 g) in glass containers with tap and stored at 5 ± 1 °C in the absence of light for 5 weeks (in triplicate). A control sample (in triplicate) with the addition of MJ (0.25 g) was also prepared. The containers were withdrawn every 7 days to determine TA and specific anthocyanin retention [21]. Yogurt with MJ (5 g) and yogurt with MJ powders (7 g) were mixed with LiChrosolv® chromatography (grade LC-MS, Merck-Milipore, Burlington, MA, USA) water (pH = 3) until reaching 10 g. Each sample was vortexed for 1 min, and centrifuged at 6,000 rpm for 20 min. The supernatant was centrifuged at 12,000 rpm for 20 min. Following previous studies, first-order degradation kinetic model ($\ln C = \ln C_0 - k(t)$) were used to fit the analytical data and to calculate degradation rate constants (k_{obs}) obtained from the slope of a plot of the natural log of the specific anthocyanin content vs. time. Retention (%) = (C_t/C_0) × 100, where C_t is the anthocyanin content at "t" time, and C_0 is the initial specific anthocyanin content. Half-life values were calculated from: ($t_{1/2} = \text{Ln}2/k_{obs}$) [10,21].

3.6. Bioaccessibility of MJ Powders Using an In Vitro Digestion Model

Samples of 0.5 g (in triplicate) of MJ powders and MJ were processed following Aravena et al. [34].

3.6.1. Mouth Digestion

Artificial saliva (9 mL) was added to each flask with sample. This solution was comprised of 14.4 mM sodium bicarbonate, 21.1 mM potassium chloride, 1.59 mM calcium chloride, and 0.2 mM magnesium chloride. The pH was adjusted to 7 with HCl (1 M). Sixty α-amylase units per milliliter of buffer were incorporated the same day the test was performed. Samples were incubated in a thermostatic bath (ZHWY-110X30, Zhicheng, Shangai, China); 37 °C for 5 min, at a shaking speed of 185 rpm.

3.6.2. Stomach Digestion

The pH of the samples was adjusted to 2.0 using HCl (1 M), then 36 mL of a pepsin solution (25 mg/mL in 0.02 M HCl) was added. Therefore, each sample (containing 9 mL of artificial saliva) was diluted 5-fold with artificial gastric juice, as it occurs in the stomach. Samples were incubated for 2 h at 37 °C with a stirring speed of 130 rpm.

3.6.3. Gut Digestion

The pH of the samples was adjusted to 6.0 with NaHCO$_3$ (1 M). Then, 0.25 mL per mL of sample of an artificial gut solution, containing pancreatin (2 g/L) and bile salts (12 g/L) dissolved in aqueous NaHCO$_3$ (0.1 M) was added. Incubation was carried out for 2 h at 37 °C and a shaking speed of 45 rpm.

Each digestion product was transferred to 50 mL Falcon tubes and the pH was adjusted (pH 3) [27]. It was then centrifuged for 10 min at 5000 rpm to recover the liquid fraction. Next, the liquid digestion product was centrifuged at 12,000 rpm before anthocyanin analysis. The measured amount of specific anthocyanin was the bioaccessible portion, which was calculated as follows:

$$\text{Bioaccessibility \%} = \frac{\text{mg anthocyanin}_i \text{ on digest product}}{\text{mg anthocyanin}_i \text{ in MJ or MJ powders}} \times 100 \tag{5}$$

where i corresponds to each anthocyanin quantified.

3.7. Statistical Analysis

The differences in MJ powders for each parameter and among specific anthocyanins were analyzed using a one-way ANOVA test for means comparison. When significant differences were found, the Tukey HSD (honest significant differences) multiple-comparison test ($p \leq 0.05$) was applied. Analyses were performed with SAS 9.2 for Microsoft Windows (2009; SAS Institute Inc., Cary, NC, USA).

4. Conclusions

The studied encapsulation methods did not generate differences on stability of 3-*O*-monoglycosylated anthocyanins in yogurt, and either on BA of anthocyanins after in vitro GI digestion. Thus the encapsulation technology using either SD or FD can be used as a protection strategy for MJ anthocyanins by producing stable anthocyanin-rich powders to be used in yogurt, also improving their BA by 9%. These results contribute to the development of new food health ingredients using maqui anthocyanins.

Author Contributions: C.F. and P.R. conceived and designed the experiments; C.B. performed the experiments; C.F., C.B. and J.P. analyzed the data; C.F., P.R. and J.P. wrote the paper.

Acknowledgments: This work was supported by FONDECYT-CONICYT project N° 3150342. The authors would like to thank Felipe Oyarzún's group for their support in HPLC analyses, Víctor Miranda for technical support, and Romina Pérez for technical assistance with in vitro digestion model tests.

Conflicts of Interest: The authors declare no conflict of interest.

Abbreviations

a_w	water activity
BA	bioaccessibility
cy-3-glu	cyanidin-3-glucoside
cy-3-sa	cyanidin-3-sambubioside
cy-3-sa-5-glu	cyanidin-3-sambubioside-5-glucoside
cy-3,5-diglu	cyanidin-3,5-diglucoside
del-3-sa-5-glu	delphinidin-3-sambubioside-5-glucoside
del-3,5-diglu	delphinidin-3,5-diglucoside
del-3-glu	delphinidin-3-glucoside
del-3-sa	delphinidin-3-sambubioside
FD	freeze-drying
GI	gastrointestinal
EA	encapsulating agents
EE	encapsulation efficiency
HPLC-DAD	high-performance liquid chromatography–photo diode array detector
MD	maltodextrin
MJ	maqui juice
RSM	response surface methodology
SD	spray-drying
SPI	soy protein isolate
TA	total anthocyanins
SEM	scanning electron microscopy

References

1. Fredes, C.; Robert, P. The powerful colour of the maqui (*Aristotelia chilensis* [Mol.] Stuntz) fruit. *J. Berry Res.* **2014**, *4*, 175–182. [CrossRef]

2. ODEPA—Oficina de Estudios y Políticas Agrarias. Boletín de frutas y hortalizas procesadas. Available online: http://www.odepa.gob.cl/wp-content/files_mf/1490990342BFrutaprocesada0317.pdf (accessed on 9 May 2017).

3. Fredes, C.; Montenegro, G.; Zoffoli, J.P.; Santander, F.; Robert, P. Comparison of total phenolic, total anthocyanin and antioxidant activity of polyphenol-rich fruits grown in Chile. *Cienc. Investig. Agrar.* **2014**, *41*, 49–60. [CrossRef]

4. Fredes, C.; Osorio, M.J.; Parada, J.; Robert, P. Stability and bioaccessibility of anthocyanins from maqui (*Aristotelia chilensis* [Mol.] Stuntz) juice microparticles. *LWT-Food Sci. Technol.* **2018**, *91*, 549–556. [CrossRef]

5. Mahdavi, S.A.; Jafari, S.M.; Ghorbani, M.; Assadpoor, E. Spray-drying microencapsulation of anthocyanins by natural biopolymers: A review. *Drying Technol.* **2014**, *32*, 509–518. [CrossRef]

6. Robert, P.; García, P.; Fredes, C. Drying and preservation of polyphenols. In *Advances in Technologies for Producing Food-Relevant Polyphenols*; Cuevas-Valenzuela, J., Vergara Salinas, J.R., Pérez-Correa, J.R., Eds.; CRC Press, Taylor and Francis Group: Boca Raton, FL, USA, 2017; pp. 281–302.

7. Desai, K.; Park, H. Recent developments in microencapsulation of food ingredients. *Drying Technol.* **2005**, *23*, 1361–1394. [CrossRef]

8. Laokuldilok, T.; Kanha, N. Effects of processing conditions on powder properties of black glutinous rice (*Oryza sativa* L.) bran anthocyanins produced by spray drying and freeze drying. *LWT Food Sci. Technol.* **2015**, *64*, 405–411. [CrossRef]

9. Darniadi, S.; Ho, P.; Murray, B.S. Comparison of blueberry powder produced via foam-mat freeze drying versus spray-drying: Evaluation of foam and powder properties. *J. Sci. Food Agric.* **2017**. [CrossRef] [PubMed]

10. Robert, P.; Gorena, T.; Romero, N.; Sepulveda, E.; Chavez, J.; Saenz, C. Encapsulation of polyphenols and anthocyanins from pomegranate (*Punica granatum*) by spray drying. *Int. J. Food Sci. Technol.* **2010**, *45*, 1386–1394. [CrossRef]

11. Fredes, C.; Yousef, G.G.; Robert, P.; Grace, M.H.; Lila, M.A.; Gómez, M.; Gebauer, M.; Montenegro, G. Anthocyanin profiling of wild maqui berries (*Aristotelia chilensis* [Mol.] Stuntz) from different geographical regions in Chile. *J. Sci. Food Agric.* **2014**, *94*, 2639–2648. [CrossRef] [PubMed]

12. Rojo, L.E.; Ribnicky, D.; Logendra, S.; Poulev, A.; Rojas-Silva, P.; Kuhn, P.; Dorna, R.; Grace, M.H.; Lila, M.A.; Raskin, I. In vitro and in vivo anti-diabetic effects of anthocyanins from maqui berry (*Aristotelia chilensis*). *Food Chem.* **2012**, *131*, 387–396. [CrossRef] [PubMed]

13. Céspedes, C.L.; Valdez-Morales, M.; Avila, J.G.; El-Hafidi, M.; Alarcón, J.; Paredes-López, O. Phytochemical profile and the antioxidant activity of Chilean wild black-berry fruits, *Aristotelia chilensis* (Mol) Stuntz (Elaeocarpaceae). *Food Chem.* **2010**, *119*, 886–895. [CrossRef]

14. Escribano-Bailón, M.T.; Alcalde-Eon, C.; Muñoz, O.; Rivas-Gonzalo, J.C.; Santos-Buelga, C. Anthocyanins in berries of Maqui (*Aristotelia chilensis* (Mol.) Stuntz). *Phytochem. Anal.* **2006**, *17*, 8–14. [CrossRef] [PubMed]

15. Gironés-Vilaplana, A.; Baenas, N.; Villaño, D.; Speisky, H.; García-Viguera, C.; Moreno, D.A. Evaluation of Latin-American fruits rich in phytochemicals with biological effects. *J. Funct. Foods.* **2014**, *7*, 599–608. [CrossRef]

16. Villacrez, J.L.; Carriazo, J.G.; Osorio, C. Microencapsulation of Andes Berry (*Rubus glaucus* Benth.) aqueous extract by spray drying. *Food Bioprocess Technol.* **2014**, *7*, 1445–1456. [CrossRef]

17. Fang, Z.; Bhandari, B. Effect of spray drying and storage on the stability of bayberry polyphenols. *Food Chem.* **2011**, *129*, 1139–1147. [CrossRef] [PubMed]

18. Santana, A.; Cano-Higuita, D.; De Oliveira, R.; Telis, V. Influence of different combinations of wall materials on the microencapsulation of jussara pulp (*Euterpe edulis*) by spray drying. *Food Chem.* **2016**, *212*, 1–9. [CrossRef] [PubMed]

19. Fang, Z.; Bhandari, B. Encapsulation of polyphenols—A review. *Trends Food Sci. Technol.* **2010**, *21*, 510–523. [CrossRef]

20. Dutra Alvim, I.; Abreu Stein, A.; Paes Koury, I.; Balardin Hellmeister Dantas, F.; de Camargo Vianna Cruz, C.L. Comparison between the spray drying and spray chilling microparticles contain ascorbic acid in a baked product application. *LWT Food Sci. Technol.* **2016**, *65*, 689–694. [CrossRef]

21. Robert, P.; Torres, V.; García, P.; Vergara, C.; Sáenz, C. The encapsulation of purple cactus pear (*Opuntia ficus-indica*) pulp by using polysaccharide-proteins as encapsulating agents. *LWT Food Sci. Technol.* **2015**, *60*, 1039–1045. [CrossRef]

22. Coisson, J.; Travaglia, F.; Piana, G.; Capasso, M.; Arlorio, M. *Euterpe oleracea* juice as a functional pigment for yogurt. *Food Res. Int.* **2005**, *38*, 893–897. [CrossRef]

23. Sung, J.-M.; Kim, Y.-B.; Kum, J.-S.; Choi, Y.-S.; Seo, D.-H.; Choi, H.-W.; Park, J.-D. Effects of freeze-dried mulberry on antioxidant activities and fermented characteristics of yogurt during refrigerated storage. *Korean J. Food Sci. Anim. Resour.* **2015**, *35*, 807–814. [CrossRef] [PubMed]

24. Ribeiro, H.L.; de Oliveira, A.V.; de Brito, E.S.; Ribeiro, P.R.B.; Souza Filho, M.M.; Azeredo, H.M.C. Stabilizing effect of montmorillonite on acerola juice anthocyanins. *Food Chem.* **2018**, *245*, 966–973. [CrossRef] [PubMed]

25. Lila, M.; Ribnicky, D.; Rojo, L.; Rojas-Silva, P.; Oren, A.; Havenaar, R.; Janle, E.; Raskin, I.; Yousef, G.; Grace, M. Complementary approaches to gauge the bioavailability and distribution of ingested berry polyphenolics. *J. Agric. Food Chem.* **2012**, *60*, 5763–5771. [CrossRef] [PubMed]

26. Lucas-Gonzalez, R.; Navarro-Coves, S.; Pérez-Álvarez, J.A.; Fernández-López, J.; Muñoz, L.A.; Viuda-Martos, M. Assessment of polyphenolic profile stability and changes in the antioxidant potential of maqui berry (*Aristotelia chilensis* (Molina) Stuntz) during in vitro gastrointestinal digestion. *Ind. Crops Prod.* **2016**, *94*, 774–782. [CrossRef]

27. Flores, F.; Singh, R.; Kerr, W.; Pegg, R.; Kong, F. Total phenolics content and antioxidant capacities of microencapsulated blueberry anthocyanins during in vitro digestion. *Food Chem.* **2014**, *153*, 272–278. [CrossRef] [PubMed]

28. Mosele, J.I.; Macià, A.; Romero, M.P.; Motilva, M.J. Stability and metabolism of *Arbutus unedo* bioactive compounds (phenolics and antioxidants) under in vitro digestion and colonic fermentation. *Food Chem.* **2016**, *201*, 120–130. [CrossRef] [PubMed]

29. Oidtmann, J.; Schantz, M.; Mäder, K.; Baum, M.; Berg, S.; Betz, M.; Kulozik, U.; Leick, S.; Rehage, H.; Schwarz, K.; et al. Preparation and comparative release characteristics of three anthocyanin encapsulation systems. *J. Agric. Food Chem.* **2012**, *60*, 844–851. [CrossRef] [PubMed]

30. McDougall, G.; Fyffe, S.; Dobson, P.; Stewart, D. Anthocyanins from red wine. Their stability under simulated gastrointestinal digestion. *Phytochemistry* **2005**, *66*, 2540–2548. [CrossRef] [PubMed]
31. AOAC—Association of Official Analytical Chemists. Fruits & Fruit Products. In *Official Methods of Analysis*; Cunniff, P., Ed.; AOAC International: Gaithersburg, MD, USA, 1996.
32. Cai, Y.Z.; Corke, H. Production and properties of spray-dried Amaranthus betacyanin pigments. *J. Food Sci.* **2000**, *65*, 1248–1252. [CrossRef]
33. Jinapong, N.; Suphantharika, M.; Jamnong, P. Production of instant soymilk powders by ultrafiltration, spray drying and fluidized bed agglomeration. *J. Food Eng.* **2008**, *84*, 194–205. [CrossRef]
34. Aravena, G.; García, O.; Muñoz, O.; Pérez-Correa, J.R.; Parada, J. The impact of cooking and delivery modes of thymol and carvacrol on retention and bioaccessibility in starchy foods. *Food Chem.* **2016**, *196*, 848–852. [CrossRef] [PubMed]

Sample Availability: Samples of the compounds are not available from the authors.

molecules

MDPI

Article

Investigating the Interaction of Ascorbic Acid with Anthocyanins and Pyranoanthocyanins

Jacob E. Farr and M. Monica Giusti *

Department of Food Science and Technology, The Ohio State University, 2015 Fyffe Ct., Columbus, OH 43210-1007, USA; farr.39@osu.edu
* Correspondence: giusti.6@osu.edu; Tel.: +1-614-247-8016

Academic Editor: Derek J. McPhee
Received: 27 February 2018; Accepted: 21 March 2018; Published: 23 March 2018

Abstract: Juices colored by anthocyanins experience color loss related to fortification with ascorbic acid (AA), thought to be the result of condensation at Carbon-4 of anthocyanins. To further understand this mechanism, pyranoanthocyanins, having a fourth-ring covalently occupying Carbon-4, were synthesized to compare its reactivity with AA against that of anthocyanins. Pyranoanthocyanins were synthesized by combining chokeberry anthocyanins with pyruvic acid. AA (250–1000 mg/L) was added to either chokeberry extract, cyanidin-3-galactoside, or 5-Carboxypyranocyanidin-3-galactoside. Samples were stored in the dark for 5 days at 25 °C and spectra (380–700 nm), color (CIE-L*c*h*), and composition changes (HPLC-MS/MS) were monitored. Extensive bleaching occurred for cyanidin-3-galactoside and chokeberry colored solutions, with a decrease in half-lives from 22.8 to 0.3 days for Cyanidin-3-galactoside when 1000 mg/L AA was added. 5-Carboxypyranocyanidin-3-galactoside solution better maintained color with limited loss in absorbance, due to the formation of colored degradation products ($\lambda_{vis-max}$ = 477 to 487 nm), and half-life decrease from 40.8 to 2.7 days, an 8–13-fold improvement compared to anthocyanins. This suggested alternative sites of reactivity with AA. Carbon-4 may be the preferred site for AA-pigment interactions, but it was not the only location. With Carbon-4 blocked, 5-Carboxypyranocyanidin-3-galactoside reacted with AA to form new pigments and reduce bleaching.

Keywords: anthocyanins; pyranoanthocyanins; ascorbic acid; bleaching; condensation

---

1. Introduction

Consumers commonly use color to make assessments on acceptance and liking, implied flavor, safety, and overall quality of food products [1]. Synthetic colorants have been used to correct the natural variation of food items, mask imperfections, as well as offer alternative product identities [1]. The innate stability of synthetic colorants over natural pigments has been a driver for their selection in coloring food products. Recently, this trend has begun to reverse as consumers have expressed concerns over the safety of synthetic colorants and preference for colorants from natural sources [2,3]. Anthocyanins are widely viewed as a natural alternative due to their wide spectrum of hue expression; however, their application has been limited due to stability [4].

Anthocyanins (ACN) are a class of water-soluble polyphenols found in many fruits and vegetables. Their color properties are greatly influenced by the substitution patterns on the aglycone structure as well as pH environment [4]. Warm hues including reds are observed at low pH but shift expression to vibrant purple-blues in more alkaline conditions. Their stability is influenced by many factors including pH, heat, enzymes, light, as well as certain bleaching agents including sulfites, hydrogen peroxide, and vitamin C (ascorbic acid, AA) [5]. The latter is of significance for the food industry, with widespread use of AA as both a fortifying agent as well as an antioxidant in many food and beverage systems [6].

It has long been known that the presence of AA in anthocyanin-colored solutions can accelerate degradation and loss of color [7]. Bisulfites, hydrogen peroxide, and ascorbic acid are electrophilic compounds and are thought to attack the same nucleophilic sites of the anthocyanin. For ascorbic acid, it has been postulated to cause mutual and irreversible destruction of both the pigment and micronutrients [8]. This is differentiated from bisulfite bleaching, which is reversible and pH-dependent [9]. This presents a major hurdle for the food industry to use ACN-based colorants, specifically in juices and beverages which are often fortified with vitamin C. Previous research [10,11] has proposed that anthocyanin bleaching is the result of the condensation of ascorbic acid, as well as other bleaching agents, at Carbon-4 (C4) of the anthocyanin (Figure 1), as this site is the most susceptible to electrophilic attack. However, there is also NMR evidence suggesting alternative sites of bisulfite addition such as Carbon-2 (C2) [12]. The proposed condensation is thought to result in the loss of conjugation in the C-ring, therefore lacking the original color expression of the pigment.

Previous work has found that anthocyanins with both 3- and 5-substitions increase pigment stability against ascorbic acid compared to just 3-substitution, likely a result of further restricting access to C4 in between. Viguera and Bridle reported that Malvidin-3,5-diglucoside experienced slower color loss as compared to Malvidin-3-glucoside. The same authors reported that direct substitution of the C4 with phenyl and methyl groups enhanced their stability against ascorbic acid color loss versus typical –H substitution [11]. Copigmentation of grape anthocyanins with rosemary polyphenols has also been shown to have a protective effect on the pigment; it is possible that the π-π interaction can limit accessibility to the chromophore [13]. Another means by which the interaction between anthocyanins and ascorbic acid could be investigated is by evaluation and comparison to pyranoanthocyanins.

Pyranoanthocyanins (Figure 1) are formed by anthocyanins undergoing heterocyclic addition of a polar carboxyl-containing compound such as pyruvic acid, acetaldehyde, or catechins, which are often byproducts of yeast fermentation [14]. This results in the formation of a fourth ring (D) that covalently occupies C4 and C5 of the pigment. Pyranoanthocyanins (PACN) are found in aged wines and have been reported in red onions and strawberries [14–16]. Previous research on pyranoanthocyanin stability has shown their enhanced resistance to bisulfite bleaching [17–19] but not to ascorbic acid. Carboxy-pyranoanthocyanins, resulting from the reaction of Malvidin-3-glucoside from grape with pyruvic acid, showed enhanced stability against bisulfite bleaching (up to 250 ppm) compared to anthocyanins [17]. Oligomeric pyranoanthocyanins were shown to exhibit complete resistance to bisulfite bleaching (up to 250 ppm) for 2 days [17]. It is possible that the oligomeric structures also block other potential sites of reaction such as C2. Acetyl-pyranoanthocyanins, synthesized with acetaldehyde, have been shown to not only overcome bisulfite bleaching, but were also reported to experience a hyperchromic shift in response to bisulfite at up to 200 ppm, an unexpected response to a common bleaching agent [19].

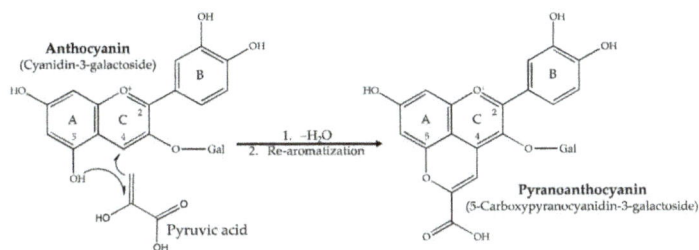

Figure 1. Formation of pyranoanthocyanin from cyanidin and pyruvic acid by heterocyclic addition.

The objective of this study was to compare the reactivity of anthocyanins and pyranoanthocyanins upon the addition of AA. If anthocyanin bleaching with AA occurs only at site C4, pyranoanthocyanins, with C4 unavailable, would not undergo bleaching. Furthermore, no changes would occur on the

PACN pigments as a results of AA addition. Our hypothesis was that C4 is not the only site of reactivity. Comparison of the reactivity of these pigments in response to ascorbic acid will aid in further understanding the deleterious interaction of this micronutrient and pigment and could allow for the better selection of anthocyanin sources in applications with ascorbic acid.

2. Results and Discussion

2.1. UV-Vis Spectrophotometry of Solutions

Anthocyanins degraded quickly in the presence of ascorbic acid, as seen in Figure 2. Chokeberry extract, containing an ACN profile which is ~70% Cyanidin-3-galactoside and with anthocyanins representing ~35% of the total area under the curve (AUC) in the max plot, showed greater resistance to bleaching compared to the purified Cyanidin-3-galactoside. This is likely the result of other chokeberry phenols playing a protective role against AA-induced degradation. Copigmentation as well as antioxidant capacity have both been demonstrated as a way by which additional polyphenols protect anthocyanins [20–22]. Other phenolic constituents of chokeberry include procyanidins, quercetin derivatives, epicatechin, and chlorogenic acid [23]. PACN (5-Carboxypyranocyanidin-3-galactoside) derived from Cyanidin-3-galactoside showed the least change in absorbance over time. Covalently occupying C4 in 5-Carboxypyranocyanidin-3-galactoside was thought to result in less change in absorbance as compared to Cyanidin-3-galactoside and chokeberry, similar to other reports of bleaching of pyranoanthocyanins observed with bisulfites [18]. All pigment solutions with ascorbic acid experienced signficant changes in maximum absorbance over time with *p*-values of less than 0.01. As AA levels increased, the loss in absorbance for each pigment over time also increased, revealing a dose-dependent effect of AA. For the 500 mg/L AA treatments over a 5-day period, 5-Carboxypyranocyanidin-3-galactoside saw a reduction of 38% reduction in maximum absorbance, chokeberry a 79% reduction, and Cyandin-3-galactoside an 88% reduction.

Changes in $\lambda_{\text{vis-max}}$ were also observed over the 5-day period. For the 500 mg/L AA level, the following hypsochromic changes in $\lambda_{\text{vis-max}}$ occurred: chokeberry, 512 to 511 nm; cyanidin-3-galactoside, 511 to 509 nm, 5-Carboxypyranocyanidin-3-galactoside, 491 to 484 nm. These shifts in $\lambda_{\text{vis-max}}$ are reflected in Figure 2. The change for 5-Carboxypyranocyanidin-3-galactoside correlated with the newly developed peaks discovered during HPLC analysis, later discussed, and resulted in the solution being more orange-red. Hypsochromic changes on ACN (chokeberry extract and cyanidin-3-galactoside) $\lambda_{\text{vis-max}}$ observed over the 5 days of the AA treatment were less than 5 nm, regardless of the levels of AA, and were more likely attributed to pigment degradation. The PACN experienced hypsochromic shifts as large as 10 nm with less loss in maximum absorbance, with shifts becoming more pronounced as AA levels increased.

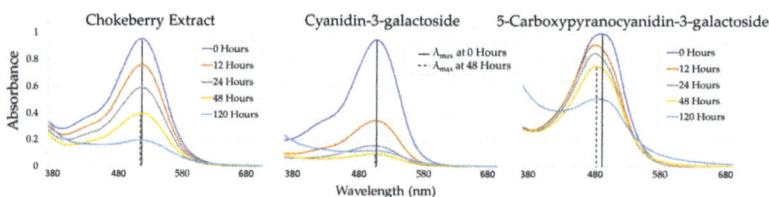

Figure 2. Spectral absorbance changes in response to 500 mg/L ascorbic acid (AA) for chokeberry, cyanidin-3-galactoside, and 5-Carboxypyranocyanidin-3-galactoside colored solutions over a 5-day period.

2.2. Kinetics of Degradation

Degradation kinetics were evaluated in terms of change in maximum absorbance of the solutions at the original $\lambda_{\text{vis-max}}$ over time. Bleaching has been previously reported as a first-order reaction, typical of ACN degradation, and was modeled as such in determining the reaction rate and half-life [24,25]. The decrease in maximum absorbance correlated with an increase in lightness (L^*) as well as the decrease in chroma (c^*) for all pigments. The reduction in maximum absorbance closely followed first-order kinetics for all treatments with an R^2 higher than 0.94 for control treatments and 0.96 for samples with added ascorbic acid. The kinetics results for each pigment and AA level can be found in Table 1. Without ascorbic acid, 5-Carboxypyranocyanidin-3-galactoside had the greatest half-life (978 h), followed by chokeberry extract (858 h), and then Cyanidin-3-galactoside (546 h). Addition of ascorbic acid dramatically reduced half-lifes for all pigments. With 1000 mg/L AA added, the 5-Carboxypyranocyanidin-3-galactoside half-life was 64 h; chokeberry extract had a half-life of 24 h; and Cyanidin-3-galactoside had a half-life of 8 h, seen in Table 1.

This order of stability was also exhibited across all AA levels. The enhanced stability and extension of half-life for 5-Carboxypyranocyanidin-3-galactoside was more evident upon the addition of ascorbic acid. Pyranoanthocyanins exhibited a half-life 8–13x higher than Cyanidin-3-galactoside in the presence of AA. Kinetics data supports the hypothesis that C4 is likely the primary or preferred site for anthocyanin-ascorbic acid interaction. It also reveals that the pyranoanthocyanin was still susceptible to ascorbic acid through alternative mechanisms or sites due to the half-life lowering in response to AA compared to its control.

The relationship between ascorbic acid level and the reaction rates was additionally assessed to observe how each of these pigments responds to ascorbic acid addition. A linear relationship was found and can be seen in Figure 3. The R^2 values for these pigments at varying AA levels are the following: chokeberry extract, 0.99; Cyanidin-3-galactoside, 0.96; and 5-carboxypyranocyanidin-3-galactoside, 0.98. Linearity is lost to some degree for Cyanidin-3-galactoside with 1000 mg/L AA addition. The slope could be effectively regarded as how responsive (or deleterious) the change in pigment solution maximum absorbance is upon AA addition, with a higher slope indicating greater susceptibility to AA. The slope of Cyanidin-3-galactoside was 3.1× that of chokeberry extract and, in comparison to 5-carboxypyranocyanidin-3-galactoside, cyanidin-3-galactoside was 8.2× and chokeberry extract 2.7× more suspectible.

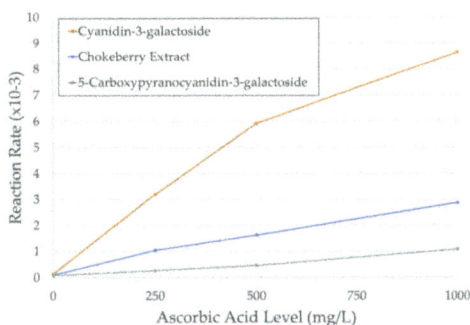

Figure 3. Reaction rates of 5-Carboxypyranocyanidin-3-galactoside, Cyanidin-3-galactoside, and chokeberry plotted against ascorbic acid level (0–1000 mg/L). Calculations are based on the changes in absorbance at the $\lambda_{\text{vis-max}}$ of the solution over time.

Table 1. Reaction rates and half-life (t$_{\frac{1}{2}}$) of solutions colored with different pigments stored at 25 °C in the dark, modeled with first-order kinetics. Calculations are based on the changes in absorbance at the $\lambda_{vis\text{-}max}$ of the solution over time.

Ascorbic Acid Level	Pigment	k (hour^{-1})	t$_{1/2}$ (h)	R^2
Control	Chokeberry Extract	8.08×10^4	858	0.947
	Cyanidin-3-galactoside	1.27×10^3	546	0.957
	5-Carboxypyranocyanidin-3-galactoside	7.08×10^4	978	0.977
250 mg/L AA	Chokeberry Extract	1.02×10^2	68	0.991
	Cyanidin-3-galactoside	3.18×10^2	22	0.996
	5-Carboxypyranocyanidin-3-galactoside	2.69×10^3	258	0.992
500 mg/L AA	Chokeberry Extract	1.60×10^2	43	0.991
	Cyanidin-3-galactoside	5.90×10^2	12	0.998
	5-Carboxypyranocyanidin-3-galactoside	4.61×10^3	150	0.965
1000 mg/L AA	Chokeberry Extract	2.85×10^2	24	0.999
	Cyanidin-3-galactoside	8.64×10^2	8	0.996
	5-Carboxypyranocyanidin-3-galactoside	1.08×10^2	64	0.998

2.3. Colorimetry

2.3.1. Lightness

Rapid color loss and extensive bleaching of pigments can be seen with CIE lightness (L^*) in Figure 4. Within 48 h exposed to 1000 mg/L AA, L^* increased from 77.2 to 96.4 (Δ19.2) for Cyanidin-3-galactoside; chokeberry, 74.4 to 89.4 (Δ15.0); and 5-Carboxypyranocyanidin-3-galactoside, 81.7 to 86.6 (Δ4.9). The presence of AA resulted in higher lightness over time, and this was dose-dependent. Pyranoanthocyanins showed the least change in L^* in reponse to AA, and Cyanidin-3-galactoside the greatest. An increase in L^* represents a lighter color expression and was most evident for chokeberry and cyanidin-3-galactoside.

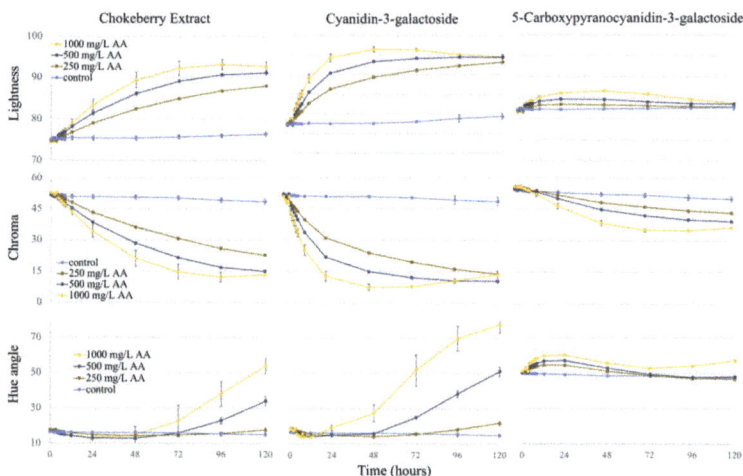

Figure 4. Colorimetric changes (CIE-L*c*h*) of change of solutions colored with chokeberry, cyanidin-3-galactoside, and 5-Carboxypyranocyanidin-3-galactoside from day 0 to day 5 for all AA levels over time. Error bars represent standard deviation.

2.3.2. Chroma

Pigment levels were standardized by absorbance at their respective $\lambda_{vis-max}$; therefore, chroma values were in close agreement at day 0. Chroma, being a measure of color intensity, is useful for determining the extent of bleaching and is reported in Figure 4. The PACN had less change in chroma compared to Cyanidin-3-galactoside and chokeberry. Chroma decreased with increasing AA levels with the exception of Cyanidin-3-galactoside after 48 h, likely the result of ascorbic acid and pigment browning playing a larger role at those times. The changes in chroma in reponse to AA addition followed: Cyanidin-3-galactoside > chokeberry > 5-Carboxypyranocyanidin-3-galactoside. All pigments (including controls) experienced a significant change in chroma over 5 days with p-values of less than 0.001.

2.3.3. Hue Angle

Synthesis of pyranoanthocyanins results in a pigment with a lower $\lambda_{vis-max}$ and a higher hue angle compared to the respective anthocyanin, having a more orange-red color expression as compared to the red color of ACN. This was clearly observed on the initial hue angle values of 5-Carboxypyranocyanidin-3-galactoside (50°), a more orange-red hue than Cyanidin-3-galactoside (16.6°), with a more red color (Figure 4). While the initial hue angle for chokeberry and Cyandin-3-galactoside started much lower and more red (<20°) than the PACN, the reaction between ACN-AA resulted in a dramatic color shift towards a yellow coloration. For the 1000 mg/L AA level, large increases in hue angle were observed from day 0 to 5 for chokeberry (17.6° to 53.8°) and cyanidin-3-galactoside (18.7° to 77.5°), while the hue angle change was much smaller for 5-Carboxypyranocyanidin-3-galactoside, changing from 51.0° to 57.1°. Changes in hue angle in the presence of AA were dose-dependent. The rapid increase in hue angle for cyanidin-3-galactoside and chokeberry was likely the result of pigment degradation and fading, whereas for 5-Carboxypyranocyanidin-3-galactoside, which better retained original chroma and lightness parameters, new pigment formation may explain hue angle changes.

2.3.4. Total Color Change (ΔE)

Total color changes (ΔE) were calculated as the color change from day 0 to 5 for each respective treatment, and are presented in Table 2. Without AA, chokeberry had the smallest ΔE, followed by 5-Carboxypyranocyanidin-3-galactoside and Cyanidin-3-galactoside. Other phenolics present in chokeberry could have enhanced color retention and stability through mechanisms such as copigmentation or radical scavenging by additional phenolics, which would not have been possible with isolated 5-Carboxypyranocyanidin-3-galactoside and cyanidin-3-galactoside. This possible explanation could be supported by previous work, where an abundance of other polyphenols has been reported in chokeberry [23]. The chokeberry fruit has previously been reported to have 89 mg/kg of quercetin, which further supports why greater stability was observed for the chokeberry treatment as compared to purified Cyanidin-3-galactoside [26]. For all levels of AA addition, chokeberry and Cyanidin-3-galactoside had a ΔE greater than 29. However, 5-Carboxypyranocyanidin-3-galactoside with AA exhibited a signficiantly smaller color shift with values of ΔE ranging from 8.7 to 10.9. This three-fold reduction in ΔE resulted in an overall better retention of color in response to AA addition.

Table 2. Days 0 and 5 colorimetric values (CIE-L*c*h*) and total color change (ΔE) of chokeberry, cyanidin-3-galactoside and 5-Carboxypyranocyanidin-3-galactoside for all AA levels over time. Numbers are means of three replications, followed by (standard deviations). ΔE: total color change of the solutions from day 0 to day 5. Values reported are means (*n* = 3) followed by standard deviation in parentheses.

Treatment		Lightness		Chroma		Hue Angle		ΔE
		Day 0	Day 5	Day 0	Day 5	Day 0	Day 5	Over 5 Days
Chokeberry Extract	Control	75.1 (0.5)	76.3 (0.5)	51.6 (0.8)	48.4 (1.2)	16.6 (0.5)	14.9 (0.8)	3.3 (0.4)
	250 mg/L AA	74.7 (0.2)	88.0 (0.2)	52.3 (0.4)	22.8 (0.3)	17.4 (0.2)	17.5 (0.5)	29.9 (0.4)
	500 mg/L AA	75.0 (0.3)	91.1 (0.6)	51.8 (0.6)	15.0 (0.4)	16.6 (0.3)	34.0 (3.1)	36.9 (0.2)
	1000 mg/L AA	74.4 (0.2)	92.7 (1.1)	52.9 (0.2)	13.3 (0.6)	17.6 (0.2)	53.8 (3.1)	42.0 (2.2)
Cyanidin-3-galactoside	Control	77.3 (0.1)	79.4 (0.8)	51.2 (0.2)	46.5 (2.1)	18.6 (0.1)	15.9 (1.0)	2.6 (1.1)
	250 mg/L AA	77.5 (0.2)	93.2 (0.3)	50.8 (0.2)	13.9 (0.7)	18.7 (0.1)	21.9 (1.2)	19.3 (0.3)
	500 mg/L AA	77.0 (0.1)	94.4 (0.6)	51.8 (0.0)	10.5 (0.4)	18.8 (0.0)	51.1 (2.7)	23.6 (0.2)
	1000 mg/L AA	77.2 (0.3)	94.5 (0.7)	50.9 (0.1)	13.7 (1.8)	18.7 (0.1)	77.5 (4.5)	27.6 (0.7)
5-Carboxypyranocyanidin-3-galactoside	Control	82.2 (0.2)	82.8 (0.6)	54.3 (0.7)	49.9 (1.3)	50.0 (0.2)	47.5 (0.6)	2.1 (0.4)
	250 mg/L AA	81.9 (0.1)	82.8 (0.3)	55.2 (0.4)	43.2 (0.2)	50.7 (0.2)	46.9 (0.3)	4.5 (0.2)
	500 mg/L AA	82.1 (0.3)	83.6 (0.4)	54.7 (0.3)	39.2 (0.3)	50.4 (0.1)	48.3 (0.4)	5.0 (0.2)
	1000 mg/L AA	81.7 (0.5)	83.7 (0.2)	55.6 (0.7)	36.2 (0.3)	51.0 (0.1)	57.1 (0.7)	5.2 (0.5)

2.4. HPLC and MS/MS Evaluation

To determine the relationship between spectral and color changes with changes in pigment composition, HPLC analysis was utilized. The initial (day 0) HPLC profile for chokeberry extract revealed the following anthocyanin profile seen in Figure 5: Cyanidin-3-galactoside (Peak #1, 70%), Cyanidin-3-glucoside (Peak #2, 3%), Cyanidin-3-arabinoside (Peak #3, 23%), and Cyanidin-3-xyloside (Peak #4, 4%), in line with the expected profile [23]. Anthocyanins contributed to 35% of the total AUC in the max plot (260–700 nm), mainly due to the presence of other polyphenols. The largest non-anthocyanin peak with a $\lambda_{\text{vis-max}}$ of 322 nm was likely chlorogenic acid, as it is reported to be present in the berry [23]. After 1 day of exposure to 1 g/L AA, all anthocyanins in the chokeberry extract decreased by 65%. The loss of individual pigments ranged from 64–68%, revealing similar degradation kinetics for all pigments present. This is not surprising considering they are all cyanidin-3-monosaccharides. By day 5, only 4% of the original pigments in the chokeberry extract had survived. The behavior of Cyanidin-3-galactoside was similar to that of chokeberry extract but was thought to experience more rapid bleaching due to the absence of other polyphenols, imparting a protective effect. The isolated anthocyanin accounted for 92% of the AUC in the day 0 maxplot. By day 1, Cyanidin-3-galactoside was reduced by 91% and day 5, >99%.

For the pyranoanthocyanin, the isolated structure accounted for 94% of the AUC from the maxplot. By day 1, this peak was reduced 93% and 99% by day 5; however, unlike the anthocyanins, where the pigments degraded into colorless forms, the PACN-AA interaction resulted in the development of new peaks in the visible range, labeled A, B, and C in Figure 5. Peak A appears to be entirely newly formed in response to AA and was not present in either the control or PACN + AA day 0 treatment. Peaks B and C were present at low levels in both the control and AA treatment at day 0 and could be colored degradation products. It appears that AA promotes the formation of these two compounds with peak B having 3.7× the AUC and peak C 15.9× AUC (470–520 nm) by day 1 as compared to the day 0 control treatment. These newly formed peaks also corroborate the spectra changes observed using the plate reader. The newly formed peaks had the following $\lambda_{\text{vis-max}}$: A, 487 nm; B, 486 nm; and C, 477 nm. The formation of these new compounds aligned with both colorimetric data (increase in hue angle) and the spectral shift (hypsochromic response) that was observed with the solutions in response to AA. The formation of the new peaks is likely how the pyranoanthocyanin solution better maintained original color expression even with the rapid loss of the parent compound. The formation of three new chromophores between the PACN-AA interaction could be the result of several different phenomena, and additional experiments were performed to include MS/MS data of the new structures.

For the three peaks produced after AA addition, it is possible that these are the result of interaction with PACN at alternative sites (not C2 or C4). With the addition of a fourth ring, ascorbic acid could have reacted with the D-ring substitution (carboxylic acid group) and produced colored byproducts. It has previously been reported that acetyl pyranoanthocyanins experience both a hyperchromic and hypsochromic shift in response to up to 200 ppm sulfites, and it was proposed this was the result of sulfite covalent linkage at the acetyl group in the D-ring, enhancing the molar absorptivity [19].

The MS/MS data generated provided valuable insight to the structures of the newly formed compounds. Peak A was the only structure that was not present in trace amounts in the control treatment. A positive ion scan revealed a parent m/z of 535. This was +18 mass units (m.u.) compared to 5-carboxypyranocyanidin-3-galactoside. A followup product ion scan revealed a daughter ion of 373 m/z, a transition of -162 m.u. from the parent ion. This is a commonly reported transition for glycosylated molecules and matches the loss of galactose from the structure [27]. This suggests that the aglycone structure is being modified and not the sugar substitution. With a parent m/z of 535, direct condensation of ascorbic or dehydroascorbic was ruled out. Ascorbic acid degradation byproducts were also considered. Ascorbic acid has previously been reported as being catalyzed by trace levels of metal (1 μM) to form hydrogen peroxide [28–30]. Ascorbic acid could form hydrogen peroxide which then could react with the pyranoanthocyanin, but the typical result would involve loss of color, and we would expect a different mass transition. Stebbins et al. [31] reported the formation of 6-hydroxy-cyanidin-3-glucoside from cyanidin-3-glucoside and ascorbic acid, the transitional m/z being +16 mass units. A different mechanism probably occurs, possibly related to the condensation of the pyranoanthocynin with other AA degradation products, and perhaps even the rearomatization of the molecule.

Peak B, under positive ionization, had a parent m/z of 519, and the respective product ion scan revealed a daughter ion at 357 mass units. Peak C had a parent m/z of 503 and the product ion scan revealed a daughter with 341 mass units The shifts for all structures was -162 m.u. from parent to daughter ion, which was again attributed to the loss of galactose. It is important to point out that the m/z of peak C was lower than that of our starting pyranoanthocyanin (5-carboxypyranocyanidin-3-galactoside, m/z 517), with a loss of 14 units. Elucidation of these chemical structures will require NMR work, currently underway.

Of the three new compounds, the structural modification induced by ascorbic acid addtion was isolated to the aglycone. Interestingly, single ion monitoring for possible direct condensation products of ascorbic or dehydroascorbic acid with 5-carboxypyranocyanidin-3-galactoside (675 and 673 m/z, respectively) did not appear in the chromatogram. This suggests that the newly formed peaks are not the result of the direct condensation of ascorbic and dehydroascorbic acids with the pyranoanthocyanin.

To further test whether the formation of the new peaks was in response to hydrogen peroxide formed as a byproduct of ascorbic acid degradation, the treatment was repeated except with hydrogen peroxide in place of ascorbic acid. The sample was monitored with 0-, 8-, and 24-h injection times and and only revealed a reduction in the original pyranoanthocyanin. Surprisingly, the three new peaks formed in response to ascorbic acid were not formed in response to direct H_2O_2 addition. No MS/MS transitions observed with PACN + AA were found by the addition of H_2O_2. With the three peaks absent in both the PDA and MS chromatogram, this theory was rejected. It was thought that ascorbic acid was degrading and that byproducts other than H_2O_2 were reacting with the pyranoanthocyanin.

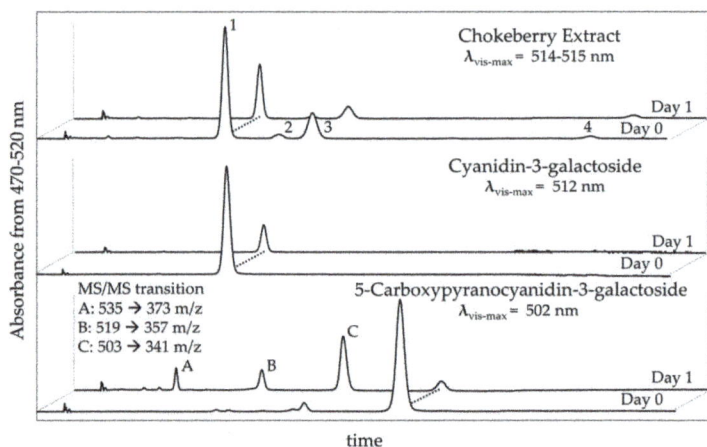

Figure 5. HPLC profiles (470–520 nm) for 5-Carboxypyranocyanidin-3-galactoside, Cyanidin-3-galactoside, and chokeberry with 1000 mg/L ascorbic acid added on days 0 and 1. MS/MS transition reported for the newly formed peaks resulting between 5-Carboxypyranocyanidin-3-galactoside and ascorbic acid interaction.

3. Materials and Methods

3.1. Materials

Powdered chokeberry fruit was provided by Artemis Inc. (Fort Wayne, IN, USA). Lab grade pyruvic acid used for the synthesis of pyranoanthocyanins was purchased from Sigma Aldrich (St. Louis, MO, USA). Analytical grade ascorbic acid (99% L-ascorbic acid) was purchased from Sigma Aldrich. HPLC grade acetonitrile and water were obtained from Fisher Scientific (Hampton, NH, USA) and HPLC grade formic acid from Sigma Aldrich.

3.2. Methods

3.2.1. Anthocyanin Semi-Purification (SPE)

Chokeberry powder was mixed with water acidified with 0.01% HCl prior to purification. The solution was loaded onto a Waters Sep-pak C18 cartridge for solid phase extraction (SPE). The column was then washed with acidified water (0.01% HCl) to remove sugars and acids, followed by washing with ethyl acetate for the removal of the more non-polar phenolics. Pigments were recovered from the cartridge with methanol acidified with 0.01% HCl, and the solvent was removed by rotary evaporation (40 °C, under vaccuum). Pigments were then solubilized and stored in acidified water for future use. This was the only preparatory step for chokeberry treatments.

3.2.2. Pyranoanthocyanin Synthesis

Pyrananthocyanins were synthesized from the semi-purified chokeberry by the addition of pyruvic acid. The extract (1000 μM cyanidin-3-glucoside equivalent) was added to a pH 2.6 citrate buffer that had 0.1% potassium sorbate and 0.1% sodium benzoate to prevent molding. A molar ratio of 1:50 (ACN: pyruvic acid) was followed as previously described [17]. The prepared anthocyanin pyruvic acid solution was stored in an incubator in the dark at 35 °C for 10 days (Isotemp, Fisher Scientific, Waltham, MA, USA). After the incubation period ended, cyanidin-3-galactoside and 5-Carboxypyranocyanidin-3-galactoside, the resulting pyranoanthocyanins from cyanidin-3-galactoside and pyruvic acid, were isolated from the solution using semipreparatory HPLC.

3.2.3. Anthocyanin and Pyranoanthocyanin Purification

A reverse phase HPLC system composed of the following modules was used: LC-6AD pumps, CBM-20A communication module, SIL-20A HT autosampler, CTO-20A column oven, and SPD-M20A Photodiode Array detector (Shimadzu, Columbia, MD, USA). The reverse-phase column selected was a 250 × 21.2 mm Luna pentafluorophenyl column with 5 μm particle size and 100 Å pore size (Phenomenex, Torrance, CA, USA). Samples were filtered prior to injection with a Phenex RC 0.45 μm, 15 mm membrane syringe filter (Phenomenex, Torrance, CA, USA). With a flow rate of 10 mL/min and a run time of 30 min, peaks were separated and collected. An isocratic system with the following solvents were used: 11:89 (Solvent A: Solvent B v/v) with Solvent A being 4.5% formic acid in HPLC grade water and Solvent B was HPLC grade acetonitrile. Elution of peaks was monitored at 500 nm. Peaks were manually collected. The two collected peaks were diluted with distilled water and again subjected to SPE semi-purification to remove formic acid and acetonitrile. Rotary evaporation was used to remove methanol (40 °C, under vaccuum), and the pigments were stored in 0.01% HCl in acidified water.

3.2.4. Anthocyanin and Pyranoanthocyanin Purity

Prior to experimentation, pigments were evaluated for purity by using an analytical HPLC only different from the previously listed one by the use of different pumps (LC-20AD, Shimadzu). Purified pigments were filtered using Phenex RC 0.45-μm membranes. A binary system with 1 mL/min flow rate was used: solvent consisted of 4.5% formic acid in HPLC grade water and solvent B consisted of HPLC grade acetonitrile. The gradient began with an isocratic flow of 6% solvent B for 17 min (elution of primary anthocyanins), increasing to 15% solvent B by 45 min (elution of primary pyranoanthocyanin), and to 40% solvent B by 50 min (wash). A 10-μL injection volume was loaded onto a Phenomenex Kinetix 5-μm EVO C18 100 A. 150 × 4.6 mm column with a Phenomenex Ultra UHPLC EVO C18 guard cartridge attached. Purity was expressed in terms of % peak area of targeted pigment as compared to the total area of all peaks present in the max plot (260–700 nm). The isolate of 5-Carboxypyranocyanidin-3-galactoside accounted for 94% of the overall area under the curve (AUC), cyanidin-3-galactoside isolate was 92% AUC, while chokeberry ACN purity was 35% AUC. Chokeberry ACN likely contained other phenols present in the source material.

3.2.5. Sample Preparation

The semi-purified chokeberry extract, the isolated cyanidin-3-galactoside, and the purified 5-Carboxypyranocyanidin-3-galactoside were diluted in pH 3.0 citrate buffer (0.1 M adjusted with HCl) until an absorbance of 1.0 at their respective $\lambda_{vis\text{-}max}$ was reached. Levels of AA of 250, 500, and 1000 mg/L were added using a concentrated ascorbic acid stock solution, and a control consisting of each pigment with the absence of AA was maintained. All samples were brought to the same final volume with additional citrate buffer. The pH of all samples were evaluted using a S220 SevenCompact pH meter (Mettler Toledo, Columbus, OH, USA) and were found to have a pH of 3.0 ± 0.05. Samples were stored in the dark at 25 °C in an incubator (listed in Section 3.2.2). UV-Vis spectrophotometry, colorimetry, and HPLC analyses were conducted over a 5-day period following the addition of ascorbic acid. UV-Vis spectral data was collected every hour for the first 8 h, and then at 12 h, and daily from that point on for 5 days. Spectra and color were eveluated with this data. HPLC analyses were conducted on days 0, 1, 3, and 5. All treatments were run in triplicate.

3.2.6. UV-Vis Spectrophotometry of Samples

A SpectraMax 190 Microplate Reader (Molecular Devices, Sunnyvale, CA, USA) with a 96-well plate (poly-D-lysine coated polystyrene) were used for the evaluation of absorbance from 380–700 nm, at 1-nm intervals. Aliquots (200 μL) of samples were loaded into individual wells, and a blank consisted of the same citrate buffer used.

3.2.7. Color Analyses of Samples

Using UV-Vis spectral data (380–700 nm, 1-nm intervals) in combination with software (*ColorBySpectra*) written for color conversion, absorbance data was translated to the CIE (Commission internationale de l'éclairage) L*c*h color space [32]. The calculations for CIE-L*c*h* implemented by the software used the following conditions: D65 standard illuminant, regular transmission, and a 10° observer angle function [33]. Color data is reported as L* (lightness), c* (chroma), and h* (hue angle).

3.2.8. HPLC Monitoring of Samples

Prepared solutions were monitored to determine the formation of potential degradation products or profile changes. Using the analytical HPLC system and conditions previously described (Section 3.2.4), chromatograms were monitored with the max plot (260–700 nm), 490 nm (near $\lambda_{vis-max}$ of 5-Carboxypyranocyanidin-3-galactoside), and 520 nm (near $\lambda_{vis-max}$ of cyanidin-3-galactoside). A max plot of 470–520 nm was later added to standardize changes in the AUC for all three pigments and their degradation products.

3.2.9. MS/MS Evaluation of Pigments

With the development of newly formed peaks for the pyranoanthocyanin, tandem mass spectrometry (MS/MS) was performed to obtain additional structural information on the novel structures. The same HPLC conditions and column mentioned in Section 3.2.4 were used on a uHPLC (iNexera) system coupled to a tandem MS unit (LCMS 8040) (Shimadzu). Electrospray ionization was used with the following conditions: 1.5 L/min nebulizing gas flow, 230 °C desolvation line temperature, 200 °C heat block temperature, and 15 L/min drying gas flow. Two total ion scans (Q3) were performed, both in positive and negative mode from 100–1500 mass unit with an event time of 0.1 s. Based on the scan results, the following events were added (positive mode) and the sample reran with secondary collisions in argon gas: product ion scan, 535 m/z; product ion scan, 519 m/z; product ion scan, 503 m/z; and precursor ion scan, 355 m/z. These secondary collision events all had an event time of 0.1 s and a collision energy of -35.0 eV. To also consider the possibility of condensation of the pyranoanthocyanin and ascorbic acid, the following single ion monitoring was also included: 675 m/z for 5-carboxypyranocyanidin-3-galactoside (517) + ascorbic acid (176) $-$ H_2O (18) and 673 m/z for dehydroscorbic acid (174).

3.2.10. Statistical Analysis of Data

Data was organized for means and standard deviations using Microsoft Excel (Redmond, Washington, DC, USA). One-way ANOVA was performed for each treatment at all time points to determine if a significant change in CIE-L*c*h and maximum absorbance occurred. One-way ANOVA (two-tailed) was also performed for each pigment (control, 250, 500, 1000 mg/L AA) at each time point to determine if and when which CIE-L*c*h and maximum absorbance became significantly different from the control. Software used for ANOVA tests was SPSS (IBM, North Castle, NY, USA).

4. Conclusions

Cyanidin-3-galactoside degraded rapidly in the presence of ascorbic acid, followed by chokeberry extract. Other phenols in chokeberry extract likely played a protective role against AA-mediated pigment bleaching. The 5-Carboxypyranocyanidin-3-galactoside colored solution exhibited the smallest change in color (ΔE) and limited bleaching in response to ascorbic acid (for 1000 mg/L AA, ΔE of 5.2 versus 27.6 for cyanidin-3-galactoside). The interaction between PACN-AA resulted in the formation of three new chromophores, as revealed by HPLC. NMR work is underway to determine the site of reaction for PACN-AA as well as the ACN-AA reactivity. The fact that PACN, with the C4 position blocked, still exhibited limited bleaching further supports the hypothesis that C4 plays an important, but not singular, role in the AA-mediated bleaching of anthocyanins. The pyranoanthocyanin better maintained absorbance and color expression in the presence of AA, not as a result of 5-Carboxypyranocyandin-3-galactoside survival, but due to the formation of colored byproducts suspected at alternative sites.

Acknowledgments: This work was supported in part by the USDA National Institute of Food and Agriculture, Hatch Project OHO01423, Accession Number 1014136.

Author Contributions: Jacob E. Farr and M. Monica Giusti designed the study and interpreted generated data. Jacob E. Farr was responsible for performing experiments and drafting the manuscript. Jacob E. Farr and M. Monica Giusti contributed to final edits and data analyses of the manuscript.

Conflicts of Interest: The authors declare no conflict of interest

References

1. Sharma, V.; McKone, H.T.; Markow, P.G. A Global Perspective on the History, Use, and Identification of Synthetic Food Dyes. *J. Chem. Educ.* **2010**, *88*, 24–28. [CrossRef]
2. Kobylewski, S.; Jacobson, M.F. Food Dyes: A Rainbow of Risks. *Decis. Sci.* **2010**, *30*, 337–360.
3. Martins, N.; Roriz, C.L.; Morales, P.; Barros, L.; Ferreira, I.C.F.R. Food colorants: Challenges, opportunities and current desires of agro-industries to ensure consumer expectations and regulatory practices. *Trends Food Sci. Technol.* **2016**, *52*, 1–15. [CrossRef]
4. Sigurdson, G.T.; Tang, P.; Giusti, M.M. Natural Colorants: Food Colorants from Natural Sources. *Annu. Rev. Food Sci. Technol.* **2017**, *8*, 261–280. [CrossRef] [PubMed]
5. Eiro, M.J.; Heinonen, M. Anthocyanin color behavior and stability during storage: Effect of intermolecular copigmentation. *J. Agric. Food Chem.* **2002**, *50*, 7461–7466. [CrossRef] [PubMed]
6. Varvara, M.; Bozzo, G.; Disanto, C.; Pagliarone, C.N.; Celano, G.V. The use of the ascorbic acid as food additive and technical-legal issues. *Ital. J. Food Saf.* **2016**, *5*. [CrossRef] [PubMed]
7. Sondheimer, E.; Kertesz, Z.I. Participation of ascorbic acid in the destruction of anthocyanin in strawberry juice and model systems. *Biol. Chem.* **1953**, *18*, 475–479. [CrossRef]
8. Shriner, R.L.; Moffett, R.B. Benzopyrylium Salts. III. Syntheses from Substituted Coumarins and Chromones. *J. Am. Chem. Soc.* **1941**, *63*, 1694–1698. [CrossRef]
9. Giusti, M.M.; Wrolstad, R.E. Characterization and Measurement of Anthocyanins by UV-visible Spectroscopy. In *Current Protocols in Food Analytical Chemistry*; John Wiley & Sons: New York City, NY, USA, 2001; Volume 5, pp. 19–31. ISBN 9780471142911.
10. Poei-Langston, M.S.; Wrolstad, R.E. Color Degradation in an Ascorbic Acid-Anthocyanin-Flavanol Model System. *J. Food Sci.* **1981**, *46*, 1218–1236. [CrossRef]
11. Garcia-Viguera, C.; Bridle, P. Influence of structure on color stability of anthocyanins and flavylium salts. *Food Chem.* **1999**, *64*, 21–26. [CrossRef]
12. Berké, B.; Chèze, C.; Vercauteren, J.; Deffieux, G. Bisulfite addition to anthocyanins: Revisited structures of colourless adducts. *Tetrahedron Lett.* **1998**, *39*, 5771–5774. [CrossRef]
13. Brenes, C.H.; Del Pozo-Insfran, D.; Talcott, S.T. Stability of copigmented anthocyanins and ascorbic acid in a grape juice model system. *J. Agric. Food Chem.* **2005**, *53*, 49–56. [CrossRef] [PubMed]
14. De Freitas, V.; Mateus, N. Formation of pyranoanthocyanins in red wines: A new and diverse class of anthocyanin derivatives. *Anal. Bioanal. Chem.* **2011**, *401*, 1467–1477. [CrossRef] [PubMed]
15. Fossen, T.; Andersen, Ø.M. Anthocyanins from red onion, Allium cepa, with novel aglycone. *Phytochemistry* **2003**, *62*, 1217–1220. [CrossRef]
16. Andersen, Ø.M.; Fossen, T.; Torskangerpoll, K.; Fossen, A.; Hauge, U. Anthocyanin from strawberry (*Fragaria ananassa*) with the novel aglycone, 5-carboxypyranopelargonidin. *Phytochemistry* **2004**, *65*, 405–410. [CrossRef] [PubMed]
17. He, J.; Carvalho, A.R.F.; Mateus, N.; De Freitas, V. Spectral Features and Stability of Oligomeric Pyranoanthocyanin-flavanol Pigments Isolated from Red Wines. *J. Agric. Food Chem* **2010**, *58*, 9249–9258. [CrossRef] [PubMed]
18. Oliveira, J.; Fernandes, V.; Miranda, C.C.C.; Santos-Buelga, C.; Silva, A.; De Freitas, V.; Mateus, N. Color properties of four cyanidin-pyruvic acid adducts. *J. Agric. Food Chem.* **2006**, *54*, 6894–6903. [CrossRef] [PubMed]

19. Gómez-Alonso, S.; Blanco-Vega, D.; Gómez, M.V.; Hermosín-Gutiérrez, I. Synthesis, isolation, structure elucidation, and color properties of 10-Acetyl-pyranoanthocyanins. *J. Agric. Food Chem.* **2012**, *60*, 12210–12223. [CrossRef] [PubMed]

20. Trouillas, P.; Sancho-García, J.C.; De Freitas, V.; Gierschner, J.; Otyepka, M.; Dangles, O. Stabilizing and Modulating Color by Copigmentation: Insights from Theory and Experiment. *Chem. Rev.* **2016**, *116*, 4937–4982. [CrossRef] [PubMed]

21. Chung, C.; Rojanasasithara, T.; Mutilangi, W.; McClements, D.J. Stabilization of natural colors and nutraceuticals: Inhibition of anthocyanin degradation in model beverages using polyphenols. *Food Chem.* **2016**, *212*, 596–603. [CrossRef] [PubMed]

22. Choe, E.; Min, D.B. Mechanisms of Antioxidants in the Oxidation of Foods. *Compr. Rev. Food Sci. Food Saf.* **2009**, *8*, 345–358. [CrossRef]

23. Kulling, S.E.; Rawel, H.M. Chokeberry (*Aronia melanocarpa*)—A review on the characteristic components and potential health effects. *Planta Med.* **2008**, *74*, 1625–1634. [CrossRef] [PubMed]

24. Sondheimer, E.; Kertesz, Z.I. The kinetics of the oxidation of strawberry anthocyanin by hydrogen peroxide. *J. Food Sci.* **1952**, *17*, 288–298. [CrossRef]

25. Ozkan, M.; Yemenicioglu, A.; Asefi, N.; Cemeroglu, B. Degradation Kinetics of Anthocyanins from Sour Cherry, Pomegranate, and Strawberry Juices by Hydrogen Peroxide. *J. Food Sci.* **2002**, *67*, 525–529. [CrossRef]

26. Häkkinen, S.H.; Kärenlampi, S.O.; Heinonen, I.M.; Mykkänen, H.M.; Törrönen, A.R. Content of the Flavonols Quercetin, Myricetin, and Kaempferol in 25 Edible Berries. *J. Agric. Food Chem.* **1999**, *47*, 2274–2279. [CrossRef] [PubMed]

27. Giusti, M.M.; Rodríguez-Saona, L.E.; Griffin, D.; Wrolstad, R.E. Electrospray and Tandem Mass Spectroscopy As Tools for Anthocyanin Characterization. *J. Agric. Food Chem.* **1999**, *47*, 4657–4664. [CrossRef] [PubMed]

28. Jansson, P.J.; Jung, H.R.; Lindqvist, C.; Nordström, T. Oxidative decomposition of vitamin C in drinking water. *Free Radic. Res.* **2004**, *38*, 855–860. [CrossRef] [PubMed]

29. Khan, M.M.T.; Martell, A.E. Metal Ion and Metal Chelate Catalyzed Oxidation of Ascorbic Acid. *J. Am. Chem. Soc.* **1967**, *89*, 4176–4185. [CrossRef] [PubMed]

30. Buettner, G.R.; Jurkiewicz, B.A. Catalytic metals, ascorbate and free radicals: Combinations to avoid. *Radiat. Res.* **1996**, *145*, 532–541. [CrossRef] [PubMed]

31. Stebbins, N.B.; Howarda, L.R.; Prior, R.L.; Brownmiller, C.; Liyanage, R.; Lay, J.O.; Yang, X.; Qian, S.Y. Ascorbic acid-catalyzed degradation of cyanidin-3-*O*-β-glucoside: Proposed mechanism and identification of a novel hydroxylated product. *J. Berry Res.* **2016**, *6*, 175–187. [CrossRef]

32. Farr, J.E.; Giusti, M.M. ColorBySpectra. Available online: https://buckeyevault.com/products/colorbyspectra (accessed on 18 October 2017).

33. CIE. *CIE 015:2004 Colorimetry*; CIE: Vienna, Austria, 2004; ISBN 978-3-901-90633-6.

Sample Availability: Samples of the pyranoanthocyanin evaluated may be requested by email contact with the corresponding author.

molecules

MDPI

Article

Cis–Trans Configuration of Coumaric Acid Acylation Affects the Spectral and Colorimetric Properties of Anthocyanins

Gregory T. Sigurdson, Peipei Tang and M. Mónica Giusti *

Department of Food Science and Technology 2015 Fyffe Ct., The Ohio State University, Columbus, OH 43210-1007, USA; sigurdson.5@osu.edu (G.T.S.); tang.451@osu.edu (P.T.)
* Correspondence: Giusti.6@osu.edu; Tel.: +1-614-247-8016

Received: 2 February 2018; Accepted: 4 March 2018; Published: 7 March 2018

Abstract: The color expression of anthocyanins can be affected by a variety of environmental factors and structural characteristics. Anthocyanin acylation (type and number of acids) is known to be key, but the influence of acyl isomers (with unique stereochemistries) remains to be explored. The objective of this study was to investigate the effects of *cis–trans* configuration of the acylating group on the spectral and colorimetric properties of anthocyanins. Petunidin-3-rutinoside-5-glucoside (Pt-3-rut-5-glu) and Delphinidin-3-rutinoside-5-glucoside (Dp-3-rut-5-glu) and their *cis* and *trans* coumaroylated derivatives were isolated from black goji and eggplant, diluted in pH 1–9 buffers, and analyzed spectrophotometrically (380–700 nm) and colorimetrically (CIELAB) during 72 h of storage (25 °C, dark). The stereochemistry of the acylating group strongly impacted the spectra, color, and stability of the Dp and Pt anthocyanins. *Cis* acylated pigments exhibited the greatest λ_{max} in all pH, as much as 66 nm greater than their *trans* counterparts, showing bluer hues. *Cis* acylation seemed to reduce hydration across pH, increasing color intensity, while *trans* acylation generally improved color retention over time. Dp-3-*cis*-*p*-cou-rut-5-glu exhibited blue hues even in pH 5 ($C^*_{ab} = 10$, $h_{ab} = 256°$) where anthocyanins are typically colorless. *Cis* or *trans* double bond configurations of the acylating group affected anthocyanin spectral and stability properties.

Keywords: delphinidin; petunidin; natural color; isomers; *cis* acylation; *trans* acylation; *Solanum melongena* L.; *Lycium ruthenicum*

1. Introduction

The color of fresh fruits and vegetables and their food products relates to their overall market sales and success. Consumers can infer a variety of attributes such as flavor, safety, nutritional value, and more due to the color of a product. Due to links between certain synthetic colorants and hyperactivity in children, allergenicity in sensitive populations, and demand for more natural products, the use of naturally derived colorants in foods has been increasing [1,2]. Colorants derived from nature now lead in usage in foods having demonstrated a 77% growth from 2009 to 2013 [3]. Colorants can be derived from many natural sources including plants, microorganisms, animals, and also minerals [4]. Of plant-derived pigments, anthocyanins comprise the largest group of water-soluble pigments; more than 700 structures have been identified [5]. They provide unique advantages and challenges to food producers as they are responsible for a wide range of colors in nature including orange, red, purple, and blue. Individual anthocyanins are also widely known to express different hues based upon structural changes that depend on environmental pH and chemical substitution patterns.

Several structural factors including chromophore methoxylation, hydroxylation, glycosylation, and acylation all affect the colors expressed by anthocyanins. The six most common anthocyanidins in edible produce differ in degree of B-ring hydroxylation and/or methoxylation, their hues typically being bluer with increasing degree of substitution [6]. Methoxylation of the chromophore, compared with hydroxylation, results in slightly redder hues; the h_{ab} values were demonstrated to be negatively correlated with not only the total anthocyanin content of processed blueberry juice but also the delphinidin/malvidin ratio [7]. Glycosylation of cyanidin at C3 typically intensified and stabilized the color of anthocyanidins in most pH; however, glycosylation of cyanidin at C3 and C5 has been reported to decrease pK_h and color stability [8–11]. Despite the decreased color stability, this glycosylation pattern also resulted in hue shifts toward purple-blue tones [12,13].

Acylation has been considered an important aspect regarding the color expression and stability of many anthocyanins [14]. The chemical mechanisms affecting colorimetric properties are believed to be complex and may include intramolecular copigmentation, steric hindrance protecting the chromophore from hydration, extension of the electron delocalization system, and further alterations in the geometric properties of the molecule [13,15,16]. Attachment of hydroxycinnamic acids typically results in an increase in pK_h and, therefore, increased color expression in comparatively higher pH, which contributes to the predominance of acylated anthocyanin colorants in the food industry [13,14]. Acylation of cyanidin derivatives with cinnamic acids has been reported to induce a hue shift towards purple. Even the location of the acyl attachment plays an important role in colorimetric properties; λ_{max} differences as large as 57 nm were found in derivatives of cyanindin-3-sinapoyl-sophoroside-5-glucoside due to attachment of the sinapoyl moiety on different locations of the sophorosoyl moiety [17]. Aromatically acylated anthocyanins are typically found esterified to *trans*-configured variants of hydroxycinnamic acids; however, few reports demonstrate the existence of *cis*-configured hydroxycinnamic acylated anthocyanins.

When found in nature, the *cis* and *trans* acylated forms are found together, but the *trans* acylated anthocyanin always predominates [18]. Only 12 naturally occurring *cis* acylated anthocyanin derivatives were identified in reviewing the current literature; however, photoirradiation was demonstrated to induce *trans* to *cis* isomerization in vitro [18–23]. Naturally occurring *cis* acylated anthocyanins are primarily reported in parts of plants that receive greater amounts of light, such as flowers, leaves, and some fruits [18–23]. Of botanical and edible fruits, the black goji (*Lycium ruthenicum*), Asian eggplant (*Solanum melongena* L.), and purple bell peppers (*Capsicum annuum* L.) have been previous identified as natural sources of *cis* coumaric acylated petunidin and delphinidin [21,24]. Therefore, these materials may serve as sources of naturally derived food colorants; however, few reports compare the spectral properties of the isolated pigments. Of those, they were characterized primarily in acidic methanol or aqueous solutions at acidic pH ($\leq$4.6) [18–20].

Structural modifications of the same chromophore result in unique colorimetric properties of each pigment, which complicates selection of pigment sources or specific pigments to obtain desired colors in food products. Therefore, there is a need to better understand how these structural components impact the color of these pigments in a wide range of conditions. Very few reports were found that compared spectral and colorimetric properties of *cis* and *trans* isomers of hydroxycinnamic acyl moieties on the same anthocyanin in acidic pH and none in alkaline pH. Therefore, the objective of this work was to investigate the impact of isomeric configurations of acyl moieties on the color expression of delphinidin and petunidin anthocyanins.

2. Results and Discussion

Despite the relative rarity of *cis* acylated anthocyanins in nature, two food sources of *cis* and *trans* acylated anthocyanins were identified for use in this study [21,22,24]. The *cis* and *trans* *p*-coumaroylated isomers of delphinidin-3-rutinoside-5-glucoside (Dp-3-rut-5-glu) were predominant in Asian eggplant (*Solanum melongena* L.) extracts, while the *cis* and *trans* *p*-coumaroylated derivatives of petunidin-3-rutinoside-5-glucoside (Pt-3-rut-5-glu) were found in black goji (*Lycium ruthenicum*). Alkaline saponification of the extracts yielded predominantly Dp-3-rut-5-glu and Pt-3-rut-5-glu, as observed in HPLC chromatograms and identified in Figure 1.

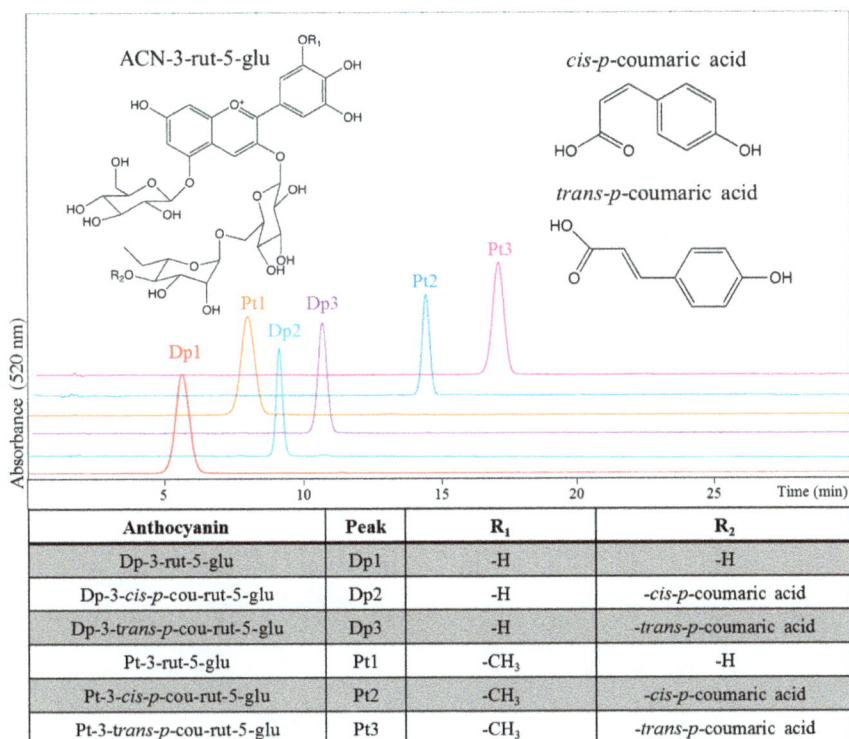

Figure 1. HPLC chromatograms (detection at 520 nm), identities, and structures of isolated delphinidin (Dp) and petunidin (Pt) derivatives.

2.1. Spectrophotometric Properties of Acylated Anthocyanin Derivatives

The differences in the acyl substitutions of these anthocyanins resulted in unique differences in visible light absorption spectra that became more pronounced as pH was increased. In very acidic pH, the spectral absorbance curves in a family of pigments were fairly similar, exhibiting similar low absorbance approaching the UV region and having little to no absorbance approaching the IR region (Figure 2). Anthocyanins glycosylated at only C3 are well documented to exhibit characteristic high proportional absorbance between 420 and 440 nm compared with the absorbance at the λ_{max} in acidic pH; however, diglycosylation at C3 and C5 results in a decrease in this absorbance between 420 and 450 nm when compared with anthocyanins glycosylated at only C3 [25]. The absorbance spectra of these Dp and Pt derivatives of this study, even the acylated ones, are consistent with these reported trends (Figure 2).

Typical of anthocyanins, the formation of predominantly colorless anthocyanin derivatives occurred in mildly acidic pH (4–6) for all derivatives (Figures 2 and 3) resulting in low absorbance and almost linear spectra. Nonacylated anthocyanin derivatives were expected to be most prone to hydration and formation of colorless structures. Dp-3-rut-5-glu showed the lowest color retention in the most pH values (Table 1); however, the *trans* acylated counterpart showed similar color bleaching. The *cis* acylated Dp isomer generally showed significantly higher color expression than the others in the series (Table 1 and Figure 3). In the case of Mv-3-*p*-cou-rut-5-malonyl-5-glu, hydroxycinnamic acyl moieties in the *cis* conformation are believed to be stereochemically nearly parallel to the anthocyanidin chromophore; while in *trans* configuration, the acyl moiety lies quasi-perpendicular to the aglycones [18]. This difference in stereochemistry may play important roles in the protection of the chromophore against hydration and, therefore, color loss. This trend was not consistent for the Pt derivatives. The *cis* acylated isomer showed slightly more color loss than did the *trans* acylated isomer in pH 6–9, but both bleached less than the nonacylated counterpart (Table 1). Perhaps the slightly bulkier methoxyl attachment on the B-ring of Pt played a role in protection from hydration.

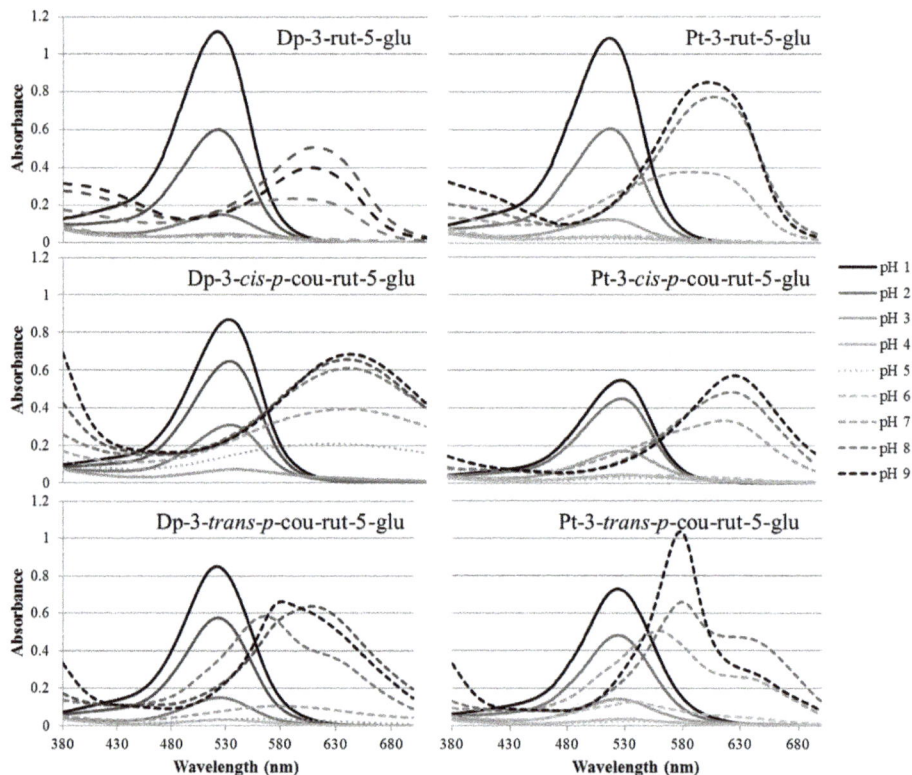

Figure 2. Visible absorbance (380–700 nm) of isolated delphinidin (Dp) and petunidin (Pt) derivatives, pH 1–9.

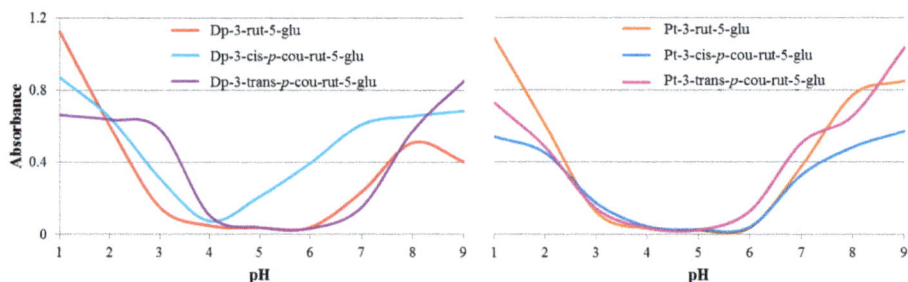

Figure 3. Absorbance of delphinidin (Dp) and petunidin (Pt) derivatives at respective λ_{max} in pH 1–9.

Table 1. λ_{max} (nm) and % absorbance retention, defined as absorbance at respective λ_{max} in pH_n/absorbance in $pH_1 \times 100$, of delphinidin (Dp) and petunidin (Pt) derivatives in pH 1–9, $n = 3$ (standard deviation in parenthesis). Different superscript letters indicate significant differences ($p < 0.05$) between derivatives of the same anthocyanidin in the same pH.

Anthocyanin	pH 1	pH 2	pH 3	pH 4	pH 5	pH 6	pH 7	pH 8	pH 9
				λ_{max} (nm)					
Dp-3-rut-5-glu	517 c (0)	518 c (0)	518 c (2)	521 c (0)	522 c (4)	450 c (1)	583 b (0)	603 c (1)	600 b (1)
Dp-3-*cis-p*-cou-rut-5-glu	528 a (1)	528 a (1)	529 a (1)	534 a (2)	617 a (1)	630 a (1)	632 a (0)	631 a (1)	632 a (1)
Dp-3-*trans*-pcou-rut-5-glu	521 b (1)	523 b (0)	523 b (1)	531 b (1)	563 b (2)	576 b (1)	566 c (0)	608 b (2)	584 c (4)
Pt-3-rut-5-glu	518 c (0)	519 c (0)	521 c (2)	519 b (6)	522 b (9)	524 c (2)	589 b (1)	610 b (1)	604 b (1)
Pt-3-*cis-p*-cou-rut-5-glu	531 a (0)	531 a (0)	532 a (1)	534 a (2)	553 a (6)	564 a (1)	621 a (1)	627 a (0)	630 a (1)
Pt-3-*trans-p*-cou-rut-5-glu	523 b (0)	524 b (1)	525 b (2)	535 a (4)	535 b (2)	538 b (0)	559 c (0)	579 c (0)	578 c (0)
				% absorbance retention					
Dp-3-rut-5-glu		53.7 c (2.4)	13.4 c (0.1)	4.3 b (0.1)	3.3 c (0.4)	3.4 c (0.0)	20.8 b (0.3)	45.2 b (1.6)	35.7 b (4.2)
Dp-3-*cis-p*-cou-rut-5-glu		77.0 a (1.1)	36.3 a (1.2)	9.0 a (0.3)	25.6 a (0.8)	45.8 a (2.5)	70.6 a (2.2)	70.6 a (2.3)	71.1 a (1.3)
Dp-3-*trans*-pcou-rut-5-glu		68.2 b (0.7)	18.6 b (0.2)	4.1 b (0.1)	4.9 b (0.2)	14.3 b (0.2)	70.5 a (2.4)	73.5 a (0.2)	68.4 a (3.0)
Pt-3-rut-5-glu		55.9 c (2.1)	11.5 c (0.2)	3.1 c (0.2)	2.1 c (0.1)	3.5 c (0.1)	34.8 c (4.1)	71.3 b (4.7)	78.3 c (3.1)
Pt-3-*cis-p*-cou-rut-5-glu		82.0 a (1.8)	31.8 a (0.4)	8.3 a (0.3)	5.1 a (0.1)	7.9 b (1.0)	60.4 b (3.3)	88.3 a (0.6)	104.4 b (0.7)
Pt-3-*trans-p*-cou-rut-5-glu		66.1 b (0.9)	19.6 b (1.1)	5.1 b (0.3)	3.5 b (0.3)	17.8 a (0.7)	69.4 a (0.8)	90.3 a (3.4)	141.8 a (4.8)

As pH increased to ≥ 6, the λ_{max} of the pigments was bathochromically shifted, and the spectral characteristics between the different species became more pronounced. Anthocyanins exist in a structurally dynamic equilibrium that is pH dependent; while in acidic pH, they exist primarily in cationic flavylium or hemiketal forms that appear red or colorless. With increasing pH to neutral and alkaline pH, quinoidal structures begin to predominate, which can become anionic, but also have greater λ_{max} and express purple and blue hues [26]. Interestingly, the absorbance spectra between the *cis* and *trans* acylated isomers of Dp-3-rut-5-glu and of Pt-3-rut-5-glu exhibited many differences (Figure 2). The *cis* acylated isomers exhibited a major peak in either acidic or alkaline pH. The peak of Dp-3-*cis-p*-cou-rut-5-glu was fairly broad in alkaline pH and was comparably sharper for Pt-3-*cis-p*-cou-rut-5-glu. In pH 7–9, the *trans* acylated isomers exhibited prominent absorbance shoulders in greater wavelengths than the λ_{max}. Interestingly, these absorbance spectra resembled mirror images of the spectra of anthocyanin-3-glucosides in acidic pH, in which they exhibit a characteristic absorbance shoulder at 420–450 nm [25]. The differences in stereochemistry between the *cis* and *trans* acylated isomers likely played a role in the unique absorbance, perhaps through intramolecular copigmentation or by structural distortions of the aglycone (stretching, bending, or torsion) that modify the π-delocalization of the chromophore by molecular substitution [16].

The λ_{max} of the different acylated derivatives also varied considerably not only across pH but also when comparing one to another in the same pH (Table 1). Generally, the λ_{max} of acylated derivatives were greater than those of the nonacylated derivatives. The impact of the different stereochemistry between the *cis* and *trans* configuration of coumaric acid on the λ_{max} of Dp and Pt derivatives was pronounced. In all pH, the λ_{max} of Dp-3-*cis-p*-cou-rut-5-glu was greater than the *trans* acylated counterpart as well as all other Dp derivatives (Table 1). This was also the case with acylated Pt except in pH 4; at which the pigments were almost completely bleached in this pH. This was consistent with previous comparisons of *cis* and *trans* isomer acylated anthocyanins in acidic pH or acidified methanol [18–20]. In pH 5, the λ_{max} and absorbance of Dp-3-*cis-p*-cou-rut-5-glu was surprisingly large (618 nm, Figure 3), 36 nm greater than Dp-3-*trans-p*-cou-rut-5-glu in the same pH. This is an atypically large λ_{max} for anthocyanins in acidic pH, especially without the presence of other cofactors such as metal ions [27]. Pt-3-*cis-p*-cou-rut-5-glu also showed rather large λ_{max} of 621–630 nm in neutral and alkaline conditions, 48–62 nm greater than the large λ_{max} of the *trans* acylated counterpart. In this pH range (6–9), Dp-3-*cis-p*-cou-rut-5-glu exhibited λ_{max} 23–66 nm greater than its *trans* acylated counterpart.

2.2. Colorimetric Properties of Acylated Anthocyanin Derivatives

The diversity in absorbance spectra of these different anthocyanins resulted in unique hues being expressed by of each of the pigments (Table 2 and Figure 4). Due to lack of molar absorptivity coefficients for each of these individual pigments, they were quantified as equivalents of Cy-3-diglucoside-5-glucoside or coumaroylated derivatives, as described in Section 3.2.4. Therefore, differences in the colorimetric data, primarily L* (luminosity/lightness) and C^*_{ab} (chroma/intensity), could be affected by the differences in concentration. L* and C^*_{ab} values were comparable between the different derivatives in the same pH and also between the derivatives of the two anthocyanidin series, displaying similar respective trends over the pH conditions tested. However, in alkaline pH, Pt derivatives showed slightly smaller L* and larger C^*_{ab} values, and slightly darker and more intense colors (Table 2). Depending on pH, differences in L* values ranged 1.2–14.4 units and ranged 2.0–19.2 for C^*_{ab} values for Dp derivatives. For Pt derivatives, differences in L* ranged 0.3–13.4 and 0.5–14.1 for C^*_{ab}. Differences were typically largest in neutral and alkaline pH and were less pronounced in acidic pH. However, in acidic pH, L* values ranged less than 10 units between different pigments in the same pH. Ranges in C^*_{ab} were greater, demonstrating the pigments showing different color intensities. In order to account for these variances, data were normalized by calculating % absorbance retention, defined as Absorbance at respective λ_{max} in pH_n / Absorbance in pH_1 × 100 (Table 1). Acylation typically worked to decrease L* and increase C^*_{ab} values, darkening and intensifying color. The effects of the acyl isomerization showed opposite effects in acidic and alkaline pH. In acidic pH, the *cis* acylated isomers generally showed lower L* values and greater C^*_{ab} values, but in alkaline pH, the *trans* acylated isomers typically expressed these traits.

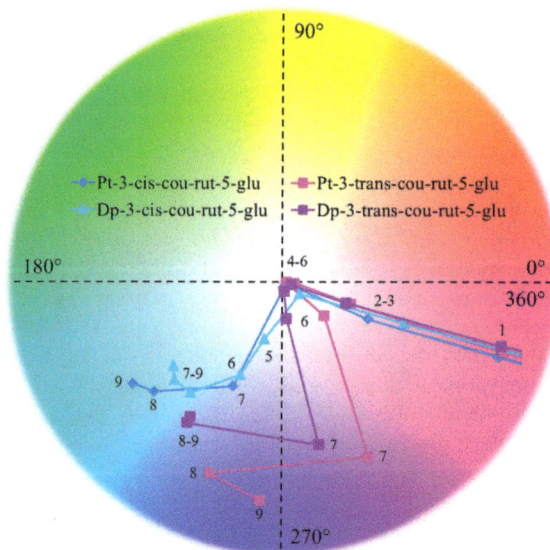

Figure 4. Hue angle (arctan b*/a*) of *cis* and *trans* acylated delphinidin (Dp) and petunidin (Pt) derivatives, pH 1–9.

Table 2. CIE-L*C*$_{ab}$h$_{ab}$ colorimetric values (standard deviation) of delphinidin (Dp) and petunidin (Pt) derivatives in pH 1–9, $n = 3$. Different superscript letters indicate significant differences ($p < 0.05$) between derivatives of the same anthocyanidin in the same pH.

Anthocyanin	pH 1	pH 2	pH 3	pH 4	pH 5	pH 6	pH 7	pH 8	pH 9
Lightness (L*)									
Dp-3-rut-5-glu	74.1 (0.1)	82.2 (0.5)	93.7 (0.1)	97.2 (0.1)	97.5 (0.2)	97.2 (0.0)	84.4 (0.3)	77.3 (0.4)	80.5 (1.5)
Dp-3-*cis*-pcou-rut-5-glu	77.7 (0.1)	81.4 (0.5)	89.8 (0.6)	95.9 (0.2)	89.0 (0.6)	82.8 (0.7)	78.7 (0.5)	79.3 (0.8)	79.4 (0.2)
Dp-3-*trans*-pcou-rut-5-glu	78.0 (0.3)	83.0 (0.4)	94.3 (0.1)	98.3 (0.1)	97.4 (0.2)	93.1 (0.1)	75.3 (1.2)	78.1 (0.2)	79.3 (1.0)
Pt-3-rut-5-glu	74.6 (0.2)	82.3 (0.3)	94.9 (0.0)	98.2 (0.1)	98.4 (0.1)	97.3 (0.1)	77.2 (2.3)	72.1 (1.1)	70.3 (1.2)
Pt-3-*cis*-pcou-rut-5-glu	81.1 (0.2)	83.7 (0.2)	92.0 (0.6)	97.3 (0.1)	98.1 (0.1)	96.9 (0.3)	83.5 (0.5)	83.3 (0.4)	83.7 (0.3)
Pt-3-*trans*-pcou-rut-5-glu	76.1 (0.3)	82.5 (0.1)	93.5 (0.5)	97.9 (0.1)	98.3 (0.1)	92.3 (0.3)	73.4 (0.3)	73.5 (0.6)	70.4 (0.3)
Chroma (C*$_{ab}$)									
Dp-3-rut-5-glu	58.3 (0.1)	41.3 (1.0)	12.2 (0.1)	3.1 (0.2)	2.0 (0.3)	2.3 (0.1)	9.0 (0.4)	22.3 (0.5)	16.2 (2.0)
Dp-3-*cis*-pcou-rut-5-glu	50.2 (0.2)	42.4 (1.1)	21.8 (1.3)	4.0 (0.3)	10.1 (0.5)	17.0 (0.8)	23.8 (0.6)	24.2 (0.8)	22.8 (0.6)
Dp-3-*trans*-pcou-rut-5-glu	48.4 (0.6)	38.2 (0.7)	11.7 (0.1)	2.0 (0.1)	1.9 (0.2)	6.4 (0.1)	28.2 (1.5)	28.4 (0.2)	35.1 (1.6)
Pt-3-rut-5-glu	58.6 (0.5)	42.5 (0.7)	10.6 (0.0)	2.4 (0.1)	1.5 (0.1)	1.9 (0.1)	22.2 (2.3)	37.5 (1.2)	39.0 (1.2)
Pt-3-*cis*-p-cou-rut-5-glu	44.5 (0.4)	38.4 (0.4)	16.0 (0.2)	3.3 (0.2)	1.2 (0.1)	1.3 (0.3)	19.2 (0.8)	27.9 (0.3)	29.9 (0.4)
Pt-3-*trans*-pcou-rut-5-glu	52.0 (0.7)	38.8 (0.4)	12.4 (0.9)	2.5 (0.1)	1.0 (0)	9.5 (0.3)	33.0 (0.3)	34.2 (0.6)	36.9 (0.3)
Hue angle (h$_{ab}$)									
Dp-3-rut-5-glu	0.3 [c] (0.1)	356.4 [a] (0.2)	359.0 [a] (0.3)	26.3 [c] (2.2)	55.5 [c] (2.6)	75.7 [c] (2.7)	272.3 [b] (1.1)	218.7 [c] (1.3)	198.4 [c] (5.3)
Dp-3-*cis*-pcou-rut-5-glu	343.9 [b] (0.2)	344.0 [c] (0.2)	341.2 [b] (0.4)	327.6 [b] (2.8)	255.5 [b] (0.4)	247.5 [b] (0.3)	231.3 [c] (0.5)	223.2 [b] (0.3)	218.6 [b] (0.5)
Dp-3-*trans*-pcou-rut-5-glu	344.5 [a] (0.1)	343.4 [b] (0.2)	341.7 [b] (0.1)	342.0 [a] (1.1)	293.9 [a] (3.1)	280.8 [a] (0.5)	283.6 [a] (1.1)	237.0 [a] (0.2)	236.8 [a] (1.7)
Pt-3-rut-5-glu	357.3 [a] (0.1)	354.2 [a] (0.1)	358.2 [a] (90.5)	34.5 [c] (2.0)	58.2 [c] (2.5)	33.7 [c] (1.5)	274.5 [b] (1.2)	231.5 [b] (0.7)	224.6 [b] (0.6)
Pt-3-*cis*-p-cou-rut-5-glu	341.2 [c] (0.1)	340.7 [c] (0.2)	336.6 [c] (3.2)	337.8 [b] (1.4)	325.0 [b] (3.6)	299.1 [b] (8.3)	246.3 [c] (0.2)	221.5 [c] (0.1)	215.0 [c] (0.9)
Pt-3-*trans*-pcou-rut-5-glu	342.8 [b] (0.1)	342.2 [b] (0.1)	342.2 [b] (0.4)	352.6 [a] (1.0)	356.3 [a] (3.9)	322.3 [c] (1.1)	296.8 [a] (0.3)	250.1 [a] (0.4)	265.0 [a] (0.2)

The hues of these derivatives varied more greatly as a function of the isomeric acylation patterns than of the type of aglycone. As expected of anthocyanins, reddish hues were expressed by these anthocyanins in acidic pH. Acylation typically altered the hue to be comparatively pinker in acidic pH, leading to hues ~340° for all acylated derivatives in pH 1–3 (Table 2). As pH was increased from 1 to 6, all the Dp and Pt derivatives became extremely faint (almost colorless), which led to some variability in hue. The hues of the nonacylated derivatives fell in the red-yellow region (26.3–75.7°) in pH 4–6 while the acylated derivatives were varied more considerably from pink to purple to blue. In pH $\geq$ 7, these Dp derivatives expressed blue colorations (hue 180–270°) while the Pt derivatives were blue and purple (hue 215–297°), suggesting that the methoxyl group on the B-ring of Pt increased the redness of the chromophore. Although the hues of the acylated isomers of the same chromophore were similar in acidic pH 1–3, they differed significantly in pH $\geq$ 4. Dp-3-*cis*-p-cou-rut-5-glu expressed blue colorations in the widest pH range of 5–9. Blue colorations expressed by anthocyanins in such acidic pH are uncommon and often require additional co-factors such as metal ions or polyacylation [28,29]. Dp-3-*trans*-p-cou-rut-5-glu, despite bearing the same acyl moiety, expressed comparatively more purple colorations (Figure 4 and Table 2). Similar observations were observed between the *cis* and *trans* acylated Pt derivatives. Hues of the *trans* acylated derivatives expressed more purple hues (hues = 250–297°) while Pt-3-*cis*-p-cou-rut-5-glu expressed blues in pH $\geq$ 7 (hues = 215–256°) (Table 2).

The *cis–trans* isomerization of acyl moieties on anthocyanins played significant roles in the color expression of the same anthocyanin, affecting both hydration (colorlessness) and hue.

2.3. Color Stability over Time

Stability of anthocyanins is inherently related to the structures of the pigments due to chemical substitution patterns of the chromophore and also the dynamic structural equilibria in which they exist as a function of pH. Table 3 displays the half-life calculations based on change in absorbance (as measured at the λ_{max} from t_0) during dark, ambient storage. As would be expected, stability of the Dp and Pt derivatives was decreased as pH was increased—in some cases, by almost a 950-fold decrease from pH 1 to pH 9 (Table 3). Dp as aglycone is known to be reactive due to the presence of reactive hydroxyl groups on the B-ring; interestingly, Dp-3-rut-5-glu and its derivatives generally exhibited greater half-lives than did Pt-3-rut-5-glu derivatives in acidic conditions.

Table 3. Half-lives (h) and rate constants (k) of delphinidin (Dp) and petunidin (Pt) derivatives in pH 1–9, $n = 3$. Different superscript letters indicate significant differences ($p < 0.05$) between derivatives of the same anthocyanidin in the same pH.

Anthocyanin	pH 1	pH 2	pH 3	pH 4	pH 5	pH 6	pH 7	pH 8	pH 9
				$t_{1/2}$ (h)					
Dp-3-rut-5-glu	675.0 [b] (9.3)	378.7 [a] (16.8)	102.7 [a] (6.2)	34.2 [b] (5.5)	* NM	* NM	1.7 [c] (0.0)	1.1 [c] (0.0)	0.7 [c] (0.0)
Dp-3-*cis-p*-cou-rut-5-glu	1077.2 [a] (21.6)	424.0 [a] (128.9)	52.9 [b] (0.6)	16.0 [c] (0.2)	13.08 [b] (0.2)	12.8 [b] (0.0)	12.6 [a] (0.1)	8.6 [b] (0.1)	7.5 [b] (0.4)
Dp-3-*trans-p*-cou-rut-5-glu	666.3 [b] (47.6)	457.5 [a] (26.2)	89.2 [a] (7.2)	53.9 [a] (1.6)	67.5 [a] (8.4)	18.3 [a] (7.7)	7.7 [b] (0.1)	13.2 [a] (0.7)	13.7 [a] (0.6)
Pt-3-rut-5-glu	183.9 [b] (26.0)	65.3 [b] (5.3)	28.5 [a] (1.9)	18.0 (1.5)	19.5 (2.5)	6.5 [a] (0.2)	1.7 [b] (0.0)	2.5 [b] (0.0)	1.1 [c] (0.0)
Pt-3-*cis-p*-cou-rut-5-glu	546.8 [b] (20.3)	137.4 [a] (19.8)	17.8 [b] (1.7)	* NM	* NM	1.9 [b] (0.1)	1.8 [b] (0.0)	3.3 [b] (0.1)	4.0 [b] (0.7)
Pt-3-*trans-p*-cou-rut-5-glu	2309.6 [a] (389.1)	109.7 [a] (11.8)	25.3 [a] (2.6)	* NM	* NM	7.1 [a] (0.8)	4.2 [a] (0.1)	7.3 [a] (0.6)	5.5 [a] (0.6)
				k					
Dp-3-rut-5-glu	0.0010 (0.0000)	0.0018 (0.0001)	0.0068 (0.0004)	0.0206 (0.0032)	* NM	* NM	0.4004 (0.0034)	0.6110 (0.0078)	1.0377 (0.0583)
Dp-3-*cis-p*-cou-rut-5-glu	0.0006 (0.0000)	0.0017 (0.0005)	0.0131 (0.0001)	0.0433 (0.0006)	0.0532 (0.0001)	0.0540 (0.0001)	0.0552 (0.0005)	0.0808 (0.0005)	0.0929 (0.0049)
Dp-3-*trans-p*-cou-rut-5-glu	0.0010 (0.0001)	0.0015 (0.0001)	0.0078 (0.0007)	0.0129 (0.0004)	0.0104 (0.0012)	0.0379 (0.0015)	0.0900 (0.0009)	0.0527 (0.0028)	0.0507 (0.0020)
Pt-3-rut-5-glu	0.0038 (0.0001)	0.0107 (0.0001)	0.0244 (0.0016)	0.0386 (0.0032)	0.0359 (0.0049)	0.1073 (0.0029)	0.4019 (0.0064)	0.2780 (0.0033)	0.6163 (0.0071)
Pt-3-*cis-p*-cou-rut-5-glu	0.0013 (0.0000)	0.0051 (0.0007)	0.0391 (0.0040)	* NM	* NM	0.3745 (0.0124)	0.3832 (0.0091)	0.2127 (0.0054)	0.1751 (0.0329)
Pt-3-*trans-p*-cou-rut-5-glu	0.0003 (0.0001)	0.0064 (0.0001)	0.0277 (0.0030)	* NM	* NM	0.0988 (0.0118)	0.1664 (0.0029)	0.0956 (0.0070)	0.1273 (0.0122)

* NM: not measurable due to lack of absorbance at initial timepoint.

Overall, acylation demonstrated stabilizing effects on Dp-3-rut-5-glu and Pt-3-rut-5-glu, increasing half-lives in most pH levels evaluated. Increases in storage stability by acylation were most pronounced in alkaline pH, leading to 19.6× and 5× increases in half-life in pH 9 when comparing Dp-3-*trans-p*-cou-rut-5-glu to Dp-3-rut-5-glu and Pt-3-*trans-p*-cou-rut-5-glu to Pt-3-rut-5-glu, respectively. The *cis* or *trans* conformation of the acyl moiety also uniquely impacted the stability for the acylated Dp-3-rut-5-glu and Pt-3-rut-5-glu derivatives. In very acidic pH 1, the half-life of Dp-3-*cis-p*-cou-rut-5-glu was greater than that of its *trans* counterpart while the half-life of Pt-3-*trans-p*-cou-rut-5-glu was greater than that of its *cis* form. However, in pH > 1, half-lives of both Dp-3-*trans-p*-cou-rut-5-glu and Pt-3-*trans-p*-cou-rut-5-glu were greater than those of the *cis* acylated counterparts. While the *cis* acylated isomers of Dp and Mv have increased color stability across pH related to protection of the chromophore against hydration [18], the retention of color over time was found to be decreased when comparing *cis* to *trans* acylated isomers. These would be important considerations in the selection of specific anthocyanins depending on the application. In low-acid (pH 4–6) products in which the pigments will be dried, such as candy panning or extruded ready-to-eat cereals, *cis* acylated anthocyanins may provide more intense coloration. However, in high moisture and neutral products such as protein or dairy beverages, anthocyanins are more labile to degradation during storage; thus, *trans* acylated anthocyanins may express colorations for longer amounts of time.

Interestingly, all the Pt derivatives (acylated or not) showed greater stability in pH 8 than in neutral pH 7 or in alkaline pH 9, which might be explained by anthocyanins' structural transformation according to the pH environment. In mildly acidic and neutral pH, the flavylium cation can be hydrated (colorless) or deprotonated and exist as different quinonoidal base forms. With subsequent increases in pH, the quinonoidal base can be ionized to have one or two negative charges [30]. The kinetic parameters pK_{a2} and pK_{a3} are used to denote the dissociation constants for the transformations from the neutral quinonoidal base to having one and two negative charges, respectively. The pK_{a2} and pK_{a3} of the Pt aglycone were found to be pH 6.99 and 8.27, respectively [31]. Thus, a large proportion of blue quinonoidal base forms with one negative charge exist, which may be more prone to degradation at pH 7, while at pH 8, a higher proportion of quinonoidal bases with 2 charges exists, further modifying the reactivity of the pigment. Nevertheless, the retention of color over time was found to be enhanced by acylation, and its orientation in space proved another important consideration.

3. Materials and Methods

3.1. Materials

Delphinidin- and petunidin-rich anthocyanin extracts were prepared from the peels of Asian eggplant varieties (*Solanum melongena* L.) and dried black goji fruits (*Lycium ruthenicum*) purchased from grocery stores (respectively, Columbus, OH, USA and Shanghai, China).

Tris(hydroxymethyl)aminomethane, 99%, was purchased from Alfa Aesar (Ward Hill, MA, USA). All other chemicals and solvents were ACS or HPLC grade and purchased from Fisher Scientific (Fair Lawn, NJ, USA).

3.2. Methods

3.2.1. Anthocyanin Preparation for Pigment Isolation

Anthocyanins were extracted from the skins of eggplants and black goji fruits with acidified 70% acetone (1.5% trifluoroacetic acid and 0.01% HCl, respectively), isolated by phase partition with water and chloroform, and purified by C_{18} solid-phase extraction with acidified water and ethyl acetate following procedures described by Rodríguez-Saona and Wrolstad [32].

To obtain nonacylated counterparts, aliquots from each extract were dissolved in 10 mL of 10% KOH for 10 min to cleave the ester bonds between the acyl and glycosyl moieties of the anthocyanins, according to procedures of Giusti and Wrolstad [33]. The extracts were again purified and concentrated by solid-phase extraction described above.

3.2.2. Anthocyanin Isolation

Dp-3-rut-5-glu and the *cis* and *trans* acyl isomers of Dp-3-*p*-cou-rut-5-glu were isolated from Asian eggplant anthocyanin extracts while Pt-3-rut-5-glu, Pt-3-*cis*-*p*-cou-rut-5-glu, and Pt-3-*trans*-*p*-cou-rut-5-glu were isolated from the black goji extracts by semipreparative reverse-phase HPLC. The system was produced by Shimadzu (Columbia, MD, USA) and composed of LC-6AD pumps, a CBM-20A communication module, an SIL-20A HT autosampler, a CTO-20A column oven, and an SPD-M20A Photodiode Array detector. LCMS Solution Software (Version 3, Shimadzu, Columbia, MD, USA) was used to monitor samples.

Anthocyanins were separated by a Luna reverse-phase pentafluorophenyl (PFP2) column with 5 μm particle size and 100 Å pore size in a 250 × 21.2 mm column (Phenomenex®, Torrance, CA, USA). The flow rate was 10.0 mL/min for a run time of 30 min. A binary gradient was used with solvents A: 4.5% formic acid in water and B: acetonitrile. For nonacylated Dp, the gradient began at 8% B and was constant for 1 min, then increased to 16.5% B over 31 min, while for acylated Dp, the gradient began at 10% B and increased to 30% by 30 min. Similarly, the gradient for nonacylated Pt began at 10% B and was constant for 1 min then increased to 15% B over 31 min. Acylated Pt was separated with a more complex gradient starting at constant 12% B for 2 min, then increasing to 21% B over 25 min and being held constant until 30 min, before finally increasing to 30% B by 50 min. Elution of anthocyanins was monitored at 520 nm, and desired peaks were collected manually. Isolates were diluted with 2–3 volumes of water and subjected to solid phase extraction to remove excess formic acid and concentrate the pigments.

3.2.3. Isolated Anthocyanin Identity—HPLC-MS

The identity of the pigments was monitored by analytical reverse-phase HPLC-MS with a similar system (Shimadzu, Columbia, MD, USA) differing by pumps LC-20AD and including an LCMS-2010 mass spectrometer. The analytical column was a Kinetix reverse-phase pentafluorophenyl (PFP2) column with 2.6 μm particle size and 100 Å pore size in a 100 × 4.6 mm column (Phenomenex®, Torrance, CA, USA). The flow rate was 0.6 mL/min for a binary gradient consisting of solvents A: 4.5% formic acid in water and B: acetonitrile. The gradient for all Dp derivatives began at 0% B, increased 0–10% from 0 to 1 min and then 10–23% B from 1 to 30 min; for Pt derivatives, the gradient began at 7% B for 2 min and up to 20% B over 30 min. Spectral data was collected at 200–700 nm. A quantity of 0.13 mL per minute was diverted to the mass spectrometer. Mass spectrometry was conducted under the positive ion mode; data were collected from m/z 200–1200. Mass spectrometry, order of elution, and comparison to literature were used to identify the anthocyanins [22,29,34,35]. The structures of the predominant and isolated anthocyanins of the extracts may be found in Figure 1.

3.2.4. Sample Preparation

The anthocyanin extracts were diluted to concentrations of 50 μM in buffers of pH 1–9 ± 0.05. Buffer systems were composed of 0.025 M KCl for pH 1–2, 0.1 M sodium acetate for pH 3–6, 0.25 M TRIS for pH 7–8, and 0.1 M sodium bicarbonate for pH 9. The pH of the buffer systems was adjusted with concentrated HCl or 10% NaOH prior to final volume adjustment. Anthocyanin concentrations of the extracts and isolates were determined by the pH differential methodology [36]. Dp-3-rut-5-glu and Pt-3-rut-5-glu isolates were expressed as cyanidin-3-diglucoside-5-glucoside equivalents, while the Dp-3-*p*-cou-rut-5-glu and Pt-3-*p*-cou-rut-5-glu isolates were expressed as Cy-3-*p*-cou-diglu-5-glu equivalents using ε reported by Ahmadiani et al [17]. After anthocyanins were diluted in buffers, the pH of every sample was verified to be ± 0.05 using a Mettler Toledo Inc. S220 SevenCompact™ pH/Ion meter (Columbus, OH, USA). Samples were equilibrated at room temperature in the dark for 30 min prior to initial analysis. Samples were then sealed and stored in the dark for 72 h at 25 °C to briefly assess the color stability of the pigments. Three replicates were prepared for each treatment.

3.2.5. Visible Spectrophotometry of Samples

After pigment dilution, samples were equilibrated at room temperature in the dark for 30 min prior to initial spectrophotometric analysis, referred to as t_0. Aliquots of 300 μL of each sample were transferred to poly-D-lysine-coated polystyrene 96-well plates and evaluated by visible absorbance (380–700 nm, 1 nm intervals) spectrophotometry with a SpectraMax 190 Microplate Reader (Molecular Devices, Sunnyvale, CA, USA). Color intensity of the pigments in pH 1–9 was also compared by % absorbance retention calculated as Absorbance in pH_n/Absorbance in pH_1 at respective λ_{max} × 100. Most anthocyanins express a high color intensity in very acidic pH ≤ 1; therefore, this proportion was used as a measure to compare the degree of color loss in different pH.

3.2.6. Colorimetry of Samples

Colorimetric data was expressed in the CIELAB communication system, the official color scale from the CIE (Commission internationale de l'éclairage), where LAB stands the color coordinates in the color space. Color parameters were calculated from spectral absorbance data (380–700 nm, 5 nm intervals) using the ColorBySpectra software [37] based on standard 1964 CIE equations, D_{65} illuminant spectral distribution, and 10° observer angle functions. Color values are expressed as a function of L* (luminosity/lightness), C^*_{ab} (chroma/intensity) and h^*_{ab} (hue angle).

3.2.7. Calculation of Sample Half-Lives

Samples in the well plates were kept for 72 h at room temperature and in the dark to compare the relative stabilities of the different spatial configurations of the acylated anthocyanins. Plates were closed with a lid and sealed with parafilm to prevent evaporation during storage. Absorbance and color measurements taken after 30 min equilibration were used as time 0, t_0. Subsequent readings were taken throughout the storage at times 2, 4, 6, 24, 48, and 72 h . Linear regressions were prepared from the natural logarithm of the absorbance (at the λ_{max} from t_0) during the time points of the study, following the formula $\ln[A_t] = -kt + \ln[A_{t0}]$. Linear regressions showed coefficient of determination values (R^2 values ≥ 0.89). The half-life ($t_{1/2}$) was calculated for each of the samples as $t_{1/2} = \ln 2/k$.

3.2.8. Statistical Evaluation of Data

Figures, means, and standard deviations were produced using Microsoft Office Excel 2010 (Office 14.0, Microsoft. Redmond, WA, USA). The λ_{max}, % absorbance retention, and half-lives of the different pigments were evaluated by one-way analysis of variance (ANOVA) (two-tailed, $\alpha = 0.05$) and post hoc Tukey's test ($\alpha = 0.05$) using Minitab 16 (Minitab Inc., State College, PA, USA). Of colorimetric data, only hue angles* were statistically compared by one-way analysis of variance (ANOVA) (two-tailed, $\alpha = 0.05$) and post hoc Tukey's test ($\alpha = 0.05$). Due to differences in exact pigments' concentrations, differences in L* and C^*_{ab} values were expected and could not be equally compared statistically; the hue angle* was considered to be less variable due to small differences in concentration.

4. Conclusions

The stereochemistry of coumaric acid acylation had a strong impact on the spectra, color, and stability of Dp and Pt anthocyanins. Pigments with *cis* isomeric acylation had the greatest λ_{max} in all pH—as much as 66 nm greater than *trans* counterparts. Therefore, *cis* acylated isomers expressed bluer hues while the *trans* isomers were comparatively redder. Acylation with *trans* isomers induced a spectral shoulder in wavelengths ~630 nm off of the peak at the λ_{max} in pH ≥ 7, while *cis* isomers were documented by a broad main peak around these wavelengths. *Cis* acylation seemed to better protect the molecule against hydration, resulting in higher color expression across pH while *trans* acylation generally improved color retention over time. Anthocyanins typically express low color intensity in the mildly acidic pH common to many foods and confectionaries; *cis* acylated anthocyanins may help to fill this industry gap. In addition, it was found that Dp-3-*cis*-*p*-cou-rut-5-glu expressed the greatest λ_{max} (617–632 nm) in the widest pH range (5–9) compared with Pt-3-*cis*-*p*-cou-rut-5-glu with λ_{max} 553–630 nm in pH 5–9. Dp-3-*cis*-*p*-cou-rut-5-glu exhibited blue hues even in pH 5 ($C^*_{ab} = 10$, $h_{ab} = 256°$) in which most anthocyanins exist in colorless hemiketal forms. Color stability across pH and over time of all derivatives was negatively impacted with increased pH; however, acylation improved stability in both of these aspects. This study has demonstrated the unique spectral properties and variety of hues of anthocyanins composed of the same chemical substitutions as to be a result of differing spatial configurations of the acyl moieties. It provides additional insight on the chemical attributes that affect the large diversity of hues expressed by anthocyanins in natural plant systems.

Acknowledgments: This work was supported in part by the USDA National Institute of Food and Agriculture, Hatch Project OHO01423, Accession Number 1014136.

Author Contributions: G.T.S., P.T., and M.M.G. conceived and designed the study and interpreted the results. G.T.S. and P.T. executed experiments and collected data and drafted the manuscript. All authors contributed to the data analyses and final edits of the manuscript.

Conflicts of Interest: The authors declare no conflict of interest.

References

1. McCann, D.; Barrett, A.; Cooper, A.; Crumpler, D.; Dalen, L.; Grimshaw, K.; Kitchin, E.; Lok, K.; Porteous, L.; Prince, E.; et al. Food additives and hyperactive behaviour in 3-year-old and 8/9-year-old children in the community: A randomised, double-blinded, placebo-controlled trial. *Lancet* **2007**, *370*, 1560–1567. [CrossRef]
2. Potera, C. The Artificial Food Dye Blues. *Environ. Health Perspect.* **2010**, *118*, 428. [CrossRef] [PubMed]
3. Dornblaser, L.; Jago, D. Colors and Flavors: The move to more natural. *Mintel* **2013**, *1*, 1–24.
4. Sigurdson, G.; Tang, P.; Giusti, M.M. Natural Colorants: Food Colorants from Natural Sources. *Annu. Rev. Food Sci. Technol.* **2017**, *8*, 261–280. [CrossRef] [PubMed]
5. Andersen, Ø.M.; Jordheim, M. Basic Anthocyanin Chemistry and Dietary Sources. In *Anthocyanins in Health and Disease*; Wallace, T.C., Giusti, M.M., Eds.; CRC Press (Taylor & Francis Group): Boca Raton, FL, USA, 2014; pp. 13–90.
6. Ananga, A.; Georgiev, V.; Ochieng, J.; Phills, B.; Tsolova, V. Production of Anthocyanins in Grape Cell Cultures: A Potential Source of Raw Material for Pharmaceutical, Food, and Cosmetic Industries. In *The Mediterranean Genetic Code-Grapevine and Olive*; Puljuha, D., Sladonja, B., Eds.; InTech: Rijeka, Croatia, 2013; pp. 247–288.
7. Cesa, S.; Carradori, S.; Bellagamba, G.; Locatelli, M.; Casadei, M.A.; Masci, A.; Paolicelli, P. Evaluation of processing effects on anthocyanin content and colour modifications of blueberry (*Vaccinium* spp.) extracts: Comparison between HPLC-DAD and CIELAB analyses. *Food Chem.* **2017**, *232*, 114–123. [CrossRef] [PubMed]
8. Rakic, V.; Skrt, M.; Miljkovic, M.; Kostic, D.; Sokolovic, D.; Poklar-Ulrih, N. Effects of pH on the stability of cyanidin and cyanidin 3-*O*-β-glucopyranoside in aqueous solution. *Hem. Ind.* **2015**, *69*, 511–522. [CrossRef]
9. Zhao, C.L.; Chen, Z.J.; Bai, X.S.; Ding, C.; Long, T.J.; Wei, F.G.; Miao, K.R. Structure-activity relationships of anthocyanidin glycosylation. *Mol. Divers.* **2014**, *18*, 687–700. [CrossRef] [PubMed]
10. Fossen, T.; Cabrita, L.; Andersen, O.M. Colour and stability of pure anthocyanins influenced by pH including the alkaline region. *Food Chem.* **1998**, *63*, 435–440. [CrossRef]
11. Mazza, G.; Brouillard, R. Recent developments in the stabilization of anthocyanins in food products. *Food Chem.* **1987**, *25*, 207–225. [CrossRef]
12. Torskangerpoll, K.; Andersen, Ø.M. Colour stability of anthocyanins in aqueous solutions at various pH values. *Food Chem.* **2005**, *89*, 427–440. [CrossRef]
13. Stintzing, F.C.; Stintzing, A.S.; Carle, R.; Frei, B.; Wrolstad, R.E. Color and antioxidant properties of cyanidin-based anthocyanin pigments. *J. Agric. Food Chem.* **2002**, *50*, 6172–6181. [CrossRef] [PubMed]
14. Giusti, M.M.; Wrolstad, R.E. Acylated anthocyanins from edible sources and their applications in food systems. *Biochem. Eng. J.* **2003**, *14*, 217–225. [CrossRef]
15. Malien-Aubert, C.; Dangles, O.; Amiot, M.J. Color stability of commercial anthocyanin-based extracts in relation to the phenolic composition. Protective effects by intra- and intermolecular copigmentation. *J. Agric. Food Chem.* **2001**, *49*, 170–176. [CrossRef] [PubMed]
16. Malcıoğlu, O.B.; Calzolari, A.; Gebauer, R.; Varsano, D.; Baroni, S. Dielectric and Thermal Effects on the Optical Properties of Natural Dyes: A Case Study on Solvated Cyanin. *J. Am. Chem. Soc.* **2011**, *133*, 15425–15433. [CrossRef] [PubMed]
17. Ahmadiani, N.; Robbins, R.J.; Collins, T.M.; Giusti, M.M. Molar absorptivity (ε) and spectral characteristics of cyanidin-based anthocyanins from red cabbage. *Food Chem.* **2016**, *197*, 900–906. [CrossRef] [PubMed]
18. George, F.; Figueiredo, P.; Toki, K.; Tatsuzawa, F.; Saito, N.; Brouillard, R. Influence of *trans-cis* isomerisation of coumaric acid substituents on colour variance and stabilisation in anthocyanins. *Phytochemistry* **2001**, *57*, 791–795. [CrossRef]

19. Yoshida, K.; Kondo, T.; Kameda, K.; Goto, T. Structure of Anthocyanins Isolated from Purple Leaves of *Perilla ocimoides* L. var. crispa Benth and Their Isomerization by Irradiation of Light. *Agric. Biol. Chem.* **1990**, *54*, 1745–1751. [CrossRef]

20. Hosokawa, K.; Fukushi, E.; Kawabata, J.; Fujii, C.; Ito, T.; Yamamura, S. Seven acylated anthocyanins in blue flowers of Hyacinthus orientalis. *Phytochemistry* **1997**, *45*, 167–171. [CrossRef]

21. Ichiyanagi, T.; Kashiwada, Y.; Shida, Y.; Ikeshiro, Y.; Kaneyuki, T.; Konishi, T. Nasunin from Eggplant Consists of *Cis−Trans* Isomers of Delphinidin 3-[4-(*p*-Coumaroyl)-L-rhamnosyl(1→6)glucopyranoside]-5-glucopyranoside. *J. Agric. Food Chem.* **2005**, *53*, 9472–9477. [CrossRef] [PubMed]

22. Zheng, J.; Ding, C.; Wang, L.; Li, G.; Shi, J.; Li, H.; Wang, H.; Suo, Y. Anthocyanins composition and antioxidant activity of wild *Lycium ruthenicum* Murr. from Qinghai-Tibet Plateau. *Food Chem.* **2011**, *126*, 859–865. [CrossRef]

23. Inami, O.; Tamura, I.; Kikuzaki, H.; Nakatami, N. Stability of anthocyanins of Sambucus canadensis and Sambucus nigra. *J. Agric. Food Chem.* **1996**, *44*, 3090–3096. [CrossRef]

24. Jin, H.; Liu, Y.; Guo, Z.; Yang, F.; Wang, J.; Li, X.; Peng, X.; Liang, X. High-performance liquid chromatography separation of *cis–trans* anthocyanin isomers from wild *Lycium ruthenicum* Murr. employing a mixed-mode reversed-phase/strong anion-exchange stationary phase. *J. Agric. Food Chem.* **2015**, *63*, 500–508. [CrossRef] [PubMed]

25. Harborne, J.B. *Comparative Biochemistry of the Flavonoids*; Academic Press: London, UK; New York, NY, USA, 1967.

26. Brouillard, R.; Delaporte, B. Chemistry of Anthocyanin Pigments. 2. * Kinetic and Thermodynamic Study of Proton Transfer, Hydration, and Tautomeric Reactions of Malvidin 3-Glucoside. *J. Am. Chem. Soc.* **1977**, *99*, 8461–8468. [CrossRef]

27. Sigurdson, G.T.; Robbins, R.J.; Collins, T.M.; Giusti, M.M. Spectral and colorimetric characteristics of metal chelates of acylated cyanidin derivatives. *Food Chem.* **2017**, *221*, 1088–1095. [CrossRef] [PubMed]

28. Yoshida, K.; Mori, M.; Kondo, T. Blue flower color development by anthocyanins: from chemical structure to cell physiology. *Nat. Prod. Rep.* **2009**, *26*, 884–915. [CrossRef] [PubMed]

29. Sigurdson, G.T.; Giusti, M.M. Bathochromic and Hyperchromic Effects of Aluminum Salt Complexation by Anthocyanins from Edible Sources for Blue Color Development. *J. Agric. Food Chem.* **2014**, *62*, 6955–6965. [CrossRef] [PubMed]

30. Asenstorfer, R.E.; Iland, P.G.; Tate, M.E.; Jones, G.P. Charge equilibria and pKa of malvidin-3-glucoside by electrophoresis. *Anal. Biochem.* **2003**, *318*, 291–299. [CrossRef]

31. León-Carmona, J.R.; Galano, A.; Alvarez-Idaboy, J.R. Deprotonation routes of anthocyanidins in aqueous solution, pKa values, and speciation under physiological conditions. *RSC Adv.* **2016**, *6*, 53421–53429. [CrossRef]

32. Rodríguez-Saona, L.E.; Wrolstad, R.E. Extraction, Isolation, and Purifification of Anthocyanins. In *Current Protocols in Food Analytical Chemistry*; Wrolstad, R.E., Acree, T.E., An, H., Decker, E.A., Penner, M.H., Reid, D.S., Schwartz, S.J., Shoemaker, C.F., Sporns, P., Eds.; John Wiley & Sons, Inc.: New York, NY, USA, 2001; pp. F1.1.1–F1.1.11.

33. Giusti, M.M.; Wrolstad, R.E. Separation and Characterization of Anthocyanins by HPLC. In *Current Protocols in Food Analytical Chemistry*; Wrolstad, R.E., Acree, T.E., An, H., Decker, E.A., Penner, M.H., Reid, D.S., Schwartz, S.J., Shoemaker, C.F., Sporns, P., Eds.; John Wiley & Sons, Inc.: New York, NY, USA, 2001; pp. F1.3.1–F1.3.13.

34. Azuma, K.; Ohyama, A.; Ippoushi, K.; Ichiyanagi, T.; Takeuchi, A.; Saito, T.; Fukuoka, H. Structures and antioxidant activity of anthocyanins in many accessions of eggplant and its related species. *J. Agric. Food Chem.* **2008**, *56*, 10154–10159. [CrossRef] [PubMed]

35. Sadilova, E.; Stintzing, F.C.; Carle, R. Anthocyanins, colour and antioxidant properties of eggplant (*Solanum melongena* L.) and violet pepper (*Capsicum annuum* L.) peel extracts. *Z. Naturforsch. Sect. C J. Biosci.* **2006**, *61*, 527–535. [CrossRef]

36. Giusti, M.M.; Wrolstad, R. Characterization and measurement of anthocyanins by UV- Visible spectroscopy. In *Current Protocols in Food Analytical Chemistry*; Wrolstad, R.E., Acree, T.E., An, H., Decker, E.A., Penner, M.H., Reid, D.S., Schwartz, S.J., Shoemaker, C.F., Sporns, P., Eds.; John Wiley & Sons, Inc.: New York, NY, USA, 2001; Volume 1, pp. F1.2.1–F1.2.13.
37. Farr, J.E.; Giusti, M.M. ColorbySpectra-Academic License. Available online: https://buckeyevault.com/products/colorbyspectra (accessed on 15 January 2018).

Sample Availability: Samples of the compounds are not available from the authors.

molecules

MDPI

Article

Theoretical Characterization by Density Functional Theory (DFT) of Delphinidin 3-*O*-Sambubioside and Its Esters Obtained by Chemical Lipophilization

Ana Selene Márquez-Rodríguez [1], Claudia Grajeda-Iglesias [2], Nora-Aydeé Sánchez-Bojorge [1], María-Cruz Figueroa-Espinoza [3], Luz-María Rodríguez-Valdez [1], María Elena Fuentes-Montero [1,*] and Erika Salas [1,*]

[1] Facultad de Ciencias Químicas, Universidad Autónoma de Chihuahua, Chihuahua 31125, Mexico; anaselene.marquez@gmail.com (A.S.M.-R.); norasanchez15@gmail.com (N.-A.S.-B.); lmrodrig@uach.mx (L.-M.R.-V.)

[2] Technion-Israel Institute of Technology, Haifa 31096, Israel; claugrajeda@gmail.com

[3] Montpellier SupAgro, 34060 Montpellier, France; maria-cruz.figueroa@cirad.fr

* Correspondence: mfuentes@uach.mx (M.E.F.-M.); esalas@uach.mx (E.S.); Tel.: +52(614)-2366000 (ext. 4279) (M.E.F.-M.); +52(614)-2366000 (ext. 4286) (E.S.)

Academic Editors: M. Monica Giusti and Gregory T. Sigurdson
Received: 1 June 2018; Accepted: 27 June 2018; Published: 29 June 2018

Abstract: Anthocyanins are water-soluble phenolic pigments. However, their poor solubility in lipidic media limits their use. This hurdle can be overcome with the lipophilization of anthocyanins, which consists of adding an aliphatic chain to a hydrophilic compound, in order to increase its solubility in lipids. Still, the unspecific chemical lipophilization of anthocyanin-esters produces molecules with different properties from their precursors. In this work, experimental changes of anthocyanin-esters obtained by chemical lipophilization are investigated *in silico* aiming specifically at observing their molecular behavior and comparing it with their anthocyanin precursor. Thus, the analysis of delphinidin 3-*O*-sambubioside and its esters employing Density Functional Theory (DFT) methods, such as the hybrid functional B3LYP in combination with the 6-31++G(d,p) Pople basis set, provides the ground state properties, the local reactivity and the molecular orbitals (MOs) of these compounds. Excited states properties were analyzed by TD-DFT with the B3LYP functional, and the M06 and M06-2X meta-GGA functionals. Local reactivity calculations showed that the electrophilic site for all the anthocyanin-esters was the same as the one for the anthocyanin precursor, however the nucleophilic site changed depending localization of the esterification. TD-DFT results indicate that the place of esterification could change the electronic transitions and the MOs spatial distribution.

Keywords: anthocyanin; lipophilization; time-depending density functional theory

1. Introduction

Anthocyanins are phenolic compounds from the flavonoids family, which are produced naturally in flowers and fruits (Figure 1). These water-soluble pigments change color in aqueous solution from red to blue depending on the pH. This spectrum responds to the change in the form of the molecule's structure when it gets protonated or hydrated, thus in aqueous solution different forms co-exist in equilibrium [1]. Besides the equilibrium states, anthocyanin studies have been performed to determine their reactivity, molecular interactions, antioxidant properties, among others [2,3].

The application of anthocyanins as industrial pigments has been limited due to their poor solubility in lipidic media. To overcome this solubility issue the lipophilization of anthocyanins has successfully been performed by enzymatic and chemical approaches [4,5]. The lipophilization reaction consists of adding an aliphatic chain to the hydrophilic anthocyanin, in order to change the polarity in the molecule and favor its solubility in non-polar matrixes. For example, a recently reported enzymatic lipophilization of malvidin 3-*O*-glucoside with oleic acid and *Candida antarctica* lipase B as catalyst generated a single anthocyanin-ester with attractive color characteristics [4]. In a different study [5], a chemical reaction was performed with delphinidin 3-*O*-sambubioside (DpS) and octanoyl chloride in an acidic medium; the reaction produced mono-, di- and tri-esters which could not be purified and fully characterized by experimental methods.

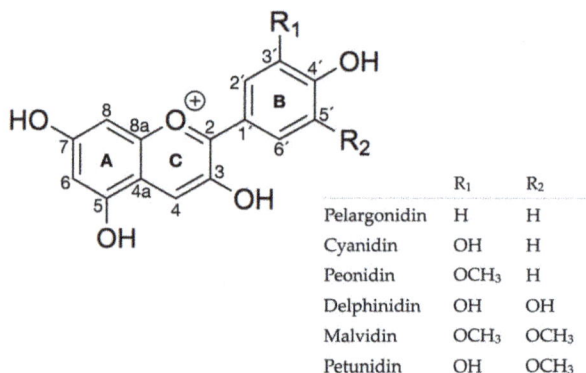

	R$_1$	R$_2$
Pelargonidin	H	H
Cyanidin	OH	H
Peonidin	OCH$_3$	H
Delphinidin	OH	OH
Malvidin	OCH$_3$	OCH$_3$
Petunidin	OH	OCH$_3$

Figure 1. Structure of anthocyanidins: pelargonidin, cyanidin, peonidin, delphinidin, malvidin and petunidin.

Some studies have been supported by theoretical research, which have incorporated geometrical analysis, molecular interactions, antioxidant properties, optical properties and solvation effects, to name a few [6–8]. These models allow the explaining and understanding of the behavior of the experimental phenomena at a molecular level, converting in silico studies into an important tool, which provides significant information about anthocyanins [9]. Particularly, the optical features of anthocyanins have been calculated with semi-empirical [10], Hartree-Fock (HF) [11], and Density Functional Theory (DFT) [12] methodologies employing several functionals and basis sets. The excited states obtained by HF have used the single excitation–configuration interaction (SE–CI) calculation to describe the electronic characteristics of anthocyanins after the geometry optimization by DFT methodology, employing the B3LYP hybrid functional and the D95 basis set [6]. Concerning semi-empirical methods, the Zerner's Intermediate Neglect of Differential Overlap (ZINDO) [13] and the Austin Model 1 (AM1) [14] showed to correctly describe the maximum wavelength (λ_{max}) of anthocyanins [15]. However, DFT is one of the electronic structure most used methods to calculate anthocyanins theoretical features, whereas Time-Depending DFT [16] has provided more accuracy to represent UV-Vis absorption spectra [6,17,18].

Specifically, the TD-DFT methodology approaches the experimental λ_{max} for flavylium cations, however, optical properties calculated employing pure DFT functionals (BLYP, BP86, PBE) underestimate the main absorption band, in accordance with Anouar et al. [18]. The hybrid functionals, which include HF exchange contributions and have given better results, are the B3LYP, B3P86, PBE0 or the meta-GGA functionals [18–20]. Moreover, diffuse functions, polarized functions for heavy atoms, and the diffuse functions for hydrogen atoms (for example the 6-31+G(d,p) Pople basis set) have been suggested when analyzing electronic properties of anthocyanins [9].

The solvation effects have been analyzed by the application of the polarizable continuum model (PCM) using the integral equation formalism polarizable variant (IEFPCM) and water as solvent, which describes the contribution of the solvent in theoretical studies of anthocyanins [21]. However, the full characterization of anthocyanin color has not yet been achieved by a theoretical methodology.

Based on a previous work [19] the B3LYP [22], M06 and M06-2X [23] were chosen for this study; these functionals have a 20%, 27% and 54% of HF exchange contribution, respectively. Therefore, the aim of this work is to explore the electronic and optical properties of the delphinidin 3-*O*-sambubioside and its chemically obtained esters, employing different DFT approaches.

2. Results and Discussion

2.1. Experimental Results

The lipophilization of anthocyanins has only been achieved recently in three studies, via chemical or enzymatic reactions [4,5,24]. The lipophilization of DpS was successfully carried out employing octanoyl chloride as reported by Grajeda-Iglesias [5]. The anthocyanins were extracted from *Hibiscus sabdariffa* calyxes and then purified; the anthocyanins were used for the lipophilization reaction that is described in the methodology section of this article. The products of this reaction were two different monoesters (E1 and E3) and a di-ester (E2). The positions where the esterification can take place are shown in Scheme 1. These structures were partially elucidated using HPLC/MS/MS. Fragmentation pathways in MS/MS spectrometry disclosed that the esterification of the anthocyanin occurred either in the sambubioside unit, in the flavylium unit or both (Figure 2). It shows the full-scan ESI in positive mode of the esters obtained by chemical lipophilization. In Figure 2a two peaks can be observed, the peak at m/z 723 corresponds to the base peak of the delphinidin 3-*O*-sambubioside monoester (E1), and the peak at m/z 429 corresponds to the flavylium fragment plus the aliphatic chain, while the loss of m/z 294 indicates the sambubioside. Thus, it could be concluded that the lipophilization took place in the flavylium core. Then, the second monoester (E3) is shown in Figure 2c, where the base peak is displayed at m/z 723, while the m/z 303 indicates the presence of the anthocyanin´s aglycon. Thereby the loss of m/z 420 corresponds to the mass of sambubioside and the octanoyl chain, which evidences that the lipophilization was carried out in the sambubioside fragment. Finally, the delphinidin 3-*O*-sambubioside di-ester (E2) can be observed at m/z 849 corresponding to its base peak, while the peak at m/z 429 implies that the mass of the flavylium core is esterified; thus, the loss of m/z 429 also denotes that the sambubioside was esterified (Figure 2b). These esters suffered a hypochromic effect (shown in the inserts of Figure 4a–c), where the more significant loss of intensity at 520 nm was in the di-ester.

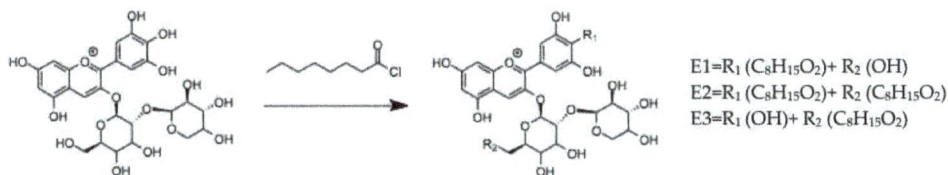

$$E1 = R_1 (C_8H_{15}O_2) + R_2 (OH)$$
$$E2 = R_1 (C_8H_{15}O_2) + R_2 (C_8H_{15}O_2)$$
$$E3 = R_1 (OH) + R_2 (C_8H_{15}O_2)$$

Scheme 1. Lipophilization reaction of DpS. Ester 1 (E1) is a monoester with the octanoyl chain joined to the B ring; Ester 2 (E2) is a di-ester with the octanoyl chains joined to the B ring and the sambubioside; Ester 3 (E3) is the monoester with the octanoyl chain joined to the sambubioside.

It has been indicated that a loss of protons in the anthocyanin structure could produce a hypochromic effect [25,26]. Besides, the change of color in anthocyanidins depends on the grade of substitution of -OH, -OCH3 and glycosylation. These phenomena have been experimentally observed and theoretically analyzed [18,27]. Thus, the union of a chain to the flavylium core could induce the hypochromic effect. Cruz et al. [4] also saw a hypochromic effect when the enzymatic lipophilization of malvidin 3-*O*-glucoside occurred in the glucose unit. The ester obtained from lipophilization of malvidin 3-*O*-glucoside with oleic acid and *Candida antarctica* lipase B had a hypochromic and bathochromic effect as well, but the new pigment preserved its attractive color, contrasting with the chemical lipophilization of DpS. This effect on the color intensity in the enzymatic produced anthocyanin ester could be due to the position of the esterification, which occurred at the glucose moiety of the malvidin. Meanwhile, the chemical lipophilization is not selective and the hypochromic effect is noticeable when the esterification occurs in the flavylium core.

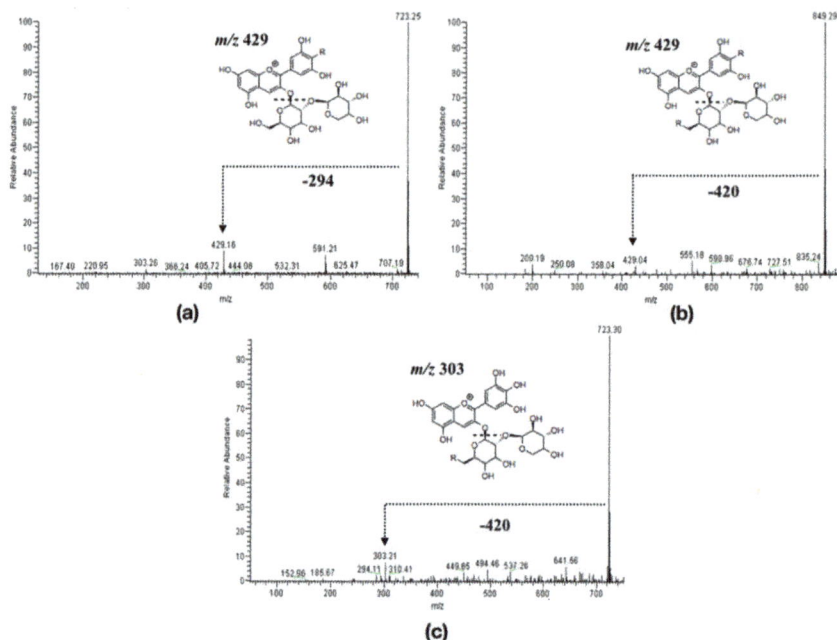

Figure 2. Full-scan (+) ESI of: (**a**) the monoester E; (**b**) di-ester E2 and (**c**) monoester E3.

2.2. Theoretical Results

2.2.1. Theoretical Methodology Selection

Only a small number of atomic structures of phenolic compounds, with exact bond length and angles, have been reported. X-Ray diffraction (XRD) data for cyanidin bromide [28] is one of them. Therefore, this experimental XRD structure was used to test the quality of the approximations for each functional. In this work, the optimized geometrical structure of Cyanidin was calculated with the DFT B3LYP [22], M06-2X and M06 [23] functionals with the 6-31G++(d,p) [29] basis set. The results from the regression analysis are shown in Table 1. The highest correlation coefficient (R^2) corresponds to the B3LYP/6-31++G(d,p) level of theory, which coincides with those reported by Sanchez-Bojorge et al. in 2015 [19].

Based on this analysis for Cyanidin, the methodology B3LYP/6-31G++(d,p) was used for the geometry optimization in DpS, E1, E2 and E3 molecules. The frequency analysis was performed to find the global minimal of the potential energy hypersurface and confirm the absence of imaginary frequencies. Thus, in order to start the Self-Consistent Field (SCF) calculations from the structural configuration closest to the global minimum energy, the ground state was obtained first in gas phase. Afterwards, the geometry optimizations and frequency analysis were done in aqueous phase, employing the IEFPCM method with the same level of theory for all the compounds. No imaginary frequencies were found in the final theoretical structures of the DpS and its esters.

Table 1. Bond lengths of cyanidin (gas phase) calculated with different methodologies and compared with experimental data reported by Ueno [28].

Distances * (Å)	Base 6-31++G(d,p)			
	Experimental	B3LYP	M06-2X	M06
C2-C1′	1.453	1.451	1.453	1.427
C4a-C5	1.432	1.432	1.429	1.411
C1′-C2′	1.409	1.417	1.411	1.413
C6-C7	1.413	1.418	1.417	1.401
C2-C3	1.396	1.408	1.402	1.410
C3′-C4′	1.400	1.415	1.411	1.394
C4-C4a	1.382	1.399	1.396	1.404
C6′-C1′	1.404	1.411	1.402	1.415
C4a-C8a	1.408	1.421	1.414	1.393
C7-C8	1.387	1.398	1.392	1.427
R^2	-	0.9329	0.9185	0.9193

* Structure is shown in Figure 1.

2.2.2. Reactivity Analyses

Chemical reactivity calculations were done in aqueous phase for all the molecules with the B3LYP/6-31++G(d,p) level of theory. To model the anthocyanin in pH < 2, the Fukui functions were calculated for the flavylium form of the DpS. The results for the DpS and its esters show the site where an electrophilic or nucleophilic attack could occur. Only one of the reactions can happen at a time. The nucleophilic attack will happen at the electrophilic site of the molecule and vice versa for the electrophilic attack. Thus, a nucleophilic molecule, like water, will attack the electrophilic site of the anthocyanin. The reaction of esterification, a form of electrophilic attack, will take place in the nucleophilic site of the anthocyanin. DpS local reactivity is located at C4 and O4′, the electrophilic and nucleophilic sites, respectively (Figure 3).

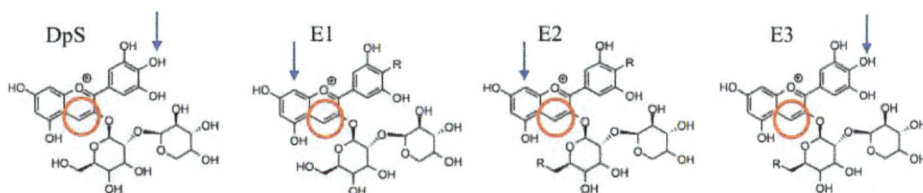

Figure 3. Local reactivity sites for anthocyanins and its esters calculated with the B3LYP/6-31++G(d,p) methodology. The red circle (C4) represents an electrophilic site and the blue arrow (O4′ or C8) represents a nucleophilic site.

The structure of E1 is based on the nucleophilic reactivity site of DpS. The esterification of DpS was done in the O4′. The reactivity calculations for E1 give the same electrophilic site (C4), but the nucleophilic site changed to C8. However, a second esterification of E1 at C8 is not possible due to the lack of oxygen, but it can take place at O7, which is the second nucleophilic site.

The geometry for E2 was chosen in accordance with experimental data obtained from LC/MS/MS (see Figure 3) and the reactivity calculations took into account the following facts. The mass spectrometry shows a di-ester: one ester where the octanoyl chain was joined to the flavylium moiety, and the other one in the sambubioside unit. The reactivity for E2 is presented in the C4 for the electrophilic site, and in C8 for the nucleophilic site. Similarly to E1, a second esterification cannot take place in the C8, instead it could take place at O7 or O5. The modeling for a di-ester with one chain joined to the B ring and another chain joined to the A ring was not calculated, in agreement with the experimental data.

The geometrical model for E3 was also chosen in accordance with mass spectrometry data, where it was observed that the octanoyl chain was joined to the sambubioside part. For this ester, the electrophilic and nucleophilic sites were in the C4 and O4', respectively. The reactivity for the E3 was equal to the native anthocyanin DpS.

The electrophilic site for the DpS, in the C4 of the C ring, agrees with the experimental results; whereas the principal nucleophilic site (O4') for the DpS does not match with previously reported data by Fulcrand [2] and de Freitas and Mateus [30] at the C6 or C8 of the A ring. In summary, the electrophilic site of all anthocyanin-esters remained in the position C4 of the C ring. Though, for E1 and E2, the nucleophilic site changed to the C8 in the A ring and for E3 it remained in the O4' of the B ring.

2.2.3. Excited States

Regarding the molecular structure modification, the change in the absorption spectra was analyzed in a theoretical study of pyranoanthocyanins [31] showing that the site of deprotonation could change the theoretical UV-Vis spectra. Thus, in DpS, the position where the hydrogen atom was substituted by the aliphatic chain also produces different spectra for the anthocyanin-esters. Thereby, the comparison of various theoretical spectra obtained with different functionals in this study will give qualitative insights into how the position of the lipophilization affects the absorbance of the molecule.

The excited states for DpS were calculated with the 6-31++G(d,p) basis set and the B3LYP, M06 and M06-2X functionals, employing the IEFPCM formalism in aqueous phase, where the calculated maximum wavelength (λ_{calc}) for each methodology were 487.7, 479.9 and 436.9 nm respectively. All the employed functionals presented a hypsochromic shift in their λ_{calc} for all the analyzed esters, when compared with the maximum absorption wavelength (λ_{calc}) of the DpS. The differences at λ_{exp} = 526 nm absorbance between the DpS and its esters encouraged the detailed comparative study of the predictive capabilities of each functional, within the TD-DFT methodology, which is presented below. A detailed description of previous functionals will be discussed in this section. The λ_{calc} closest to the experimental λ_{exp} value registered at 526 nm for the DpS were obtained with the B3LYP functional, which has a 20% HF exchange. In this manner, Table 2 shows the most important theoretical results for electronic transitions and their percentage of contribution, the corresponding vertical absorption wavelengths (λ_{calc}) and oscillator strengths (f) of the four analyzed molecules calculated particularly with the DFT: B3LYP/6-31++G(d,p) methodology.

Table 2. Vertical absorption wavelength (λ_{calc}), oscillator strength (f) and main electronic transitions for DpS, E1, E2 and E3, calculated with the B3LYP/6-31++G(d,p) level of theory in aqueous phase, employing the IEFPCM model [a].

Molecule		λ_{calc} (nm)	Oscillator Strength (f)	Electronic Transitions		
	1	487.7	0.5409	H-0→L+0 (+92%)	H-2→L+0 (+7%)	
DpS	2	463.0	0.0273	H-1→L+0 (+98%)		
	3	422.0	0.1237	H-2→L+0 (+89%)	H-0→L+0 (+8%)	
	1	484.0	0.0417	H-1→L+0 (+85%)	H-0→L+0 (+14%)	
E1	2	478.1	0.3598	H-0→L+0 (+76%)	H-1→L+0 (+15%)	H-2→L+0 (+8%)
	3	410.7	0.2969	H-2→L+0 (+89%)	H-0→L+0 (+9%)	
	1	485.4	0.0265	H-1→L+0 (+90%)	H-0→L+0 (+9%)	
E2	2	477.6	0.3769	H-0→L+0 (+81%)	H-1→L+0 (+10%)	H-2→L+0 (+8%)
	3	411.2	0.2943	H-2→L+0 (+89%)	H-0→L+0 (+9%)	
	1	486.9	0.5391	H-0→L+0 (+91%)	H-2→L+0 (+7%)	
E3	2	463.5	0.0280	H-1→L+0 (+99%)		
	3	422.4	0.1212	H-2→L+0 (+89%)	H-0→L+0 (+8%)	

[a] Delphinidin 3-O-sambubioside λ_{exp} at 526 nm.

Figure 4 shows the experimental and theoretical absorption spectra of DpS compared with E1, E2 and E3 at B3LYP/6-31++G(d,p) level of theory. It can be seen, that the normalized TD-DFT response of the oscillators of the native anthocyanin and its esters show UV-Vis differences; which are similar to the experimental response, for example, the absorbance intensity is diminished when an esterification has occurred. Although, the difference in the relative absorbances cannot be quantified, as the experimental UV-Vis spectra of DpS and its different esters were analyzed at different concentrations.

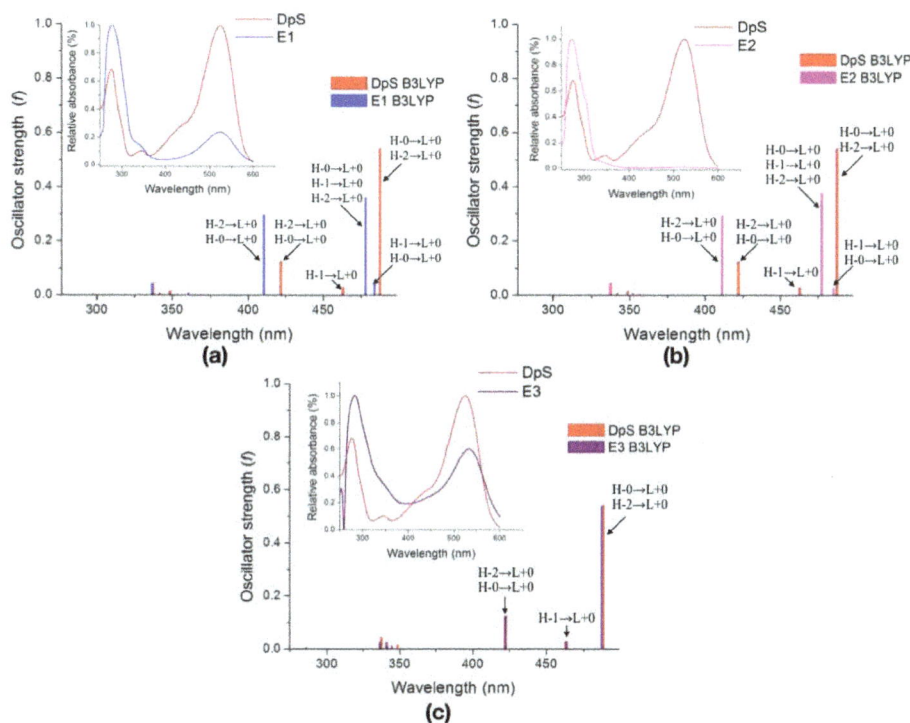

Figure 4. UV-Vis absorption spectra calculated with TD-DFT: B3LYP/6-31++G(d,p) of the DpS compared with the: (**a**) monoester E1; (**b**) di-ester E2; (**c**) monoester E3. [The inserts present the experimental UV-Vis spectra of the DpS contrasted with E1 (**a**), E2 (**b**) and E3 (**c**)].

The main experimental absorption band for DpS, which is around 520 nm, was assigned to the first excited state that corresponds to the theoretical H-0→L+0 and H-2→L+0 electronic transitions at 487.7 nm (f = 0.5409) (see Figure 4). The experimental UV-Vis spectrum of DpS presented a shoulder at 420 nm; it was assigned to the third excited state at 422 nm (f = 0.5391) and related to H-2→L+0 and H-0→L+0 electronic transitions.

When a lipophilization occurs in the sambubioside fragment (E3), the behavior in the theoretical UV-Vis spectrum is similar to the DpS spectrum, however, experimentally a minor hypochromic effect is observed in the anthocyanin-ester (See insert in Figure 4c). As mentioned before, a decrease in intensity of the absorbance was observed in the Malvidin 3-*O*-glucoside oleic acid conjugate reported by Cruz et al. in 2016 [4], with no loss of the color attractiveness.

E1 and E2 molecules display a different behavior compared with DpS and E3, although theoretically similar between them. It is noticed that for E1 the experimental band at 520 nm corresponds to two excited states, where the first state is mainly described by the H-1→L+0 electronic transition at 484 nm (f = 0.0417). Meanwhile, for E2, the same experimental band disappears from the spectrum (see insert in Figure 4b), but the theoretical calculations showed an excited state at 485.4 nm

(f = 0.0265) that mainly corresponds to the H-1→L+0 electronic transition. In the second excited state for E1 and E2, electronic transitions are mainly given by H-0→L+0 at 478.1 nm (f = 0.3598) and 477.6 nm (f = 0.3769) respectively. While, for E1 and E2, the excited states corresponding to the absorbance at 420 nm are related to the H-2→L+0 and H-0→L+0 electronic transitions at 410 (f = 0.2969) and 411.2 nm (f = 0.2943), respectively.

For the DpS and E3, the TD-DFT results show that the H-0→L+0 transition is the main contribution of the first excited state, in accordance with Woodford [32]. Nevertheless, using the B3LYP functional for the E1 and E2 a different configuration arises: the first excited state is represented mainly by the H-1→L+0 electronic transition, and the second excited state, is given by H-0→L+0 (the main electronic transition), which corresponds to the strongest oscillator force. This inversion could be explained when lipophilization occurs in the flavylium core, causing a hypochromic shift. Moreover, E1 and E2 second excited states present an increment from two to three electronic transitions, while E3 maintained only two transitions, as happened with DpS.

Other two functionals were tested in order to corroborate the trend of the obtained electronic transitions calculated by B3LYP. The M06 functional was one of them. It presented the first excited state for the DpS at 479.9 nm, which is 46 nm below the λ_{exp} = 526 nm (Table 3). The first excited state is associated with the H-0→L+0 and H-2→L+0 electronic transitions, with an oscillator strength value of 0.5886. Second and third excited states show equal electronic transitions than the ones obtained when the B3LYP functional was employed.

Table 3. Vertical absorption wavelength (λ_{calc}), oscillator strength (f) and main electronic transitions for DpS, E1, E2 and E3, calculated with the M06/6-31++G(d,p) level of theory in aqueous phase, employing the IEFPCM model [a].

Molecule		λ_{calc} (nm)	Oscillator Strength (f)	Electronic Transition	
	1	479.9	0.5886	H-0→L+0 (+93%)	H-2→L+0 (+7%)
DpS	2	442.9	0.0137	H-1→L+0 (+99%)	
	3	407.8	0.1136	H-2→L+0 (+89%)	H-0→L+0 (+7%)
	1	470.4	0.4380	H-0→L+0 (+92%)	H-2→L+0 (+7%)
E1	2	458.7	0.0151	H-1→L+0 (+99%)	
	3	398.1	0.2720	H-2→L+0 (+89%)	H-0→L+0 (+8%)
	1	469.9	0.4380	H-0→L+0 (+92%)	H-2→L+0 (+7%)
E2	2	460.1	0.0167	H-1→L+0 (+99%)	
	3	398.5	0.2719	H-2→L+0 (+89%)	H-0→L+0 (+8%)
	1	469.9	0.4750	H-0→L+0 (+91%)	H-2→L+0 (+8%)
E3	2	465.5	0.0094	H-1→L+0 (+99%)	
	3	402.7	0.1793	H-2→L+0 (+87%)	H-0→L+0 (+8%)

[a] Delphinidin 3-*O*-sambubioside λ_{exp} at 526 nm.

The M06 functional, with a 27% of HF exchange, does not show the inversion in the excited states presented by B3LYP functional between the first and second excited states for E1 and E2. Thus, the M06 functional presents only a hypsochromic shift in the λ_{calc} when the esterification occurs either in the flavylium moiety or in the sambubioside unit, without any other outstanding changes in the electronic transitions (Figure 5). The other functional used for comparison with B3LYP was M06-2X. Table 4 shows the relevant theoretical information about the electronic transitions, the calculated vertical absorption wavelengths and the oscillator strength at M06-2X/6-31++G(d,p).

The first excited state for DpS calculated with M06-2X/6-31++G(d,p) level of theory is assigned to H-0→L+0 and H-2→L+0 electronic transitions at 436.9 nm (f = 0.6474), which means a severe shift from λ_{exp} = 526 nm (Figure 6). This difference in maximum absorption wavelength can be attributed to the HF exchange of the M06-2X functional (54%), which is twice the amount compared to the B3LYP or the M06 functionals. The second excited state, corresponding to the experimental band at 420 nm, is assigned to H-1→L+0 and H-2→L+0 electronic transitions, displayed at 357.2 nm (f = 0.1095), and it also still shows a shift from the experimental value.

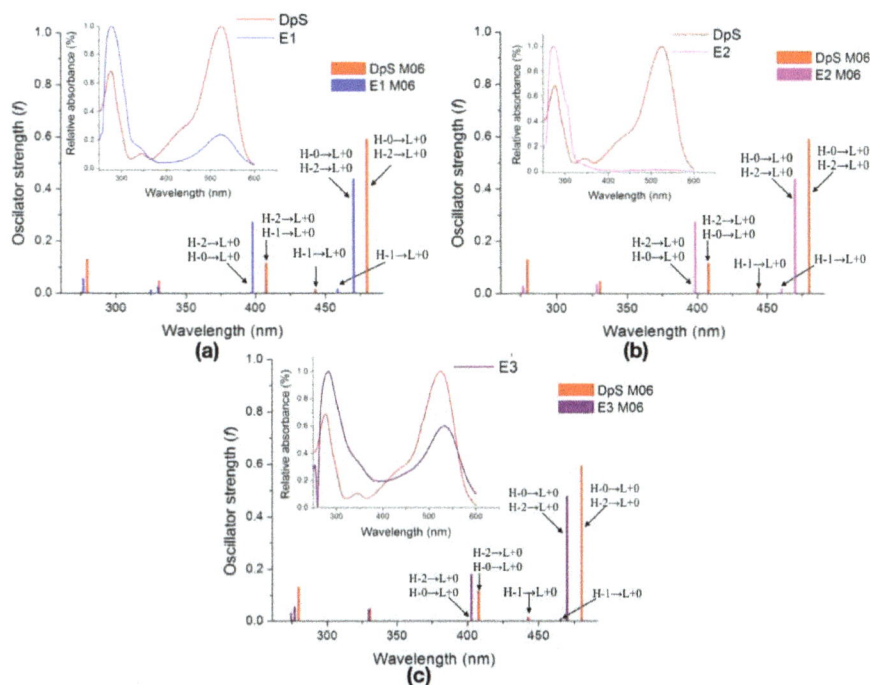

Figure 5. UV-Vis absorption spectra calculated with TD-DFT: M06/6-31++G(d,p) of the DpS compared with the: (**a**) monoester E1; (**b**) di-ester E2; (**c**) the monoester E3. [The inserts present the experimental UV-Vis spectra of the DpS contrasted with E1 (**a**), E2 (**b**) and E3 (**c**)].

Table 4. Vertical absorption wavelength (λ_{calc}), oscillator strength (f) and main electronic transitions for DpS, E1, E2 and E3, calculated with the M06-2X/6-31++G(d,p) level of theory in aqueous phase, employing the IEFPCM model [a].

Molecule		λ_{calc} (nm)	Oscillator Strength (f)	Electronic Transition		
DpS	1	436.9	0.6474	H-0→L+0 (+92%)	H-2→L+0 (+6%)	
	2	357.2	0.1095	H-1→L+0 (+50%)	H-2→L+0 (+39%)	
	3	350.3	0.0532	H-2→L+0 (+49%)	H-1→L+0 (+46%)	
E1	1	429.5	0.5450	H-0→L+0 (+93%)	H-2→L+0 (+6%)	
	2	373.5	0.0188	H-1→L+0 (+97%)		
	3	349.7	0.2484	H-2→L+0 (+86%)	H-0→L+0 (+5%)	
E2	1	428.9	0.5452	H-0→L+0 (+93%)	H-2→L+0 (+6%)	
	2	374.0	0.0182	H-1→L+0 (+97%)		
	3	349.8	0.2515	H-2→L+0 (+85%)	H-0→L+0 (+5%)	
E3	1	429.8	0.5582	H-0→L+0 (+91%)	H-2→L+0 (+7%)	
	2	378.1	0.0161	H-1→L+0 (+97%)		
	3	352.5	0.1833	H-2→L+0 (+83%)	H-0→L+0 (+7%)	H-4→L+0 (+5%)

[a] Delphinidin 3-*O*-sambubioside λ_{exp} at 526 nm.

Based on the results reported in Table 4, when the anthocyanin-ester presents an esterification, either in the flavylium core (E1), in the sambubioside (E3) or both (flavylium core and sambubioside) (E2), the first excited state preserves the same electronic transition as the DpS; while the oscillator strength displays a diminished value compared with the DpS at same level of theory. The second excited state for E1, E2 and E3 presents only the H-1→L+0 electronic transition at 373.5 (f = 0.0188), 374 (f = 0.0182) and 378.1 nm (f = 0.0161) respectively.

The diminishing in the oscillator strength of the anthocyanin-esters is noticeable for all functionals, compared with the oscillator strength of the DpS. This marked effect is not observed for E3 when it is calculated with B3LYP/6-31++G(d,p) level of theory, where the oscillator strength and the λ_{calc} remained similar to the values of DpS.

In summary, the results of M06-2X are similar to the ones of M06 functional, and the results of both meta-GGA are different from B3LYP, which presented an inversion of excited states. In addition, the B3LYP functional has been reported [19] to correlate better with the experimental data for the absorption properties.

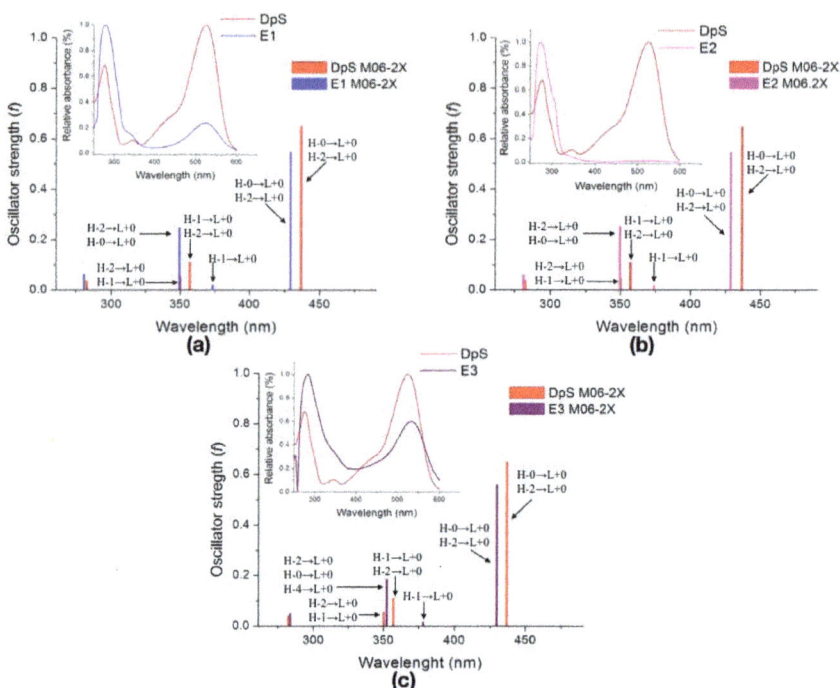

Figure 6. UV-Vis absorption spectra calculated with TD-DFT: M06-2X/6-31++G(d,p) of the DpS compared with the: (**a**) monoester E1; (**b**) di-ester E2; (**c**) the monoester E3. [The inserts present the experimental UV-Vis spectra of the DpS contrasted with E1 (**a**), E2 (**b**) and E3 (**c**)].

The B3LYP functional explained better the UV-Vis spectra than Minnesota functionals, since the HF contribution is only 20% for B3LYP [33]. This implies that the LYP correlation energy considered in the B3LYP functional, which is partially responsible for the covalent and conjugated nature of our molecules, adjusts better than the one from the M06 family (where HF exchange are between 27% and 54%) [23]. This result coincides with the one reported by Anouar [18], where the B3P86 and B3LYP functionals show the best performance for phenolic compounds.

2.2.4. Molecular Orbitals

In order to gain insight into the origin of the differences between the esterified molecules, the B3LYP/6-31++G(d,p) methodology (the one which gave a structure nearest to experimental XRD data) was employed to perform the following molecular orbitals (MOs) calculations.

Highest occupied molecular orbital (HOMO) and lowest unoccupied molecular orbital (LUMO) maps corresponding to the principal electronic transition (H-0→L+0) were estimated through a single point calculation. Computed MOs are all distributed along the anthocyanin structures, as shown in Figure 7. It seems there are no significant differences between MOs of the DpS molecule and its esters. However, there is a slight dissimilarity in their spatial distributions, whereas DpS and E3 exhibit some electronic density at O5′, E1 and E2 do not (Figure 7). The HOMO-LUMO gap for the DpS (2.852 eV) is similar to the one mentioned by Rustioni [34] (2.87 eV).

The HOMO-LUMO gaps are also almost equal in all the analyzed structures. Although, when the esterification occurs in the sambubioside unit (E3), its HOMO-LUMO energy gap is close to the one of DpS.

Comparing these findings with the previous reactivity results, it can be added that the electrophilic sites for all the studied molecules in this work correspond to the site of the largest LUMO density, and the nucleophilic sites are related to the largest HOMO density. Thus, although no significant differences are shown in the calculated MOs for DpS and E3, these present different nucleophilic behaviors than E1 or E2, which were esterified in the B ring. Chemical reactivity plus MOs analysis might clarify the fact that, any insignificant change occurring in the principal electronic transitions could modify the macroscopic properties of the molecules.

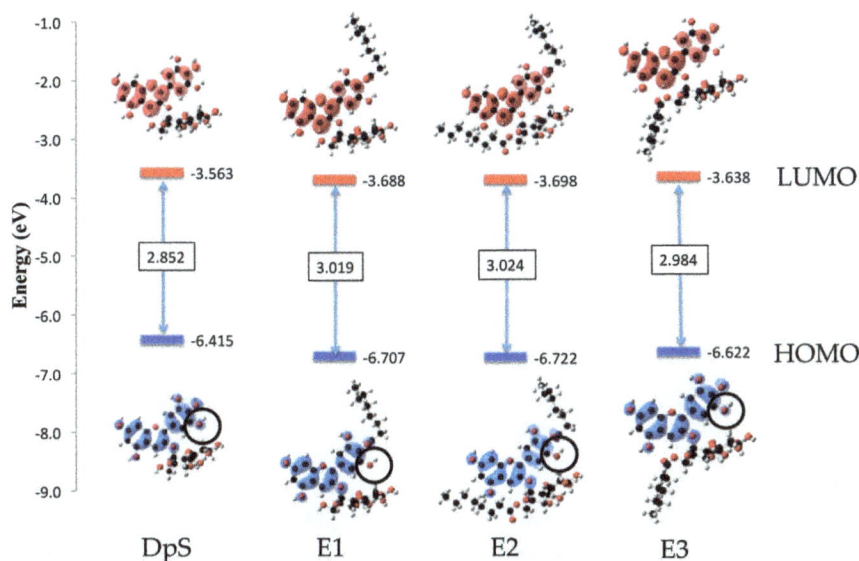

Figure 7. Molecular orbitals and HOMO-LUMO gaps for delphinidin 3-*O*-sambubioside and its esters. The black circles highlight the position where the HOMO is affected by the esterification at ring B.

3. Materials and Methods

3.1. Experimental Methodology

3.1.1. Anthocyanin Extraction and Purification

The purification of the hibiscus anthocyanins was described by Grajeda-Iglesias [35]. Briefly, 100 g of grounded powder of *Hibiscus sabdariffa* L. was mixed with 250 mL of acidified water-ethanol (80–20, *v/v*) and then placed in a cold ultrasonic bath for 30 min. The obtained extract was then filtered, freeze-dried and stored at −20 °C. The fractionation of the phenolic extract was carried out by solid phase extraction (SPE) using a Sep-Pak® cartridges (35 cc vac, 10 g, 55–105 µm; Millipore Waters, Milford, MA, USA);

200 mg of the hibiscus extract was dissolved in 2 mL of acidified water (acetic acid 5%, v/v) and then placed on a cartridge previously activated with methanol and equilibrated with acidified water. The elution was done by increasing the percentage of methanol in an alcohol-water mixture. Two main fractions were obtained from the SPE, the fraction 2 (F2) containing DpS and fraction 3 (F3) containing cyanidin 3-*O*-sambubioside. Each fraction obtained was evaporated under vacuum at 32 °C and freeze-dried for further purification. Semipreparative HPLC was used to purify delphinidin 3-*O*-sambubioside. ^{1}H-NMR (400 MHz) and ^{13}C-NMR analysis were performed to confirm structure and purity [35].

3.1.2. Anthocyanins Lipophilization

Lipophilization of the purified DpS was performed following the experimental procedure described in [5]. Briefly, a purified fraction of DpS (0.0168 mmol equiv) was completely dissolved in 1 mL of DMF. Ten equivalents of octanoyl chloride were incorporated into the anthocyanin solution; the reaction was left for 24 h at room temperature in an inert atmosphere in the dark. The reaction progress was monitored on HPLC (HPLC-ESI-MS, Thermo Fisher, San Jose, CA, USA), and it was stopped adding the solution into a Sep-Pak C18, previously activated with methanol and equilibrated with acidified water. Three different fractions were obtained by sequentially adding acidified water, ethyl acetate and methanol. The fractions were rotaevaporated until dryness and recovered in methanol-acidified water (50–50 v/v).

3.2. Theoretical Methodology

3.2.1. Methodology Selection

The hybrid DFT [12] functionals B3LYP [22,33,36], M06 and M06-2X [23] with the 6-31G++(d,p) [29] basis set were employed to find the best correlation between theoretical data of cyanidin and X-ray diffraction (XRD) data of cyanidin bromide [28]. The best correlation between experimental and theoretical methodologies was used to select the method for obtaining the geometrical structure of the DpS. The selected method B3LYP/6-31++G(d,p) was employed for the geometrical structure optimization of delphinidin 3-*O*-Sambubioside (DpS), delphinidin 3-*O*-sambubioside monoester (E1), delphinidin 3-*O*-sambubioside di-ester (E2) and delphinidin 3-*O*-sambubioside monoester (E3), shown in Figure 1. The optimized parameters of ground state structure of each compound were used to calculate analytical frequencies at the same level of theory, where was confirmed the absence of imaginary frequencies.

3.2.2. Local Reactivity

Fukui functions $f_k(r)$ given in Equation (1) [37] were employed to estimate the local reactivity in the DpS, E1, E2 and E3 molecules. Fukui functions for specific atom k are computed by:

$$f_k(r) = f_k^+(r) + f_k^-(r), \tag{1}$$

$$f_k^+(r) = q_k(N+1) - q_k(N), \tag{2}$$

$$f_k^-(r) = q_k(N) - q_k(N-1), \tag{3}$$

where: $f_k^+(r)$ is the electrophilic contribution (Equation (2)) and $f_k^-(r)$ is the nucleophilic contribution (Equation (3)).

B3LYP/6-31++G(d,p) level of theory was employed with the IEFPCM [38] in aqueous phase to obtain the Hirshfeld population analysis and finding the electrophilic and nucleophilic sites of all molecules.

3.2.3. Excited States

Geometry structure of anthocyanins and its esters were obtained employing the B3LYP/l6-31++G(d,p) methodology, calculated first in gas phase, through the approximation of an isolated molecule; while the solvent effect was determined with the polarizable continuum model (PCM), using the IEFPCM formalism in aqueous phase. Then, the excited states of anthocyanins and its esters were obtained by TD-DFT [16], using the B3LYP, M06 and M06-2X functionals with the 6-31++G(d,p) basis set, in order to calculate the UV-Vis spectra. The time-dependent equations were solved for ten excited states and employing the IEFPCM model in water to find the solvent effect for each molecule.

3.2.4. Molecular Orbitals

The HOMO (High occupied molecular orbital) energy and LUMO (Lower unoccupied molecular orbital) energy and the HOMO-LUMO energy gap for each pigment were obtained from a single point calculation for the ground state, employing the IEFPCM model in aqueous phase. B3LYP/6-31++G(d,p) level of theory was used for MOs calculations.

Gaussian16 program was employed to compute all the theoretical calculations [39]. The complete methodology is summarized in Figure 8.

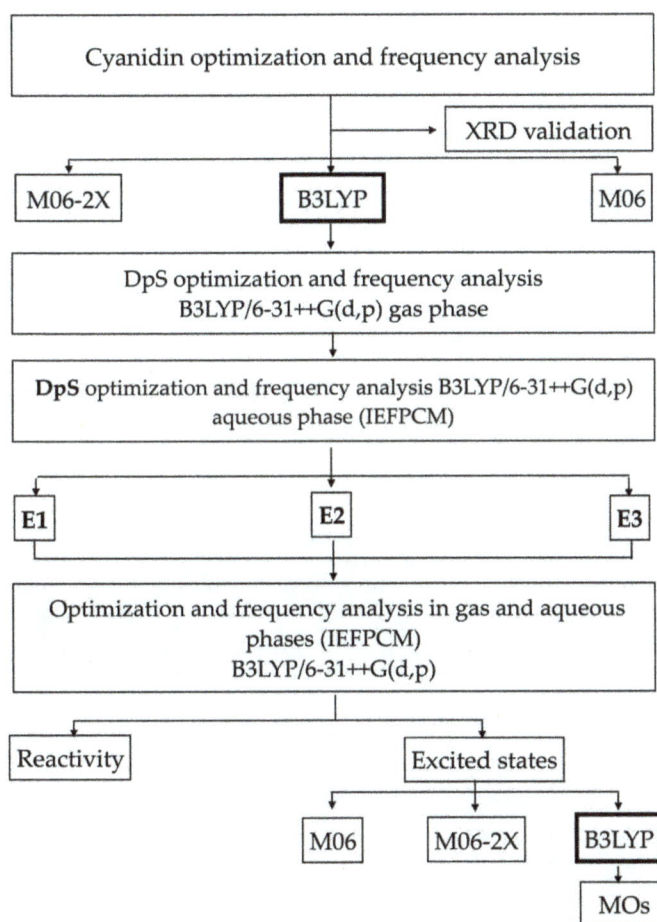

Figure 8. DTF and TD-DFT calculation scheme used to characterize DpS and its esters. Functional showing the best correlation is in a bold chart.

Molecules **2018**, *23*, 1587

4. Conclusions

Density Functional Theory gives a useful model when trying to obtain information about molecular behavior, which cannot be fully characterized by experimental procedures. Particularly in this work, DFT: B3LYP/6-31++G(d,p) presents the best correlation for the geometrical structure parameters, while the TD-DFT: B3LYP/6-31++G(d,p) methodology gives a view of the quantic landscape of the electronic behavior of lipophilized anthocyanins.

B3LYP is a high correlation functional, which describes better the MOs involved in the UV-Vis spectra of the anthocyanin-esters due to the conjugated nature of the flavylium moiety. For instance, it shows that neither of the esters contribute to the HOMO and LUMO. Although, the esterification of the anthocyanins may slightly compromise the electron resonance along the flavylium core, changing the transition at the λ_{calc} and the number of transitions in the excited state with the strongest oscillator strength.

The reactivity in anthocyanin-esters could change from the anthocyanin precursor if the lipophilization occurs in the flavylium core. E3 maintains the same excited states and electronic transitions as the DpS. From all tested methodologies, theoretical results of the E3 at TD-DFT: B3LYP/6-31++G(d,p) demonstrated that the lipophilization in the sambubioside fragment prevents the loss of absorbance intensity in the anthocyanin-esters; since its excited states remained the same as its precursor DpS. For this reason, the lipophilization of anthocyanins in the sugar moiety could improve the solubility in lipidic systems without significantly diminishing their color.

Although DFT is a reliable method, the exploration of the potential energy hypersurface is a complex numerical problem. The challenge involved in the ground state calculation is further incremented when adding the complexity of the resonant structure of the flavylium cation. Also, TD-DFT is known to have a good response when the excited structure does not deviate far from the ground state structure, and the chosen functional is adequate for the system under study. These considerations contribute to understanding the result that B3LYP functional is the one that better reproduces the experimental maximum wavelength λ_{exp}.

Another strength of the study, is the use of water as solvent in the calculations in order to mimic real experimental conditions. Lastly, one of the main contributions of this work is that experimental data was used to assist in the building of the molecular model, using DFT based quantum-chemical methods, meaning that mass spectrometry fragmentation data gave the information of the possible position of the esterification.

Author Contributions: M.-C.F.-E., L.-M.R.-V., M.E.F.-M. and E.S. designed the experiments; C.G.-I. performed the experiments; A.S.M.-R. performed the theoretical calculations; N.-A.S.-B., L.M.R.-V. and M.E.F.-M. designed the theoretical analysis; and A.S.M.-R., N.-A.S.-B., L.-M.R.-V., M.E.F.-M. and E.S. wrote the manuscript.

Funding: This research received no external funding

Acknowledgments: Authors wish to thank the Mexican Council for Science and Technology (CONACYT) for financial support of a doctoral scholarship and UACH for financial support.

Conflicts of Interest: The authors declare no conflicts of interest.

References

1. Brouillard, R. Chemical structure of anthocyanins. In *Anthocyanins as Food Colors*, 1st ed.; Markakis, P., Ed.; Academic Press: New York, NY, USA, 1982; pp. 1–40.
2. Fulcrand, H.; Atanasova, V.; Salas, E.; Cheynier, V. The fate of anthocyanins in wine: Are there determining factors? *Red Wine Color Reveal. Myster.* **2004**, *886*, 68–88. [CrossRef]
3. Castañeda-Ovando, A.; de Lourdes Pacheco-Hernández, M.; Páez-Hernández, M.E.; Rodríguez, J.A.; Galán-Vidal, C.A. Chemical studies of anthocyanins: A review. *Food Chem.* **2009**, *113*, 859–871. [CrossRef]
4. Cruz, L.; Fernandes, I.; Guimarães, M.; de Freitas, V.; Mateus, N. Enzymatic synthesis, structural characterization and antioxidant capacity assessment of a new lipophilic malvidin-3-glucoside–oleic acid conjugate. *Food Funct.* **2016**, *7*, 2754–2762. [CrossRef] [PubMed]

5. Grajeda-Iglesias, C.; Salas, E.; Barouh, N.; Baréa, B.; Figueroa-Espinoza, M.C. Lipophilization and MS characterization of the main anthocyanins purified from hibiscus flowers. *Food Chem.* **2017**, *230*, 189–194. [CrossRef] [PubMed]

6. Sakata, K.; Saito, N.; Honda, T. Ab initio study of molecular structures and excited states in anthocyanidins. *Tetrahedron* **2006**, *62*, 3721–3731. [CrossRef]

7. Di Meo, F.; Sancho Garcia, J.C.; Dangles, O.; Trouillas, P. Highlights on anthocyanin pigmentation and copigmentation: A matter of flavonoid π-stacking complexation to be described by DFT-D. *J. Chem. Theory Comput.* **2012**, *8*, 2034–2043. [CrossRef] [PubMed]

8. Chassaing, S.; Lefeuvre, D.; Jacquet, R.; Jourdes, M.; Teissedre, P.; Dangles, O.; Quideau, S. Physicochemical Studies of New Anthocyano-Ellagitannin Hybrid Pigments: About the Origin of the Influence of Oak C-Glycosidic Ellagitannins on Wine. *Eur. J. Org. Chem.* **2010**, *2010*, 55–63. [CrossRef]

9. Trouillas, P.; Di Meo, F.; Gierschner, J.; Linares, M.; Sancho-García, J.C.; Otyepka, M. Optical properties of wine pigments: theoretical guidelines with new methodological perspectives. *Tetrahedron* **2015**, *71*, 3079–3088. [CrossRef]

10. Kurtin, W.E.; Song, P.S. Electronic structures and spectra of some natural products of theoretical interest-I. Molecular orbital studies of anthocyanidins. *Tetrahedron* **1968**, *24*, 2255–2267. [CrossRef]

11. Roothaan, C.C.J. New developments in molecular orbital theory. *Rev. Mod. Phys.* **1951**, *23*, 69–89. [CrossRef]

12. Hohenberg, P.; Kohn, W. Inhomogeneous electron gas. *Phys. Rev.* **1964**, *136*, B864–B871. [CrossRef]

13. Ridley, J.; Zerner, M. An intermediate neglect of differential overlap technique for spectroscopy: Pyrrole and the azines. *Theor. Chim. Acta* **1973**, *32*, 111–134. [CrossRef]

14. Dewar, M.J.S.; Zoebisch, E.G.; Healy, E.F.; Stewart, J.J.P. Development and use of quantum mechanical molecular models. 76. AM1: A new general purpose quantum mechanical molecular model. *J. Am. Chem. Soc.* **1985**, *107*, 3902–3909. [CrossRef]

15. Freitas, A.A.; Shimizu, K.; Quina, F.H. A Computational study of substituted flavylium salts and their quinonoidal conjugate-bases: S0 → S1 electronic transition, absolute pK. *J. Braz. Chem. Soc.* **2007**, *18*, 1537–1546. [CrossRef]

16. Runge, E.; Gross, E.K.U. Density-functional theory for time-dependent systems. *Phys. Rev. Lett.* **1984**, *52*, 997–1000. [CrossRef]

17. Quartarolo, A.D.; Russo, N. A computational study (TDDFT and RICC2) of the electronic spectra of pyranoanthocyanins in the gas phase and solution. *J. Chem. Theory Comput.* **2011**, *7*, 1073–1081. [CrossRef] [PubMed]

18. Anouar, E.H.; Gierschner, J.; Duroux, J.L.; Trouillas, P. UV/Visible spectra of natural polyphenols: A time-dependent density functional theory study. *Food Chem.* **2012**, *131*, 79–89. [CrossRef]

19. Sanchez-Bojorge, N.A.; Rodriguez-Valdez, L.M.; Glossman-Mitnik, D.; Flores-Holguin, N. Theoretical calculation of the maximum absorption wavelength for Cyanidin molecules with several methodologies. *Comput. Theory Chem.* **2015**, *1067*, 129–134. [CrossRef]

20. Mohr, T.; Aroulmoji, V.; Ravindran, R.S.; Müller, M.; Ranjitha, S.; Rajarajan, G.; Anbarasan, P.M. DFT and TD-DFT study on geometries, electronic structures and electronic absorption of some metal free dye sensitizers for dye sensitized solar cells. *Spectrochim. Acta Part A Mol. Biomol. Spectrosc.* **2015**, *135*, 1066–1073. [CrossRef] [PubMed]

21. Tomasi, J.; Mennucci, B.; Cammi, R. Quantum mechanical continuum solvation models. *Chem. Rev.* **2005**, *105*, 2999–3093. [CrossRef] [PubMed]

22. Becke, A.D. Density-functional thermochemistry. III. The role of exact exchange. *J. Chem. Phys.* **1993**, *98*, 5648–5652. [CrossRef]

23. Zhao, Y.; Truhlar, D.G. The M06 suite of density functionals for main group thermochemistry, thermochemical kinetics, noncovalent interactions, excited states, and transition elements: Two new functionals and systematic testing of four M06-class functionals and 12 other function. *Theor. Chem. Acc.* **2008**, *120*, 215–241. [CrossRef]

24. De Castro, V.C.; da Silva, P.H.A.; de Oliveira, E.B.; Desobry, S.; Humeau, C. Extraction, identification and enzymatic synthesis of acylated derivatives of anthocyanins from jaboticaba (Myrciaria cauliflora) fruits. *Int. J. Food Sci. Technol.* **2014**, *49*, 196–204. [CrossRef]

25. Trouillas, P.; Sancho-García, J.C.; de Freitas, V.; Gierschner, J.; Otyepka, M.; Dangles, O. Stabilizing and Modulating Color by Copigmentation: Insights from Theory and Experiment. *Chem. Rev.* **2016**, *116*, 4937–4982. [CrossRef] [PubMed]

26. Nave, F.; Petrov, V.; Pina, F.; Teixeira, N.; Mateus, N.; de Freitas, V. Thermodynamic and kinetic properties of a red wine pigment: Catechin-(4,8)-malvidin-3-*O*-glucoside. *J. Phys. Chem. B* **2010**, *114*, 13487–13496. [CrossRef] [PubMed]

27. Borkowski, T.; Szymusiak, H.; Gliszczynska-Swiglo, A.; Tyrakowska, B. The effect of 3-*O*-β-glucosylation on structural transformations of anthocyanidins. *Food Res. Int.* **2005**, *38*, 1031–1037. [CrossRef]

28. Ueno, K. Cyanidinin bromide monohydrate (3,5,7,3',4'-pentahydrxyflavylium bromide monohydrate). *Acta Crystallogr.* **1977**, *114*, 114–116. [CrossRef]

29. Ditchfield, R.; Hehre, W.J.; Pople, J.A. Self-Consistent Molecular-Orbital Methods. IX. An Extended Gaussian-Type Basis for Molecular-Orbital Studies of Organic Molecules. *J. Chem. Phys.* **1971**, *54*, 724–728. [CrossRef]

30. De Freitas, V.; Mateus, N. Chemical transformations of anthocyanins yielding a variety of colours (Review). *Environ. Chem. Lett.* **2006**, *4*, 175–183. [CrossRef]

31. Vallverdú-Queralt, A.; Biler, M.; Meudec, E.; le Guernevé, C.; Vernhet, A.; Mazauric, J.; Legras, J.; Loonis, M.; Trouillas, P.; Cheynier, V.; et al. p-Hydroxyphenyl-pyranoanthocyanins: An experimental and theoretical investigation of their acid-base properties and molecular interactions. *Int. J. Mol. Sci.* **2016**, *17*, 1842. [CrossRef] [PubMed]

32. Woodford, J.N. A DFT investigation of anthocyanidins. *Chem. Phys. Lett.* **2005**, *410*, 182–187. [CrossRef]

33. Lee, C.; Yang, W.; Parr, R.G. Development of the Colle-Salvetti correlation-energy formula into a functional of the electron density. *Phys. Rev. B* **1988**, *37*, 785–789. [CrossRef]

34. Rustioni, L.; di Meo, F.; Guillaume, M.; Failla, O.; Trouillas, P. Tuning color variation in grape anthocyanins at the molecular scale. *Food Chem.* **2013**, *141*, 4349–4357. [CrossRef] [PubMed]

35. Grajeda-Iglesias, C.; Figueroa-Espinoza, M.C.; Barouh, N.; Baréa, B.; Fernandes, A.; de Freitas, V.; Salas, E. Isolation and Characterization of Anthocyanins from Hibiscus sabdariffa Flowers. *J. Nat. Prod.* **2016**, *79*, 1709–1718. [CrossRef] [PubMed]

36. Miehlich, B.; Savin, A.; Stoll, H.; Preuss, H. Results obtained with the correlation energy density functionals of becke and Lee, Yang and Parr. *Chem. Phys. Lett.* **1989**, *157*, 200–206. [CrossRef]

37. Yang, W.; Mortier, W.J. The Use of Global and Local Molecular Parameters for the Analysis of the Gas-Phase Basicity of Amines. *J. Am. Chem. Soc.* **1986**, *108*, 5708–5711. [CrossRef] [PubMed]

38. Cancès, E.; Mennucci, B.; Tomasi, J. A new integral equation formalism for the polarizable continuum model: Theoretical background and applications to Isotropic and anisotropic dielectrics. *J. Chem. Phys.* **1997**, *107*, 3032–3041. [CrossRef]

39. Frisch, M.J.; Trucks, G.W.; Schlegel, H.B.; Scuseria, G.E.; Robb, M.A.; Cheeseman, J.R.; Scalmani, G.; Barone, V.; Petersson, G.A.; Nakatsuji, H.; et al. *Gaussian 16, Revision B.01*; Gaussian, Inc.: Wallingford, CT, USA, 2016.

Sample Availability: Samples of the compounds are not available from the authors.

molecules

MDPI

Article

Bioactive Phytochemicals and Antioxidant Properties of the Grains and Sprouts of Colored Wheat Genotypes

Oksana Sytar [1,2,*], Paulina Bośko [3], Marek Živčák [1], Marian Brestic [1] and Iryna Smetanska [4]

[1] Department of Plant Physiology, Slovak University of Agriculture in Nitra, A. Hlinku 2,
    949 76 Nitra, Slovakia; marek.zivcak@uniag.sk (M.Z.); marian.brestic@uniag.sk (M.B.)
[2] Department of Plant Biology, Educational and Scientific Center "Institute of Biology and Medicine", Taras
    Shevchenko National University of Kyiv, Hlushkova Avenue, 2, 03127 Kyiv, Ukraine
[3] Department of Pig Breeding, Animal Nutrition and Food, West Pomeranian University of Technology in
    Szczecin, Klemensa Janickiego 29, 71-270 Szczecin, Poland; paulina.bosko@zut.edu.pl
[4] Plant Production and Processing, University of Applied Sciences Weihenstephan-Triesdorf, Markgrafenstr
    16, 91746 Weidenbach, Germany; iryna.smetanska@hswt.de
*   Correspondence: oksana.sytar@gmail.com; Tel.: +421-373414822

Academic Editors: Monica Giusti and Gregory T. Sigurdson
Received: 3 August 2018; Accepted: 6 September 2018; Published: 6 September 2018

Abstract: The grains and sprouts of colored wheat genotypes (having blue, purple and yellow colored grains) contain specific anthocyanidins, such as pelargonidin and cyanidin derivatives, that produce beneficial health effects. The objective of the presented study is to compare the antioxidant capacity and contents of bioactive phytochemicals in grains and sprouts of wheat genotypes that differ in grain color. The methods α, α-diphenyl-β-picrylhydrazyl (DPPH) and 2,2′-azino-bis (3-ethylbenzothiazoline-6-sulphonic acid) (ABTS) scavenging activities, together with spectrophotometrical and high-performance thin-layer chromatography (HPTLC) methods, were used to study the presence of total phenolics, flavonoids, anthocyanins and anthocyanidins (pelargonidin, peonidin, cyanidin, delphinidin) content. It was predicted that the sprouts of all colored wheat genotypes would have significantly higher total flavonoids, total phenolics, anthocyanidin levels and antioxidant activity than the grains. The correlation results between antioxidant activity and contents of bioactive phytochemicals in grains and sprouts of colored wheat genotypes have shown a high correlation for cyanidin and pelargonidin, especially in grains, as well as quercetin in sprouts. It was found that total anthocyanin, quercetin and pelargonidin contents were significantly higher in the sprouts of the purple wheat genotypes than in the blue or yellow wheat genotypes. Delphinidin was detected at a higher level in the grains than in the sprouts of the blue wheat genotypes. Peonidin was present at very low quantities in the grains of all colored wheat genotypes. The sprouts of the purple wheat genotypes, among the colored wheat genotypes, had the highest pelargonidin, cyanidin and quercetin contents and, therefore, can be a promising source for functional food use.

Keywords: wheat; anthocyanidins; pelargonidin; cyanidin; antioxidants; antioxidant activity

---

1. Introduction

Increasing interest in the health benefits of whole wheat grain has encouraged breeders to further enhance the antioxidant contents of cereal plants [1–3]. The phenolic compounds are of considerable interest due to their antioxidant properties, including anthocyanins as an important subgroup of phenolic antioxidants. Phenolic compounds from purple corn are known to have anti-inflammatory, anti-carcinogenic and anti-angiogenesis properties and have great potential as food colorants [1,4,5]. Natural anthocyanin antioxidants, due to strong antioxidant capacities, have been found to mitigate lifestyle diseases, such as diabetes, obesity, hyperglycemia, hypertension and cardiovascular diseases [6,7].

Recently, the development of colored wheat genotypes among different cereal genetic resources has begun due to their high anthocyanin contents. The presence of secondary metabolites, such as tannins, polyphenols, carotenoids and anthocyanins, in wheat grains leads to a variety of grain colors. The presence of anthocyanins is determined by the *Ba* and *Pp* genes, which are responsible for the blue aleurone and purple pericarp grain colors, respectively. The yellow endosperm (*Psy* genes) of wheat grain has been shown to be regulated by the presence of carotenoids [8]. The use of pigmented wheat grains support the enrichment of total dietary fiber, phenolic acids, anthocyanins and β-glucans in the wheat-bran fraction. The presence of higher levels of secondary metabolites in the wheat-bran fraction induces higher total antioxidant activity than observed in refined flour. For example, lutein is the most abundant carotenoid in refined flour [9]. A high content of biologically active compounds, which correlates to high antioxidant activity in colored wheat genotypes, may support their use as stress-tolerant genotypes [10] or in functional food production.

The *Pp* genes acquired from the tetraploid wheat *Triticum turgidum* L. subsp. *abyssinicum* Vavilov (origin-Abyssinian region, Ethiopia) are responsible for the purple color of wheat grain, which is characterized by a high content of anthocyanins in pericarp. The cyanidins 3-glucoside, cyanidin 3-rutinoside and cyanidin 3-(6"-succinyl-glucoside) are the main anthocyanidin-3-*O*-glycosides in purple wheat grains [11,12]. The difference between purple and blue grains is due to differences in the anthocyanin composition, which is especially visible in the cross sections of kernels [11]. The dominant anthocyanins for blue aleurone wheat are delphinidin 3-rutinoside and delphinidin 3-glucoside. Cyanidin 3-glucoside and cyanidin 3-rutinoside are present in blue aleurone wheat grains, but in smaller quantities than in the purple grains [12].

The edible part of the grain of cereals is the caryopsis [3], which, in the commercially available purple wheat exhibited exceptional antioxidant capacities based on scavenging of DPPH, ABTS activities and peroxyl radicals [13]. Anthocyanin-3-*O*-glycosides-rich biscuits from purple wheat showed a high level of cyanidin 3-*O*-glucoside and exhibited high antioxidant activity [5,14]. There is currently great interest in the application of germination processes that can significantly enhance the dietary and health benefits of grains via increasing the content of bioactive phytochemicals in the sprouts [15]. Moderately hydrophobic polyphenolic fractions and hydrophilic peptidic fractions were found in wheat sprouts [16]. The presence of total phenolic compound content is correlated with high antioxidant activity [17]. The high antioxidant capacity of wheat sprouts can support their use as food supplements to act against diseases induced by free radicals [18].

The grains and sprouts of colored wheat genotypes may be potential sources of natural anthocyanin antioxidants. Research on grain composition is more available than research on sprouts. Therefore, the objective of the presented experimental work was to compare antioxidant capacity and contents of bioactive phytochemicals in grains and sprouts of wheat genotypes differing in grain color.

2. Results

2.1. Antioxidant Activity and Total Phenolic, Flavonoid and Anthocyanin Contents in Grains and Sprouts

Antioxidant activity in the grains and sprouts of colored wheat genotypes has been studied in the presented experimental work (Figure 1).

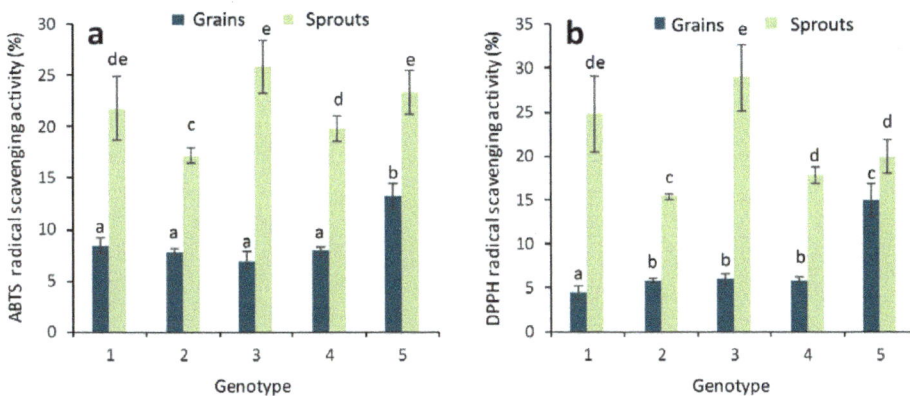

Figure 1. Antioxidant activity (ABTS test) (**a**) and (DPPH test) (**b**) in the grains and sprouts of colored wheat genotypes. 1—Citrus yellow, 2—KM 53-14 Blue, 3—KM 178-14 purple, 4—Skorpion Blue aleurone, 5—PS Karkulka purple. The columns represent the mean values ± S.E. for five replicates. Statistically significant differences among treatments at each time are indicated by different lowercase letters (Duncan test; $p < 0.05$).

The grain sample from the genotype PS Karkulka purple was shown to have the highest antioxidant activity among the studied grains of colored wheat genotypes (ABTS and DPPH radical scavenging capacity parameters). The seedlings of the genotype PS Karkulka purple had increased antioxidant activity compared to that of the grain (18% higher) (Figure 1a,b). Three- or four-fold higher antioxidant activity was found in the seedlings of other experimental wheat genotypes.

A significant increase in the total phenolic and flavonoid contents was observed in all studied sprouts of colored wheat genotypes (Figures 2 and 3), with a similar trend of increased antioxidant activity in the sprouts of colored wheat genotypes.

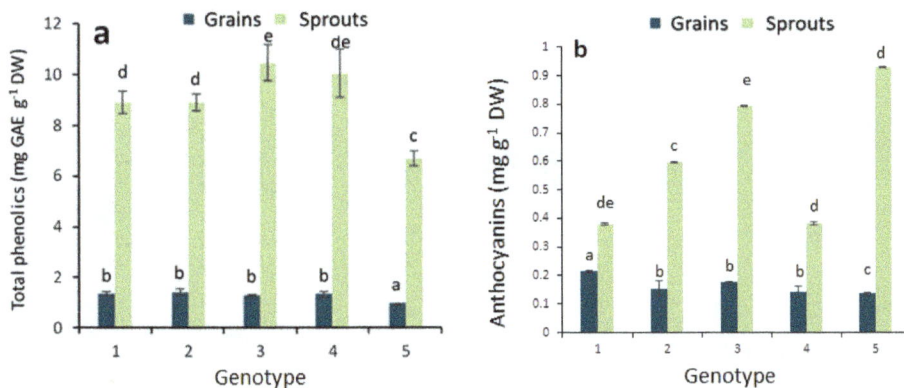

Figure 2. Total phenolics (**a**) and total anthocyanins (**b**) in the grains and sprouts of colored wheat genotypes. 1—Citrus yellow, 2—KM 53-14 Blue, 3—KM 178-14 purple, 4—Skorpion Blue aleurone, 5—PS Karkulka purple. The columns represent the mean values ± S.E. for five replicates. Statistically significant differences among treatments at each time are indicated by different lowercase letters (Duncan test; $p < 0.05$).

There was a significantly higher anthocyanin content in the sprouts of the purple wheat genotypes than in blue or yellow wheat genotypes (Figure 2b). In addition, an increased total anthocyanin content was found in the sprouts of the yellow wheat genotype. Among the grains of the colored wheat genotypes, the lowest total anthocyanin content was measured in the Citrus yellow.

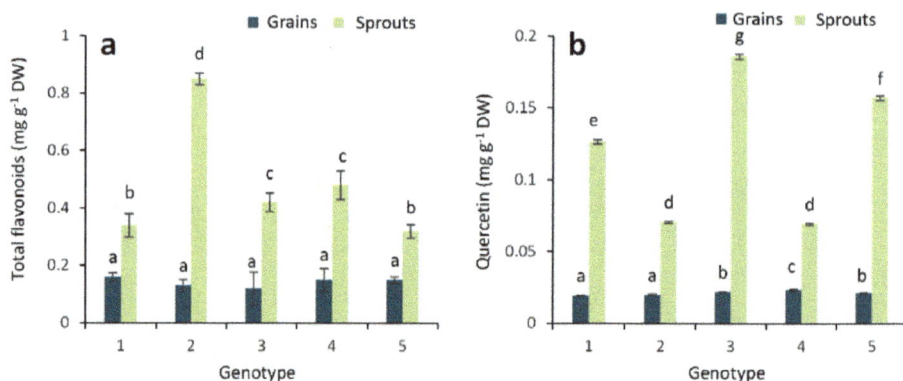

Figure 3. Total flavonoids (**a**) and quercetin (**b**) in the grains and sprouts of colored wheat genotypes. 1—Citrus yellow, 2—KM 53-14 Blue, 3—KM 178-14 purple, 4—Skorpion Blue aleurone, 5—PS Karkulka purple. The columns represent the mean values ± S.E. for five replicates. Statistically significant differences among treatments at each time are indicated by different lowercase letters (Duncan test; $p < 0.05$).

2.2. Anthocyanin Composition in the Grains and Sprouts of Colored Wheat Genotypes

We found pelargonidin in all the experimental grains of colored wheat genotypes (Figure 4), and the sprouts of the colored wheat genotypes exhibited significantly higher pelargonidin contents than the grains. The highest pelargonidin content was measured in the purple genotype PS Karkulka purple.

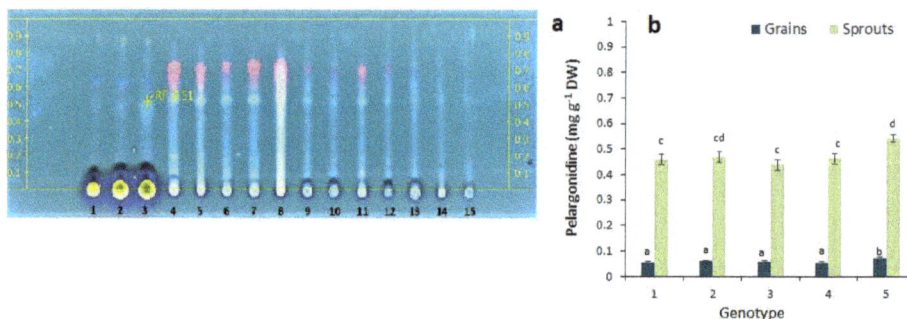

Figure 4. (**a**) HPTLC chromatogram of pelargonidin separation for the sprouts (tracks 4–8: track 4—Citrus yellow, track 5—KM 53-14 Blue, track 6—KM 178-14 purple, track 7—Skorpion Blue aleurone, track 8—PS Karkulka purple) and grains (tracks 9–13: track 9—Citrus yellow, track 10—KM 53-14 Blue, track 11—KM 178-14 purple, track 12—Skorpion Blue aleurone, track 13—PS Karkulka purple) extracts from colored wheat genotypes; standard mixture (tracks 1–3) (RF 0.51). (**b**) Pelargonidin contents in the grains and sprouts of colored wheat genotypes: 1—Citrus yellow, 2—KM 53-14 Blue, 3—KM 178-14 purple, 4—Skorpion Blue aleurone, and 5—PS Karkulka purple. The columns represent the mean values ± S.E. for five replicates. Statistically significant differences among treatments at each time are indicated by different lowercase letters (Duncan test; $p < 0.05$).

Our study revealed the presence of cyanidin in the blue genotypes, with higher cyanidin contents in the sprouts (Figure 5a). We suggest that a more detailed HPLC analysis of anthocyanins for further development of this research topic would be useful as well.

In our HPTLC analysis, delphinidin was identified in the grains and sprouts of the colored wheat genotypes (Figure 5b). The highest content of delphinidin was found in the blue wheat genotypes (Skorpion blue aleurone and KM 53-14 blue). The sprouts of the colored wheat genotypes did not show increased delphinidin contents. Peonidin was present at very low quantities in the grains of the colored wheat genotypes and was not identified in the sprouts (data not shown).

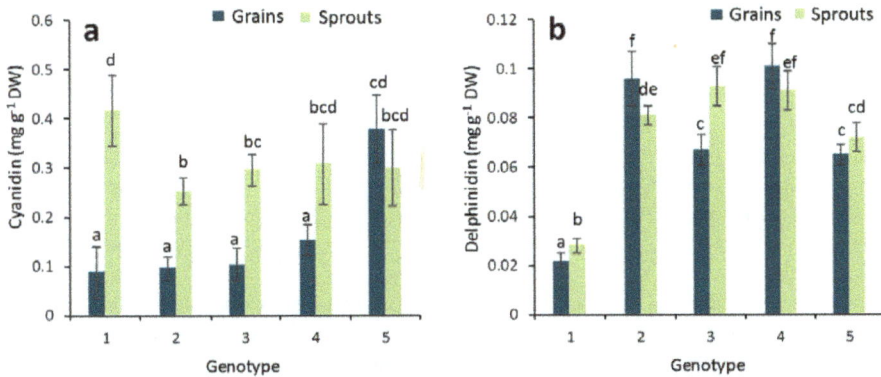

Figure 5. Cyanidin (**a**) and delphinidin (**b**) in the grains and sprouts of colored wheat genotypes. The columns represent the mean values ± S.E. for five replicates. 1—Citrus yellow, 2—KM 53-14 Blue, 3—KM 178-14 purple, 4—Skorpion Blue aleurone, 5—PS Karkulka purple. Statistically significant differences among treatments at each time are indicated by different lowercase letters (Duncan test; $p < 0.05$).

The correlation results between the antioxidant activity (DPPH and ABTS tests) and contents of bioactive phytochemicals in grains and sprouts of colored wheat genotypes (Table 1) has shown a high correlation for cyanidin and pelargonidin, especially in grains, as well as quercetin in sprouts. The positive correlation was also found in total phenolics, anthocyanins and flavonoids, but the level of correlation was generally moderate or low. The correlation between antioxidant activity and the delphinidin content was low in all cases.

Table 1. Results of the correlation between antioxidant activity and contents of bioactive phytochemicals in grains and sprouts of colored wheat genotypes.

Sample	Antiox Activ.	Total Phenolics	Total Anthocyan	Total Flavonoids	Quercetin	Pelargonidin	Cyanidin	Delphinidin
Grains	ABTS	0.51 *	0.29	0.43 *	0.08	0.68 *	0.96 *	0.04
	DPPH	0.62 *	0.38	0.22	0.20	0.59 *	0.96 *	0.09
Sprouts	ABTS	0.34	0.48	0.16	0.69 *	0.55 *	0.63 *	0.22
	DPPH	0.48	0.21	0.14	0.67 *	0.26	0.60 *	0.02

* Correlations are significant at $p < 0.05000$.

3. Discussion

A significant increase in total phenolic content was observed in all studied sprouts of the colored wheat genotypes, which agrees with the similar trend of increased antioxidant activity in the sprouts of the colored wheat genotypes. There was a significantly higher anthocyanin content in the sprouts of the purple wheat genotypes than in the blue or yellow wheat genotypes. In addition, an increased total anthocyanin content was found in the sprouts of the yellow wheat genotype. Among the grains of the colored wheat genotypes, the lowest total anthocyanin content was measured in the yellow wheat genotype. This result was expected because grains of the yellow wheat genotype have yellow endosperms (*Psy* genes), which are characterized by high carotenoid levels but not high anthocyanin levels [6]. The grains of the blue and purple wheat genotypes were characterized by a higher total anthocyanin content than that in the grains of the yellow genotype. This data confirmed previous results regarding the presence of total anthocyanins in whole meal flour and in the bran of purple and blue grains [19].

Sprouts of all the colored wheat genotypes showed significant increases in total flavonoid, total phenolic contents and antioxidant activity compared to those in the grains. The novel literature shows that the sprouts of *Triticum* species may be valuable for the development of functional foods due to increased total polyphenol and free phenolic acid contents [20,21].

Anthocyanin is in the form of glycoside while anthocyanidin is known as the aglycone. The most common types of anthocyanidins are cyanidin, delphinidin, pelargonidin, peonidin, petunidin and malvidin [22]. In the HPTLC analysis we analyzed the cyanidin, delphinidin, pelargonidin and peonidin. The special interest regarding these compounds was based on their possible role in color formation of the experimental colored wheat genotypes and the health effects of specific anthocyanidins.

Anthocyanins and their aglycone forms (anthocyanidins—malvidin, cyanidin, peonidin, and delphinidin) are flavonoids which are present in notable concentrations in berries (blueberries, bilberries, cranberries, elderberries, raspberry seeds and strawberries) [23]. Consumer interest in biologically active compounds also from wheat plants that produce health effects is increasing. The presence of anthocyanidin delphinidin in the grains of blue wheat genotypes, as well as cyanidin 3-glucoside, cyanidin 3-rutinoside and succinyl glucoside in purple grains has been reported previously [12].

In the current study, we observed the presence of anthocyanidins in wheat sprouts. Specifically, anthocyanidins that are reported to be useful for prophylaxis of some diseases. Moreover, there is evidence that biologically active substances from wheat sprouts may be partially absorbed during digestion [24]. We hypothesized that grains and sprouts of specific colored wheat genotypes, in particular, can have high healthy antioxidant contents and can therefore be a promising source for functional foods.

It is known that pelargonidin chloride is present in greater amounts in the sprouts of colored wheat genotypes than in the grains. The sprouts of radish plants have been shown to have a high pelargonidin content [25]. Pink and purple pigmentation of potato sprouts (*Tuberosum* sp.) is also caused by derivatives of pelargonidin [26]. Pelargonidin chloride from plants has anti-inflammatory effects [27]. Decreasing NO production is related to the inhibition of iNOS protein and mRNA expression, which is affected by pelargonidin in a dose-dependent manner [28]. Pelargonidin chloride also has antidiabetic effects. In vitro studies have shown that insulin secretion by beta-cells increases more in the presence of a pelargonidin derivative than in the presence of a leucocyanidin derivative, which is reported to be a good anti-diabetic agent [29].

Cyanidin has been found in the grains of colored wheat genotypes and at two- to three-fold higher levels in the corresponding sprouts. Anthocyanin glucosides, namely cyanidin 3-*O*-rutinoside, cyanidin 3-*O*-glucoside, cyanidin 3-*O*-galactoside and cyanidin 3-*O*-galactopyranosyl-rhamnoside, were detected in common buckwheat [30]. Knievel et al. (2009) [12] showed that in purple grains, cyanidin 3-glucoside and cyanidin 3-rutinoside are the main anthocyanidin-3-*O*-glycosides, but their contents are lower in blue grains due to the presence of specific blue color genes (*Ba* genes) and purple pericarp color genes (*Pp* genes) [8]. Our study has demonstrated the presence of cyanidin in the grains of blue genotypes of wheat, with higher cyanidin content in their sprouts.

The anthocyanidins cyanidin and delphinidin possess high radical scavenging activity, suppress cell proliferation and increase the apoptosis of MCF7 breast cancer cells [31]. In the previous HPTLC analysis, the 3-glucosides of delphinidin, cyanidin, malvidin and peonidin, further cyanidin glycosides and respective anthocyanidins were found in powdered berry extracts [32]. In the present experiment, the highest content of delphinidin was observed in the blue wheat genotypes (Skorpion blue aleurone and KM 53-14 blue). The sprouts of the purple and yellow wheat genotypes did not show increased delphinidin contents. Delphinidin and its derivatives were also found in the sprouts of radish and buckwheat plants [25,33]. The effects of hydroxycinnamic acids on blue color expression of cyanidin derivatives have been studied, and it was shown that interactions with phenolic compounds can play important roles in color expression [34].

The study of colored wheat genotype sprouts showed a similar trend of increased total flavonoids content [35]. The presence of quercetin in wheat sprouts has been previously demonstrated [36]. It was confirmed that the nutritional promise of wheat sprouts was due to high catalase and peroxidase activity, together with a high content of organic phosphates [23]. The health effects of quercetin are well studied, especially quercetin's ability to scavenge hydroxyl radical, peroxynitrite and other free radicals [36]. The possible use of cyanidin derivatives as nutraceuticals has also been discussed [36].

The methods used in this study did not cover the full spectrum of anthocyanins, which, in turn, provides potential for future research focused on analyzing the full profile of anthocyanidins and quercetin compounds. Moreover, the presented correlation analysis between the individual phytochemicals and the antioxidant capacity is informative for further investigations. In addition, the link between the presence of anthocyanidins and cyanidin derivatives in grains and sprouts, and their role in plant stress resistance could be interesting, as it was previously tested in salt stress experiments [37]. Thus, the colored wheat genotypes are becoming an increasingly attractive target for biological and food research.

4. Materials and Methods

4.1. Reagents and Chemicals

All anthocyanin standards were obtained as HPLC-grade chloride salts. Cyanidin (cn; 3,30,40,5,7-pentahydroxyflavylium), delphinidin (dp; 3,30,40,5,50,7-hexahydroxyflavylium) and peonidin (pn; 3,40,5,7-tetrahydroxy-30-methoxyflavylium) were provided by Cayman chemistry (Hamburg, Germany). Pelargonidin (pg; 3,40,5,7-tetrahydroxyflavylium) was provided by Sigma–Aldrich (Darmstadt, Germany). For the mobile phases, ethyl acetate and toluene (Merck, Darmstadt, Germany), as well as formic acid (Sigma–Aldrich, Darmstadt, Germany), were purchased at HPTLC grade. Double-distilled water was prepared using a Heraeus Destamat Bi-18E (Thermo Fisher Scientific, Schwerte, Germany). Methanol (gradient grade and LC-MS Chromasolv), hydrochloric acid (37% and 25%, reagent grade) and 2,2-diphenyl-1-picrylhydrazyl radical (DPPH) were ordered from Sigma–Aldrich, Darmstadt, Germany. All chemicals used for the bioassay, the HPTLC silica gel plates (60 F254, 20–10 cm) and the potassium acetate used for humidity control during plate development were purchased from Merck. The HPTLC plates were pre-washed with methanol to the upper plate edge and dried at 120 °C for 15 min on a thin-layer chromatography (TLC) plate heater (CAMAG, Muttenz, Switzerland). The plates were cooled to room temperature in a desiccator with phosphorus pentoxide (P99%, Sigma–Aldrich, Darmstadt, Germany) and temporarily stored while protected by a clean glass plate and wrapped in aluminum foil.

4.2. Plant Objects

Seeds of wheat genotypes (*Triticum* sp.) with different pigments—Citrus yellow, KM 53-14 Blue, KM 178-14 purple, Skorpion Blue aleurone and PS Karkulka purple—were provided by the Agricultural Research Institute Kromeriz, Kromeriz, Czech Republic. The pigmented seeds were characterized based on visual pink, blue and yellow color assessment (Figure 6).

Figure 6. Experimental differently pigmented wheat grains (**A**) and sprouts (**B**). 1—Citrus yellow, 2—KM 53-14 Blue, 3—KM 178-14 purple, 4—Skorpion Blue aleurone, 5—PS Karkulka purple.

Grains were washed extensively with distilled water and sterilized with 5% sodium hypochlorite for 5 min. Then, seeds were sown in petri dishes with absorbent pads and directly watered with 3 mL of 1/4 strength Hoagland's nutrient solution [38]. The germination process proceeded under controlled conditions in a growth chamber with the following parameters: Relative humidity of 60–70% and light/dark regime of 16/8 h at 25/20 °C. The length of germination process was 10 days. Then, 5 cm sprouts were harvested and frozen in liquid nitrogen to prevent phenolic compounds volatization. Samples were taken for analysis after the freeze-drying process was complete, where the material was ground by a flint mill (20,000× *g*, 2 min).

4.3. Determination of DPPH Radical Scavenging Capacity

The DPPH assay previously described [39] was used with some modifications. The stock reagent solution (1×10^{-3} M) was prepared by dissolving 22 mg of DPPH in 50 mL methanol. The stock solution was stored at 20 °C until use. The weight of the samples was 0.02 g, and all samples were assayed six times. The extraction was performed in two steps: First, 0.02 g of dry material was placed in an Eppendorf tube, and 1 mL of distilled water was added to the tube. The samples were heated for 15 min at 95 °C. Then, the material was centrifuged for 5 min (12,000 rpm, 25 °C). The extract was added to a new tube.

The supernatant was diluted with 1 mL of distilled water, reheated for 10 min at 95 °C and then spun again (12,000 rpm, 25 °C, 5 min). The extract was added to a new tube. The working solution (6×10^{-5} M) was prepared by mixing 6 mL of the stock solution with 100 mL of methanol. The optical absorbance was measured at 515 nm with a Jenway UV/Vis 6405 spectrophotometer (Jenway, Chelmsford, UK). Then, 0.1 mL of each experimental extract was mixed with 3.9 mL of the DPPH solution to react, followed by vortexing for 30 s and a further reaction time of 30 min. The absorbance was measured at 515 nm. A sample with no added extract was used as a control. The DPPH scavenging capacity was determined based on the following formula:

$$\text{DPPH scavenging capacity (\%)} = [(A_{control} - A_{sample})/A_{control}] \times 100 \quad (1)$$

A = absorbance at 515 nm.

4.4. ABTS Radical Scavenging Assay

For the ABTS assay, the method of Re et al. (1999) was adopted [40]. The stock solutions included 7 mM ABTS solution and 2.4 mM potassium persulfate solution. The working solution was prepared by mixing equal quantities of the two stock solutions and allowing them to react for 12–16 h at room temperature in the dark. The mixture was diluted by mixing 1 mL ABTS solution with 60 mL of methanol to obtain an absorbance of 0.706 ± 0.001 units at 734 nm using the spectrophotometer (Jenway 6505 UV/Vis). The ABTS solution was freshly prepared for each assay. The aqueous extraction of samples was undertaken. A total of 0.02 g of dry material was placed in an Eppendorf tube, and 1 mL of distilled water was added to the tube. The samples were heated for 15 min at 95 °C. Then, the material was centrifuged for 5 min (12,000 rpm, 25 °C). The procedure was repeated twice, with supernatants collected in the separate Eppendorf tube. Sample extracts (1 mL of each) could react with 1 mL of the ABTS solution, and the absorbance was taken at 734 nm after 7 min using the spectrophotometer. The ABTS + scavenging capacity of the extract was calculated as percentage inhibition.

$$\text{ABTS radical scavenging activity (\%)} = [(\text{Abs}_{\text{control}} - \text{Abs}_{\text{sample}})]/(\text{Abs}_{\text{control}})] \times 100 \qquad (2)$$

where $\text{Abs}_{\text{control}}$ is the absorbance of ABTS radical + methanol; $\text{Abs}_{\text{sample}}$ is the absorbance of ABTS radical + sample extract.

4.5. Determination of Total Phenolics

The total phenolic content was estimated using the Folin–Ciocalteu reagent [41]. Twenty milligrams of freeze-dried samples in the form of powder were mixed with 500 μL of 70% methanol (HPLC-Gradient grade, Sigma–Aldrich, Darmstadt, Germany) at 70 °C for 10 min. The extracts were centrifuged for 10 min at $3500 \times g$. The supernatants were collected in individual tubes. The pellets were re-extracted under the same conditions. The supernatants were combined and used to estimate the total phenolic content, and 20 μL of extract was dissolved into 2 mL of distilled water for the total phenolics analysis. Folin–Ciocalteu reagent, previously diluted ten-fold with distilled water and kept at 25 °C for 3–8 min, was used for the analysis; 200 μL of dissolved extract was mixed with 0.8 mL of sodium bicarbonate (75 g L^{-1}) solution, and 1 mL of Folin–Ciocalteu reagent was added to the mixture. The mixture was left to react for 60 min at 25 °C. Samples of sprouts were taken after finishing the freeze-drying process where the material was ground by a flint mill ($20,000 \times g$, 2 min). The absorbance for total phenolics was detected at 765 nm with a Jenway UV/Vis 6405 spectrophotometer (Jenway, Chelmsford, UK). The results are described as gallic acid equivalents (GAE/g sample).

4.6. Detection of Total Flavonoid Content

The total flavonoid content was detected using the aluminum chloride colorimetric method. Samples weighing 0.1 g were used for flavonoid extraction with 7 mL of 95% ethanol for 16–18 h. Then, 0.5 mL of the extract stock solution was reacted with 1.5 mL of 95% ethanol, 0.1 mL of 10% aluminum chloride, 0.1 mL of 1 M potassium acetate, and 2.8 mL of distilled water. Aluminum chloride solution was replaced with the same amount of distilled water in the blank. After the mixture was incubated for 30 min at room temperature, the absorbance of the reaction mixture was measured at $\lambda = 415$ nm. Quercetin was used as a standard for the calibration curves. Ten milligrams of quercetin were dissolved in ethanol, and then, the solution was diluted to 25, 50 and 100 g mL^{-1}. A calibration curve was created by measuring the absorbance of the dilutions at 415 nm (λ_{max} of quercetin) with a Jenway 6405 UV/Vis spectrophotometer.

4.7. Estimation of Anthocyanins

Total anthocyanins were estimated by a pH-differential method [42]. A known weight of samples was soaked in 3 mL of acidified methanol (1% *v/v* HCl) for 24 h in darkness at 4 °C with occasional shaking. Approximately 2 mL of distilled water and 4.8 mL of chloroform were mixed and added to the extract. The mixture was centrifuged for 15 min at $5000 \times g$. The absorbance of the upper phase was determined at 530 and 657 nm [43]. The concentrations of the anthocyanins as mg g^{-1} dry weight of the differently treated plants was determined using the following equation:

$$\text{Total Anthocyanins} = [\text{OD530} - 0.25 \text{ OD657}] \times \text{TV}/[\text{DW} \times 1000] \qquad (3)$$

OD = optical density; TV = total volume of the extract (mL); DW = weight of the dry leaf tissue (g). The anthocyanins content was finally expressed as mg cyanidin-3,5-diglucoside equivalents per g DW.

4.8. Stock Solutions and Sample Preparation for HPTLC

Anthocyanin stock solutions of 1 mg mL^{-1} each were individually prepared in a mixture of methanol and hydrochloric acid 25% (4:1, v/v). For the anthocyanin stock solutions, cyanidin chloride, pelargonidin chloride, peonidin chloride and delphinidin chloride (each 1 mg/mL) were individually dissolved in acidified methanol, and the same preparation was used for quercetin.

For sample preparation, 0.2 g of wheat DW of each sample was dissolved in 2 mL of acidified methanol (methanol and hydrochloric acid 25% mixture, 4:1, v/v). After sonication for 30 min at room temperature, the solutions were filtered through 0.45 μm cellulose filters and stored at −20 °C.

4.9. TLC Conditions

TLC aluminum oxide 60 F$_{254}$ plates (10 × 10 cm, Merck, Darmstadt, Germany) were used for the TLC analysis. Standards with different volumes (1, 2, 4, 6 μL) and samples in 4–6 μL volumes were applied. The prepared stationary phase was applied using a Linomat Vapplicator (CAMAG, Muttenz, Switzerland) under the next plate's condition with 10-mm-wide bands at 10 mm from the edge, with 10 mm from the plate bottom and a distance of 8 mm between the bands.

A mixture of chloroform/methanol/acetone/ammonia 25% (10:22:53:0.2, $v/v/v/v$) was used for the mobile phase. Chromatograms were prepared in glass chromatographic chambers (17.5 × 16 × 8.2 cm in size; Sigma–Aldrich, Darmstadt, Germany). The conditions for developing the chromatogram were room temperature with 9 cm over 40 min. Saturation was established for 15 min with vapor of the mobile phase. The plates were scanned after they were air-dried in the dark. The resulting spots were analyzed at 254 and 366 nm using a TLC Scanner 3 (CAMAG, Muttenz, Switzerland).

4.10. High-Performance Thin-Layer Chromatography

HPTLC methods for analysis of anthocyanidins in powdered, freeze-dried wheat samples were not found in the literature, except for the analysis of anthocyanidins in powdered berry samples [32]. Samples of sprouts were taken after the freeze-drying process was completed, where the material was ground by a flint mill (20,000× g, 2 min). Up to 18 tracks can be applied on one 20–10 cm HPTLC plate. For four-point-calibration, volumes of 2, 4, 6, and 8 μL of the standard mixture solution were applied at 20–250 ng/band, depending on the substance. Sample volumes of 2, 4 or 6 μL were used for the anthocyanidin analyses, depending on the expected anthocyanin concentration. For unknown concentrations, various volumes were applied, and the volume showing the best separation performance (not overloaded but above LOQ) was applied three-fold and used for quantitation. This volume was usually 2 or 4 μL for the given sample preparation. The plate activity was adjusted at a relative air humidity for 4 min at 25 ± 2% in the humidity control unit with a saturated potassium acetate solution (260 g/100 g water). The migration distance was 70 mm from the lower plate edge, and the migration time was 30 min. After development, the plate was automatically dried in a stream of cold air for 3 min.

For anthocyanidins separation, the plate was cut below the co-eluting anthocyanidin fraction (below solvent front) of the first solvent system using the TLC Plate Cutter (CAMAG, Muttenz, Switzerland). The upper plate part was developed with a mixture of ethyl acetate:toluene:formic acid:water (10:3:1.2:0.8; $v/v/v/v$) to a migration distance of 45 mm from the lower plate edge (almost to the plate top) in a twin-trough chamber (20–10 cm, CAMAG, Muttenz, Switzerland). Images of the plates were recorded with a TLC Visualizer documentation system equipped with a high-resolution 12-bit CCD digital camera (CAMAG, Muttenz, Switzerland). As the content of anthocyanidins is lower than that of anthocyanins, higher sample volumes had to be applied, which led to an overloaded separation of the anthocyanins [32]. The same sample should be applied at a low and high volume, e.g., 2 and 10 µL. This combined 2-step method was recommended by Krüger et al. (2013) [44].

The images were taken under white light illumination in the transmission mode with an exposure time of 35 ms. Spectra of the corresponding zones were used for the determination of wavelengths to measure individual anthocyanidins. A TLC Scanner 4 (CAMAG, Muttenz, Switzerland) was used for anthocyanidin detection. Densitometric evaluation was completed with a halogen tungsten lamp. The multi-wavelength scan was used to measure absorbance at 555 nm for delphinidin, at 530 nm for cyanidin and at 520 nm for pelargonidin and peonidin. The measuring scanning speed was 20 mm/s and the slit size was 6 mm–0.2 mm. The data were processed with the software winCATS, Version 1.4.7.2018 (CAMAG, Muttenz, Switzerland).

4.11. Statistical Analysis

Statistical analyses were performed using a two-way (genotype × sample type) analysis of variance (ANOVA) and Duncan's multiple range test performed at $p = 0.05$ (STATISTICA 10, StatSoft, Tulsa, OK, USA). Mean values were calculated from five replicates per cultivar for each type of sample (sprouts and grains). Data are presented as mean ± standard error from five replicates (SE). The correlation analysis between the antioxidative activity results and the phytochemicals contents was analyzed using the Statistica software, providing a correlation index and statistical significance ($p < 0.05$) for each correlation.

5. Conclusions

The comparative study of the grains and sprouts of colored wheat genotypes showed that sprouts have higher contents of biologically active compounds of a polyphenolic nature and antioxidant activities than grains. This result indicates that wheat sprouts from colored wheat genotypes are a valuable source for further food or pharmaceutical use. The significantly high content of pelargonidin and cyanidin derivatives in the sprouts of purple wheat genotypes, together with their high antioxidant activity, makes them valid candidates for functional food production. The genetic variability of colored wheat genotypes supports the availability of specific anthocyanidins from wheat sprouts: Blue wheat genotypes contain mostly delphinidin, and purple genotypes have a high quantity of pelargonidin. Further studies are needed to qualitatively analyze colored wheat sprouts, especially HPLC analyses of flavonoids and anthocyanins in the form of glycoside compositions.

Author Contributions: O.S. and P.B. conceived and designed the experiments; M.B. and I.S. performed the experiments; M.Z. and O.S. wrote the paper; M.B. and I.S. reviewed and corrected the document.

Funding: This work was supported by DAAD scholarship and research project of the Scientific Grant Agency of Slovak Republic APVV-15-0721.

Acknowledgments: We would like to thank Martinek P. for the given colored wheat grains collection and Bruckova K. for her technical assistance.

Conflicts of Interest: The authors declare no conflict of interest. The funding sponsors had no role in the design of the study; in the collection, analyses, or interpretation of data; in the writing of the manuscript, and in the decision to publish the results.

References

1.   Lao, F.; Sigurdson, G.T.; Giusti, M.M. Health benefits of purple corn (*Zea mays* L.) phenolic compounds. *Compr. Rev. Food Sci. Food Saf.* **2017**, *16*, 234–246. [CrossRef]

2. Havrlentová, M.; Pšenáková, I.; Žofajová, A.; Rückschloss, l.; Kraic, J. Anthocyanins in wheat seed—A mini review. *Nova Biotechnol. Chim.* **2014**, *13*, 1–12. [CrossRef]

3. Zykin, P.A.; Andreeva, E.A.; Lykholay, A.N.; Tsvetkova, N.V.; Voylokov, A.V. Anthocyanin composition and content in rye plants with different grain color. *Molecules* **2018**, *23*, 948. [CrossRef] [PubMed]

4. Lao, F.; Giusti, M.M. Quantification of purple corn (*Zea mays* L.) anthocyanins using spectrophotometric and HPLC approaches: Method comparison and correlation. *Food Anal. Methods* **2016**, *9*, 1367. [CrossRef]

5. Yu, L.; Beta, T. Identification and antioxidant properties of phenolic compounds during production of bread from purple wheat grains. *Molecules* **2015**, *20*, 15525–15549. [CrossRef] [PubMed]

6. Jing, P.; Giusti, M.M. Contribution of berry anthocyanins to their chemopreventive properties. In *Berries and Cancer Prevention*; Seeram, N., Stoner, G., Eds.; Springer: New York, NY, USA, 2011; pp. 3–40.

7. Putta, S.; Yarla, N.S.; Kumar, E.K.; Lakkappa, D.B.; Kamal, M.A.; Scotti, L.; Scotti, M.T.; Ashraf, G.M.; Barreto, G.E.; Rao, B.S.B.; et al. Preventive and therapeutic potentials of anthocyanins in diabetes and associated complications. *Curr. Med. Chem.* **2017**. [CrossRef] [PubMed]

8. Martinek, P.; Jirsa, O.; Vaculová, K.; Chrpová, J.; Watanabe, N.; Burešová, V.; Kopecký, D.; Štiasna, K.; Vyhnánek, T.; Trojan, V. Use of wheat gene resources with different grain colour in breeding. In Proceedings of the Tagungsband der 64. Jahrestagung der Vereinigung der Pflanzenzüchter und Saatgutkaufleute Österreichs, Raumberg-Gumpenstein, Austria, 25–26 November 2013; pp. 75–78.

9. Giordano, D.; Locatelli, M.; Travaglia, F.; Bordiga, M.; Reyneri, A.; Coïsson, J.D.; Blandino, M. Bioactive compound and antioxidant activity distribution in roller-milled and pearled fractions of conventional and pigmented wheat varieties. *Food Chem.* **2017**, *233*, 483–491. [CrossRef] [PubMed]

10. Mamoucha, S.; Tsafantakis, N.; Fokialakis, N.; Christodoulakis, N.S. A two-season impact study on *Globularia alypum*: Adaptive leaf structures and secondary metabolite variations. *Plant Biosyst.* **2018**, *152*, 1118–1127. [CrossRef]

11. Abdel-Aal, E.S.M.; Hucl, P. Composition and stability of anthocyanins in blue-grained wheat. *J. Agric. Food Chem.* **2003**, *51*, 2174–2180. [CrossRef] [PubMed]

12. Knievel, D.C.; Abdel-Aal, E.S.M.; Rabalski, I.; Nakamura, T.; Hucl, P. Grain color development and the inheritance of high anthocyanin blue aleurone and purple pericarp in spring wheat (*Triticum aestivum* L.). *J. Cereal Sci.* **2009**, *50*, 113–120. [CrossRef]

13. Abdel-Aal, E.S.M.; Hucl, P.; Rabalski, I. Compositional and antioxidant properties of anthocyanin-rich products prepared from purple wheat. *Food Chem.* **2018**, *254*, 13–19. [CrossRef] [PubMed]

14. Pasqualone, A.; Bianco, A.M.; Paradiso, V.M.; Summo, C.; Gambacorta, G.; Caponio, F.; Blanco, A. Production and characterization of functional biscuits obtained from purple wheat. *Food Chem.* **2015**, *180*, 64–70. [CrossRef] [PubMed]

15. Nelson, K.; Stojanovska, L.; Vasiljevic, T.; Mathai, M. Germinated grains: A superior whole grain functional food? *Can. J. Physiol. Pharmacol.* **2015**, *91*, 429–441. [CrossRef] [PubMed]

16. Perni, S.; Calzuola, I.; Caprara, G.A.; Gianfranceschi, G.L.; Marsili, V. Natural antioxidants in wheat sprout extracts. *Curr. Org. Chem.* **2014**, *18*, 2950–2960. [CrossRef]

17. Farasat, M.; Khavari-Nejad, R.-A.; Nabavi, S.M.B.; Namjooyan, F. antioxidant activity, total phenolics and flavonoid contents of some edible green seaweeds from northern coasts of the persian gulf. *IJPR* **2014**, *13*, 163–170. [PubMed]

18. Ravikumar, P.; Shalini, G.; Jeyam, M. Wheat seedlings as food supplement to combat free radicals: An in vitro approach. *Indian J. Pharm. Sci.* **2015**, *77*, 592–598. [PubMed]

19. Abdel-Aal, E.S.M.; Hucl, P. A rapid method for quantifying total anthocyanins in blue aleurone and purple pericarp wheats. *Cereal Chem.* **1999**, *76*, 350–354. [CrossRef]

20. Cevallos-Casals, B.A.; Cisneros-Zevallos, L. Impact of germination on phenolic content and antioxidant activity of 13 edible seed species. *Food Chem.* **2010**, *119*, 1485–1490. [CrossRef]

21. Benincasa, P.; Galieni, A.; Manetta, A.C.; Pace, R.; Guiducci, M.; Pisante, M.; Stagnari, F. Phenolic compounds in grains, sprouts and wheatgrass of hulled and non-hulled wheat species. *J. Sci. Food Agric.* **2015**, *95*, 1795–1803. [CrossRef] [PubMed]

22. Khoo, H.E.; Azlan, A.; Tang, S.T.; Lim, S.M. Anthocyanidins and anthocyanins: Colored pigments as food, pharmaceutical ingredients, and the potential health benefits. *Food Nutr. Res.* **2017**, *61*, 1361779. [CrossRef] [PubMed]

23. Mecocci, P.; Tinarelli, C.; Schulz, R.J.; Polidori, M.C. Nutraceuticals in cognitive impairment and Alzheimer's disease. *Front. Pharmacol.* **2014**, *5*, 147. [CrossRef] [PubMed]

24. Marsili, V.; Calzuola, I.; Gianfranceschi, G.L. Nutritional relevance of wheat sprouts containing high levels of organic phosphates and antioxidant compounds. *J. Clin. Gastroenterol.* **2004**, *38*, 123–126. [CrossRef]

25. Hanlon, P.R.; Barnes, D.M. Phytochemical composition and biological activity of 8 varieties of radish (*Raphanus sativus* L.) sprouts and mature taproots. *Food Sci.* **2011**, *76*, 185–192. [CrossRef] [PubMed]

26. Howard, H.W.; Kukimura, H.; Whitmore, E.T. The anthocyanin pigments of the tubers and sprouts of Tuberosum potatoes. *Potato Res.* **1970**, *13*, 142–145. [CrossRef]

27. Amini, A.M.; Muzs, K.; Spencer, J.P.; Yaqoob, P. Pelargonidin-3-*O*-glucoside and its metabolites have modest anti-inflammatory effects in human whole blood cultures. *Nutr. Res.* **2017**, *46*, 88–95. [CrossRef] [PubMed]

28. Hämäläinen, M.; Nieminen, R.; Vuorela, P.; Heinonen, M.; Moilanen, E. Anti-inflammatory effects of flavonoids: Genistein, kaempferol, quercetin, and daidzein inhibit STAT-1 and NF-κB Activations, whereas flavone, isorhamnetin, naringenin, and pelargonidin inhibit only NF-κB activation along with their inhibitory effect on iNOS expression and NO production in activated macrophages. *Mediat. Inflamm.* **2007**, *2007*, 45673.

29. Cherian, S.; Kumar, R.V.; Augusti, K.T.; Kidwai, J.R. Antidiabetic effect of a glycoside of pelargonidin isolated from the bark of *Ficus bengalensis* Linn. *Indian J. Biochem. Biophys.* **1992**, *29*, 380–382. [PubMed]

30. Kim, S.J.; Maeda, T.; Sarker, M.Z.; Takigawa, S.; Matsuura-Endo, C.; Yamauchi, H.; Mukasa, Y.; Saito, K.; Hashimoto, N.; Noda, T.; et al. Identification of anthocyanins in the sprouts of buckwheat. *J. Agric. Food Chem.* **2007**, *55*, 6314–6318. [CrossRef] [PubMed]

31. Tang, J.; Oroudjev, E.; Wilson, L.; Ayoub, G. Delphinidin and cyanidin exhibit antiproliferative and apoptotic effects in MCF7 human breast cancer cells. *Integr. Cancer Sci. Ther.* **2015**, *2*, 82–86.

32. Cretu, G.C.; Morlock, G.E. Analysis of anthocyanins in powdered berry extracts by planar chromatography linked with bioassay and mass spectrometry. *Food Chem.* **2014**, *146*, 104–112. [CrossRef] [PubMed]

33. Li, X.; Thwe, A.A.; Park, N.I.; Suzuki, T.; Kim, S.J.; Park, S.U. Accumulation of phenylpropanoids and correlated gene expression during the development of tartary buckwheat sprouts. *J. Agric. Food Chem.* **2012**, *60*, 5629–5635. [CrossRef] [PubMed]

34. Sigurdson, G.T.; Robbins, R.J.; Collins, T.M.; Giusti, M.M. Effects of hydroxycinnamic acids on blue color expression of cyanidin derivatives and their metal chelates. *Food Chem.* **2017**, *234*, 131–138. [CrossRef] [PubMed]

35. Calzuola, I.; Marsili, V.; Gianfranceschi, G.L. Synthesis of antioxidants in wheat sprouts. *J. Agric. Food Chem.* **2004**, *52*, 5201–5206. [CrossRef] [PubMed]

36. Boots, A.W.; Haenen, G.R.; Bast, A. Health effects of quercetin: From antioxidant to nutraceutical. *Eur. J. Pharmacol.* **2008**, *585*, 325–337. [CrossRef] [PubMed]

37. Mbarki, S.; Sytar, O.; Zivcak, M.; Abdelly, C.; Cerda, A.; Brestic, M. Anthocyanins of Coloured Wheat Genotypes in Specific Response to Salt Stress. *Molecules* **2018**, *23*, 1518. [CrossRef] [PubMed]

38. Epstein, E.; Bloom, A.J. *Mineral Nutrition of Plants: Principles and Perspectives*; Sinauer Associates: Sunderland, MA, USA, 1972; 412p.

39. Molyneux, P. The use of the stable free radical diphenylpicryl-hydrazyl (DPPH) for estimating antioxidant activity. *Songklanakarin J. Sci. Technol.* **2004**, *26*, 211–219.

40. Re, R.; Pellegrini, N.; Proteggente, A.; Pannala, A.; Yang, M.; Rice-Evans, C. Antioxidant activity applying an improved ABTS radical cation decolorization assay. *Free Radic. Biol. Med.* **1999**, *26*, 1231–1237. [CrossRef]

41. Singleton, V.L.; Rossi, J.A., Jr. Colorimentry of total phenolics with phosphomolybdic-phosphotungstic acid reagents. *Am. J. Enol. Viticult.* **1965**, *16*, 144–158.

42. Giusti, M.M.; Wrolstad, R.E. Characterization and measurement of anthocyanins with UV-visible spectroscopy. In *Current Protocols in Food Analytical Chemistry Banner*; Wiley: New York, NY, USA, 2001; pp. F1.2.1–F1.2.13.

43. Mirecki, R.M.; Teramura, A.H. Effects of ultraviolet-B irradiance on soybean. V. The dependence of plant sensitivity on the photosynthetic photon flux density during and after leaf expansion. *Plant Physiol.* **1984**, *74*, 475–480. [CrossRef] [PubMed]

44. Krüger, S.; Urmann, O.; Morlock, G. Development of a planar chromatographic method for quantitation of anthocyanins in pomace, feed, juice and wine. *J. Chromatogr. A* **2013**, *1289*, 105–118. [CrossRef] [PubMed]

Sample Availability: Samples of the compounds pelargonidin, peonidin, delphinidin, cyanidin are not available from the authors.

molecules

MDPI

Article

Phytoestrogenic Activity of Blackcurrant Anthocyanins Is Partially Mediated through Estrogen Receptor Beta

Naoki Nanashima [1,*], Kayo Horie [1] and Hayato Maeda [2]

[1]   Department of Bioscience and Laboratory Medicine, Hirosaki University Graduate School of Health Sciences, 66-1 Hon-cho, Hirosaki, Aomori 036-8564, Japan; k-horie@hirosaki-u.ac.jp

[2]   Faculty of Agriculture and Life Science, Hirosaki University, 3 Bunkyo-cho, Hirosaki, Aomori 036-8561, Japan; hayatosp@hirosaki-u.ac.jp

*   Correspondence: nnaoki@hirosaki-u.ac.jp; Tel.: +81-172-5968

Received: 27 November 2017; Accepted: 27 December 2017; Published: 29 December 2017

Abstract: Phytoestrogens are plant compounds with estrogenic effects found in many foods. We have previously reported phytoestrogen activity of blackcurrant anthocyanins (cyanidin-3-glucoside, cyanidin-3-rutinoside, delphinidin-3-glucoside, and delphinidin-3-rutinoside) via the estrogen receptor (ER)α. In this study, we investigated the participation of ERβ in the phytoestrogen activity of these anthocyanins. Blackcurrant anthocyanin induced ERβ-mediated transcriptional activity, and the IC_{50} of ERβ was lower than that of ERα, indicating that blackcurrant anthocyanins have a higher binding affinity to ERβ. In silico docking analysis of cyanidin and delphinidin, the core portions of the compound that fits within the ligand-binding pocket of ERβ, showed that similarly to 17β-estradiol, hydrogen bonds formed with the ERβ residues Glu305, Arg346, and His475. No fitting placement of glucoside or rutinoside sugar chains within the ligand-binding pocket of ERβ-estradiol complex was detected. However, as the conformation of helices 3 and 12 in ERβ varies depending on the ligand, we suggest that the surrounding structure, including these helices, adopts a conformation capable of accommodating glucoside or rutinoside. Comparison of ERα and ERβ docking structures revealed that the selectivity for ERβ is higher than that for ERα, similar to genistein. These results show that blackcurrant anthocyanins exert phytoestrogen activity via ERβ.

Keywords: anthocyanin; blackcurrant; estrogen receptor β; phytoestrogen

1. Introduction

Estrogens affect the functions of organs and tissues such as bones, blood vessels, skin, and brain, and participate in the underlying mechanisms of diseases such as metabolic syndrome [1–4]. The estrogen receptor (ER) has two subtypes, ERα and ERβ. ERα is mainly present in female reproductive organs such as mammary gland and uterus, whereas ERβ is found all over the body regardless of sex. The ERβ gene was cloned in 1996 [5], and the receptor is known to be involved in several diseases such as osteoporosis [6], breast cancer [1,7,8], and obesity [9], although many functions remain unclear. Although estrogen promotes the proliferation of breast cancer cells via ERα, ERβ inhibits cell proliferation. Thus, it is known that ERβ inhibits the activity of ERα [8,10].

Blackcurrants (*Ribes nigrum* L.) contain high levels of flavonoids, a group of polyphenolic compounds that includes anthocyanins and flavonols. Blackcurrants are reported to contain four anthocyanins: cyanidin-3-glucoside (C3G), cyanidin-3-rutinoside (C3R), delphinidin-3-glucoside (D3G), and delphinidin-3-rutinoside (D3R) (Figure 1). D3R and C3R are anthocyanins specific to blackcurrant [11]. Blackcurrant anthocyanins are known to have some health benefits such as amelioration of obesity and inflammation and prevention of breast cancer [12–14].

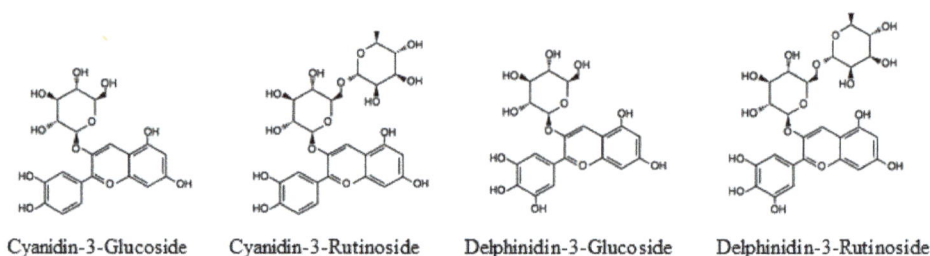

Cyanidin-3-Glucoside   Cyanidin-3-Rutinoside   Delphinidin-3-Glucoside   Delphinidin-3-Rutinoside

Figure 1. Chemical structures of anthocyanins derived from blackcurrant.

Phytoestrogens are a chemically diverse group of plant compounds with estrogenic effects in animals and include isoflavones, lignans, coumestans, and flavonoids; they are found in many foods [15–18]. The structure of anthocyanins is similar to that of flavanones and isoflavones. Although many health benefits of blackcurrant phytochemicals have been reported, no studies have addressed their phytoestrogenic activity. In contrast, phytoestrogen activity of the anthocyanins cyanidin and delphinidin has been reported by Schmitt et al. [19]. Recently, we have reported that these anthocyanins have a phytoestrogenic effect via ERα [20], but the participation of ERβ is unknown. Liquiritigenin, genistein, and S-equol are natural ligands of ERβ [21–23], and are known to inhibit the proliferation of breast, prostate, and colon cancers [24]. It is becoming clear that ERβ is involved in various diseases, and it is becoming the target of pharmacological studies [25].

To improve menopause-associated symptoms, postmenopausal women may undergo hormone replacement therapy. However, when using estrogen preparations, the risk of venous thrombosis and breast cancer must also be considered. In contrast, no association of phytoestrogens with venous thrombosis has been reported, and these compounds may suppress breast cancer. Thus, phytoestrogen is considered an important alternative to estrogen preparations [26,27].

The objective of this study was to investigate whether an anthocyanin-rich blackcurrant extract (BCE) and four blackcurrant anthocyanins exert phytoestrogenic activity via ERβ. We investigated ERβ-mediated transactivation by blackcurrant anthocyanins. In addition, the binding ability of black currant anthocyanins to ERβ was determined using competition binding assays and in silico analysis of the docking of four anthocyanins to the ERβ-17β-estradiol (E2) complex. The affinity of E2 to ERβ is very similar to that of ERα, but affinity to phytoestrogens such as genistein and S-equol is high [21,23]. Therefore, based on the interaction between genistein and ERα or ERβ [28], the interaction of cyanidin with ERα or ERβ was evaluated in silico.

2. Results and Discussion

2.1. ERβ Transactivation Activity of Blackcurrant Anthocyanins

Blackcurrant anthocyanins exhibited estrogenic activity in human ERβ reporter assays at 50.0 μM ($p < 0.05$), whereas BCE exhibited estrogenic activity at 10.0 μg/mL ($p < 0.05$), but not at 1.0 μg/mL (Figure 2a). BCE- and anthocyanin-mediated induction of estrogen response element-dependent luciferase activity was inhibited by co-treatment with 1 μM fulvestrant (Figure 2b), indicating that these effects are ERβ-mediated. These results suggest that blackcurrant anthocyanins and BCE have phytoestrogenic activity mediated via ERβ signaling.

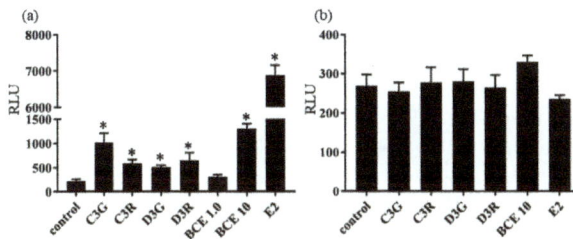

Figure 2. ERβ reporter assay of cells treated with 50 μM anthocyanins and 1.0 or 10.0 μg/mL blackcurrant extracts (BCE) or 100 pM 17β-estradiol (E2) in the absence (**a**) or presence (**b**) of 1.0 μM fulvestrant for 24 h. RLU, relative light units. Data are shown as the mean ± standard error of the mean of at least three independent experiments. * $p < 0.05$ vs. control.

2.2. Binding of Blackcurrant Anthocyanins to ERβ

We next investigated whether the phytoestrogenic activity of blackcurrant anthocyanins in vitro resulted from binding to ERβ using PolarScreen assays, and we calculated the approximate IC_{50} values. The IC_{50} of E2, BCE, C3G, C3R, D3G, and C3R was 3.2 nM, 3.5 μg/mL, 2.8 μM, 9.6 μM, 9.7 μM, and 2.3 μM, respectively (Figure 3). BCE and the four blackcurrant anthocyanins exhibited the ability to bind to ERβ. The IC_{50} of each anthocyanin was approximately 1/1000 of that of E2, which is consistent with the reported much weaker effect of phytoestrogens compared to endogenous estrogen [19,20,29]. These results suggest that blackcurrant anthocyanins have a high affinity for ERβ, similar to genistein, because the ERβ IC_{50} was lower than the ERα IC_{50} determined in our previous study [20].

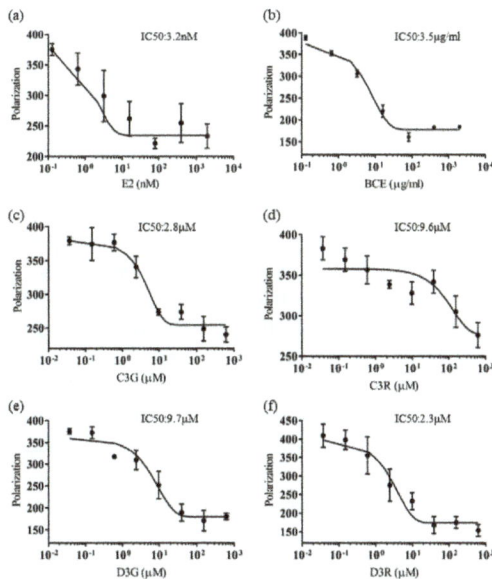

Figure 3. Competitive binding curves of blackcurrant anthocyanin-induced displacement of fluorescein-labeled 17β-estradiol (E2) from human ERβ. ERβ and fluorescein-labeled estradiol were incubated for 2 h with a serial dilution of (**a**) E2; (**b**) blackcurrant extract; (**c**) cyanidin-3-glucoside (C3G); (**d**) cyanidin-3-rutinoside (C3R); (**e**) delphinidin-3-glucoside (D3G); and (**f**) delphinidin-3-rutinoside (D3R) at least in triplicate. IC_{50} corresponds to the concentration of test compound inhibiting 50% of binding of 4.5 nM Fluormone ES2 Green to ERβ. Error bars represent the standard error of the mean.

2.3. In Silico Docking Analysis of Estradiol and ERβ

The ligand-binding domain of ERβ formed a homodimer similar to that of ERα, and estradiol bound inside the ligand-binding pocket of ERβ. In the state with bound estradiol, helix 12 (green) was positioned in such a way as to close the ligand-binding pocket (Figure 4). Because the amino acid residues involved in the binding of estradiol to ERβ were not described by Mocklinghoff [30], the residues forming a hydrogen bond with estradiol were determined using the Swiss-PDB Viewer [31]. Like ERα, residues Glu305, Arg346, and His475 within the binding pocket formed a hydrogen bond with estradiol in the stereostructure (PDB ID: 3OLS) of the ERβ/estradiol complex (Figure 4).

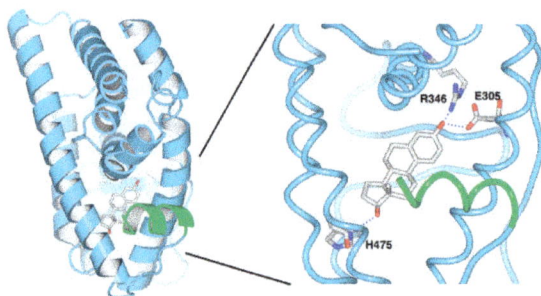

Figure 4. Ligand-binding pocket of the active ERβ conformation (PDB ID: 3OLS) showing interactions with 17β-estradiol (E2). Helix 12 is colored green.

2.4. In Silico Docking Analysis of C3G, C3R, D3G, or D3R and ERβ

In the docking model, cyanidin and delphinidin did not collide with the amino acid residues and atoms of ERβ, and fit within the internal pocket space (Figure 5a,b). Like estradiol, the hydroxyl group at position 4 of the phenyl group of cyanidin and delphinidin formed hydrogen bonds with Glu305 and Arg346 of ERβ, and the hydroxyl group at position 5 of the benzopyrylium group formed a hydrogen bond with His475 of ERβ (Figure 5a,b). These results suggest that cyanidin and delphinidin bind inside the binding pocket of ERβ in the same arrangement as estradiol.

Based on the docking analysis of the cyanidin and delphinidin skeletons, C3G, C3R, D3G, and D3R were placed in ERβ, and the space where the glucose or rutinose at position 3 fits was investigated by rotating the bond with glucoside or rutinoside. Glucose or rutinose collided with amino acid residues present in helices 3 and 12, and an arrangement in which sugar chains fit in the space was not found (Figures 5c–f and 6). These results suggest that there is not enough space inside the pocket of the ERβ-estradiol complex to bind sugar chains, which is in agreement with the report by Fan et al. [32]. However, helices 3 and 12 are known to change conformation depending on the type of ligand [33–35]. Therefore, if these helices have a conformation somewhat different from that of the ERβ-estradiol complex, which provides a space for accommodating sugar chains, glucoside or rutinoside may also be able to bind. Similarly, we were unable to find, using in silico docking analysis of ERα, an arrangement in which sugar chains of glucose and rutinose bind without steric hindrance, although we have previously reported that these four anthocyanins act as agonists [20]. In ERβ, it was also suggested that helices 3 and 12 form an appropriate conformation for four kinds of anthocyanins, thereby indicating that helix 12 adopts an agonist-like arrangement.

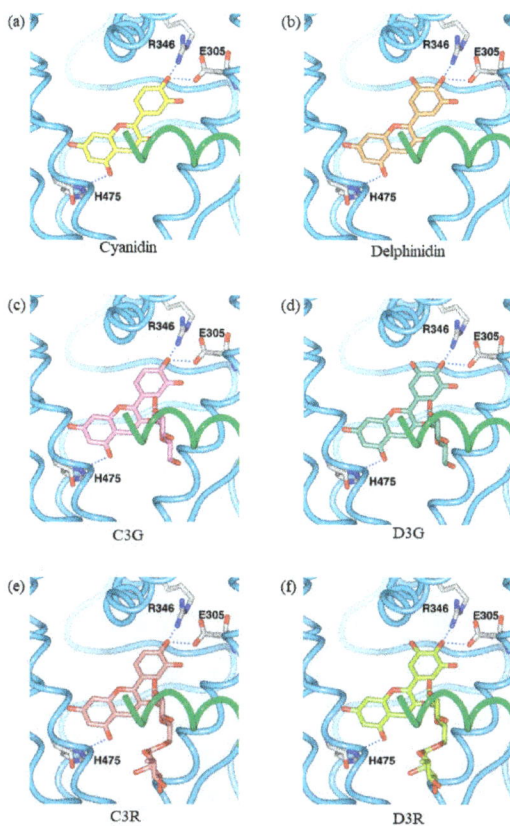

Figure 5. Ligand-binding pocket of the active ERβ conformation (PDB ID: 3OLS) showing interactions with (**a**) cyanidin; (**b**) delphinidin; (**c**) cyanidin-3-glucoside (C3G); (**d**) delphinidin-3-glucoside (D3G); (**e**) cyanidin-3-rutinoside (C3R) and (**f**) delphinidin-3-rutinoside (D3R). Helix 12 is colored green.

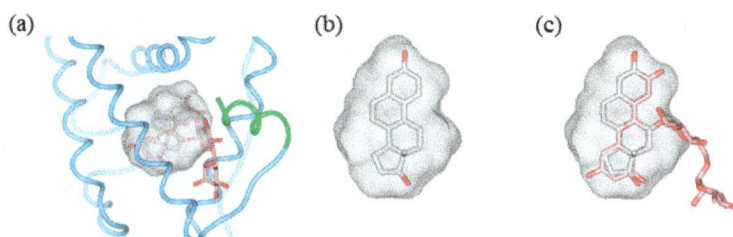

Figure 6. Interaction between the ligand-binding pocket of the ERβ and 17β-estradiol (E2) complex (PDB ID: 3OLS) and the sugar chain of cyanidin-3-rutinoside (C3R). (**a**) Docking model of C3R (light red) to the ERβ and estradiol complex (gray); (**b**) Surface shape of the binding site and appearance of E2; (**c**) Overlapping E2 and C3R; the sugar chain of C3R does not fit.

2.5. Differences in Anthocyanin Binding to ERα and ERβ

Manas et al. have determined the conformation of the genistein/ERα and genistein/ERβ complexes (PDB ID: 1X7R and 1X7J) and reported the selectivity factor of genistein to ERβ [28]. To investigate the differences in anthocyanin interaction with ERα and ERβ, we used ERα/cyanidin and ERβ/cyanidin complex models, and each ER residue located within 5.0 Å from each atom of cyanidin was determined using the Waals software. Nineteen residues were detected, and the only two residues different between ERα and ERβ were ERα Leu384 and ERβ Met336, and ERα Met421 and ERβ Ile373 (Figure 7a,b and Table 1). These differences are consistent with those reported in the genistein and ERα and ERβ binding pockets [28]. Hydrogen bonds form between cyanidin and Glu305, Arg346, and His475 of ERβ, and these residues are conserved in ERα (Figure 7a,b and Table 1).

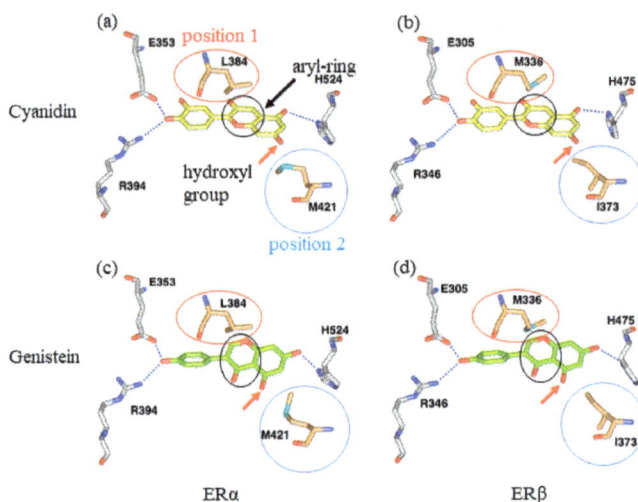

Figure 7. Difference in binding affinity of cyanidin to ERα and ERβ. Hydrogen bonds of each compound and ERs are indicated as blue dotted lines. Red, blue, and black circles indicate position 1, position 2, and aryl ring, respectively. Red arrows indicate hydroxyl groups. (**a**) Interaction of cyanidin with ERα; (**b**) cyanidin with ERβ; (**c**) genistein with ERα; and (**d**) genistein with ERβ.

Table 1. Comparison of predicted interactions between cyanidin and ERα or ERβ.

Amino Acid Residue		Interaction with Cyanidin	
ERα	ERβ	Common to ERα & ERβ	ERβ only
Ala350	Ala302	hydrophobic interaction	
Glu353	Glu305	hydrogen bond	-
Leu384	Met336		interaction with aryl ring (position 1)
Arg394	Arg346	hydrogen bond	-
Phe404	Phe356	hydrophobic interaction	-
Met421	Ile373	-	hydrophobic interaction interaction with hydroxyl group (position 2)
His524	His475	hydrogen bond	-

Each ER amino acid residue is shown located within 5.0 Å from each atom of cyanidin. -: none.

Hydrophobic interactions of Ile373 in ERβ, in addition to those of Ala302 and Phe356, corresponded to the interactions of Ala350 and Phe404 in ERα, which was inferred from the complex containing cyanidin (Table 1). In this study, the positions of ERα Leu384 and ERβ Met336 were named position 1, and the positions of ERα Met421 and ERβ Ile373 were named position 2 (Figure 7).

We observed stabilization of the protein structure in an interaction between methionine and aromatic rings, called the methionine-aromatic interaction, and selectivity to ERβ in compounds having an aryl aromatic ring positioned in the B-ring of genistein (Figure 7c,d), and thus ERβ Met336 is estimated to have a more favorable interaction with the aryl group of genistein than ERα Leu384 [28]. Therefore, this interaction is considered to underlie the selectivity of genistein for ERβ rather than ERα. Cyanidin and delphinidin have aryl groups at positions corresponding to the B-ring of genistein (Figures 1 and 7a,b). Based on the report of Manas et al. we suggest that cyanidin and delphinidin can also interact more favorably with ERβ Met336 compared to ERα Leu384, similar to genistein [28].

There is a hydroxyl group (5-OH) at position 5 of genistein near position 2 (Figure 7c,d). The side chain of Met421 in the ERα/genistein complex (PDB ID: 1X7R) adopts a rotamer whose lone pair of sulfur atoms avoids the oxygen atom of 5-OH of genistein. Furthermore, it is different from the rotamer of the side chain of Met421 of the ERα/estradiol complex (PDB ID: 1ERE) [30]. It is also known that dimethylsulfide clearly repels hydroxyl groups and that propane attracts weakly at an angle in which lone pairs of electrons face each other [28].

The hydroxyl group of position 7 of cyanidin is in the vicinity of ERα Met421 and ERβ Ile373 (Figure 7a,b). The analysis of the genistein complex suggests that the hydroxyl group at this position may repel ERα Met421 when binding to ERα (Figure 7c). In contrast, we suggest that ERβ does not repel ERβ Ile373, and the side chain of Ile373 and the carbon atoms at positions 6 and 7 may form a hydrophobic interaction (Figure 7a,b and Table 1). Therefore, ERβ Ile373 seems to be more accommodating to cyanidin and delphinidin structures than ERα Met421.

Given that estrogen levels decrease after menopause, dietary phytoestrogen may alleviate postmenopausal health concerns related to skin, bone, and cardiovascular heath [36–39]. In addition, it is known that E2 also affects male diseases such as benign prostatic hyperplasia and prostate cancer [40,41]. In particular, ERβ is expressed regardless of sex, and thus it is important to consider this receptor as a therapeutic target [25]. Furthermore, we previously orally administered BCE to female rats aged 3 weeks, and showed that BCE also had phytoestrogenic activity also in vivo [20]. We thus predict that as phytoestrogens, blackcurrant anthocyanins have many biological activities.

3. Materials and Methods

3.1. Materials

The BCE powder, CaNZac-35, was purchased from Koyo Mercantile (Tokyo, Japan). BCE contains high concentrations of polyphenols (37.6 g/100 g BCE) and anthocyanins (38.0 g/100 g BCE) [20]. C3G, C3R, D3G, and D3R (see Figure 1 for chemical structures) were purchased from Nagara Science (Gifu, Japan). E2 and fulvestrant (ICI 182,780) were purchased from Sigma-Aldrich (St. Louis, MO, USA).

3.2. ER Transactivation Assays

To assess the activation of human ERβ, nuclear receptor transactivation assay kits were obtained from Indigo Biosciences (State College, PA, USA). Briefly, the test compounds were prepared and diluted in a medium provided by the manufacturer. The cell recovery medium provided in the assay kit was thawed, warmed to 37 °C, and added to the frozen reporter cells. The cell suspension (100 µL) was dispensed into the wells of a 96-well assay plate and the test compounds (100 µL) were added to the cells at the indicated concentrations and incubated for 24 h. Luciferase activity was quantified using a TriStar LB941 multimode plate-reader (Berthold Technologies, Bad Wildbad, Germany).

3.3. Competitive Binding Assays

Competitive binding assays were performed using the PolarScreen ERβ Competitive Binding Assay Kit Green (Life Technologies, Carlsbad, CA, USA) according to the manufacturer's protocol. Recombinant human ERβ (23 nM) and 4.5 nM Fluormone ES2 Green (fluorescently labeled estradiol) were incubated for 2 h with the test compounds. Fluorescence polarization was measured using a Flex Station 3 (Molecular Devices, Sunnyvale, CA, USA). Approximate IC_{50} values, which indicate the ligand concentration that yields 50% inhibition of Fluormone ES2 Green, were determined from competitive binding curves generated using GraphPad Prism ver. 7.03 for Windows (GraphPad Software, San Diego, CA, USA).

3.4. Molecular Docking Simulations

In silico docking analysis was performed to investigate the interactions between blackcurrant anthocyanins and ERβ. The interaction between E2 and ERβ was used as positive control. The steric structures of C3G and C3R were obtained from the ZINC (http://zinc.docking.org) compound database (AC4097706 and AC4097715, respectively). D3G and D3R steric structure models were constructed using MarvinSketch (ChemAxon http://www.chemaxon.com/products/marvin/) based on the structures of C3G and C3R, respectively. Docking models based on the X-ray crystal structure of human ERβ with E2 were obtained from the Protein Data Bank (PDB) (http://www.rcsb.org/pdb/) (PDB ID: 3OLS) [35], which enabled analysis of docking to the active type (with E2) of ER. The steric structures of anthocyanins were fitted to the ER steric structure by superimposition on the molecular frame structure of E2 using Waals (Altif Laboratories, Tokyo, Japan). Hydrogen bonding and atomic interactions were determined using Swiss-Pdb Viewer programs available at http://swissmodel.expasy.org/. These analyses were performed at Altif Labs.

3.5. Statistical Analysis

Results are expressed as the mean ± standard error of the mean (SEM) of at least three independent experiments. Statistical analyses were performed using BellCurve for Excel ver. 2.13 software (Social Survey Research Information, Tokyo, Japan) and Kruskal-Wallis analysis with the Steel post-hoc test; $p < 0.05$ was considered to indicate statistical significance.

4. Conclusions

We investigated the possibility of blackcurrant anthocyanins binding to ERβ. The results show that these anthocyanins induced ERβ transcriptional activity, and that the IC_{50} was smaller for ERβ than for ERα. Consistent with these results, the affinity for ERβ was higher than that for ERα. In the structure of the ERβ/estradiol complex, some steric hindrance was found between sugar chain atoms and helices 3 and 12. However, as the conformation of these helices varies dynamically, we suggest that when each of the four blackcurrant anthocyanins bind to ERβ, they adopt a conformation suitable for accommodating glucoside or rutinoside. These results reveal that blackcurrant anthocyanins have phytoestrogen activity via ERβ. Therefore, blackcurrant anthocyanins may be effective for improvement of various senile-stage disorders known to be associated with ERβ, such as menopausal disorder and breast cancer.

Acknowledgments: The authors thank Fumiko Matsuzawa (Altif Laboratories) for her support. We would like to thank Editage (www.editage.jp) for English language editing. This research is partially supported by Japan Society for the Promotion of Science KAKENHI Grant Number 16K00844, the COI Next Generation Researchers Collaborative Research Fund from Japan Science and Technology Agency, Initiative for Realizing Diversity in the Research Environment, Funds for the Development of Human Resources in Science and Technology, Ministry of Education, Culture, Sports, Science and Technology and Hirosaki University Institutional Research Grant for Young Investigators.

Author Contributions: Naoki Nanashima designed the study, performed the experiments, analyzed the data, and wrote the manuscript; Kayo Horie and Hayato Maeda contributed to analysis of the data.

Conflicts of Interest: The authors declare no conflict of interest.

References

1. Jia, M.; Dahlman-Wright, K.; Gustafsson, J.A. Estrogen receptor alpha and beta in health and disease. *Best Pract. Res. Clin. Endocrinol. Metab.* **2015**, *29*, 557–568. [CrossRef] [PubMed]
2. Lobo, R.A. Metabolic syndrome after menopause and the role of hormones. *Maturitas* **2008**, *60*, 10–18. [CrossRef] [PubMed]

3. Prisby, R.D. Mechanical, hormonal and metabolic influences on blood vessels, blood flow and bone. *J. Endocrinol.* **2017**, *235*, R77–R100. [CrossRef] [PubMed]

4. Thornton, M.J. Estrogens and aging skin. *Dermatoendocrinology.* **2013**, *5*, 264–270. [CrossRef] [PubMed]

5. Kuiper, G.G.; Enmark, E.; Pelto-Huikko, M.; Nilsson, S.; Gustafsson, J.A. Cloning of a novel receptor expressed in rat prostate and ovary. *Proc. Natl. Acad. Sci. USA* **1996**, *93*, 5925–5930. [CrossRef] [PubMed]

6. Shearman, A.M.; Karasik, D.; Gruenthal, K.M.; Demissie, S.; Cupples, L.A.; Housman, D.E.; Kiel, D.P. Estrogen receptor beta polymorphisms are associated with bone mass in women and men: The Framingham Study. *J. Bone Miner. Res.* **2004**, *19*, 773–781. [CrossRef] [PubMed]

7. Hayashi, S.I.; Eguchi, H.; Tanimoto, K.; Yoshida, T.; Omoto, Y.; Inoue, A.; Yoshida, N.; Yamaguchi, Y. The expression and function of estrogen receptor alpha and beta in human breast cancer and its clinical application. *Endocr. Relat. Cancer* **2003**, *10*, 193–202. [CrossRef] [PubMed]

8. Lazennec, G.; Bresson, D.; Lucas, A.; Chauveau, C.; Vignon, F. ER beta inhibits proliferation and invasion of breast cancer cells. *Endocrinology* **2001**, *142*, 4120–4130. [CrossRef] [PubMed]

9. Yepuru, M.; Eswaraka, J.; Kearbey, J.D.; Barrett, C.M.; Raghow, S.; Veverka, K.A.; Miller, D.D.; Dalton, J.T.; Narayanan, R. Estrogen receptor-{beta}-selective ligands alleviate high-fat diet- and ovariectomy-induced obesity in mice. *J. Biol. Chem.* **2010**, *285*, 31292–31303. [CrossRef] [PubMed]

10. Williams, C.; Edvardsson, K.; Lewandowski, S.A.; Strom, A.; Gustafsson, J.A. A genome-wide study of the repressive effects of estrogen receptor beta on estrogen receptor alpha signaling in breast cancer cells. *Oncogene* **2008**, *27*, 1019–1032. [CrossRef] [PubMed]

11. Gopalan, A.; Reuben, S.C.; Ahmed, S.; Darvesh, A.S.; Hohmann, J.; Bishayee, A. The health benefits of blackcurrants. *Food Funct.* **2012**, *3*, 795–809. [CrossRef] [PubMed]

12. Lee, Y.M.; Yoon, Y.; Yoon, H.; Park, H.M.; Song, S.; Yeum, K.J. Dietary anthocyanins against obesity and inflammation. *Nutrients* **2017**, *9*, 1089. [CrossRef] [PubMed]

13. Nanashima, N.; Horie, K.; Chiba, M.; Nakano, M.; Maeda, H.; Nakamura, T. Anthocyaninrich blackcurrant extract inhibits proliferation of the MCF10A healthy human breast epithelial cell line through induction of G0/G1 arrest and apoptosis. *Mol. Med. Rep.* **2017**, *16*, 6134–6141. [PubMed]

14. Shaw, O.M.; Nyanhanda, T.; McGhie, T.K.; Harper, J.L.; Hurst, R.D. Blackcurrant anthocyanins modulate CCL11 secretion and suppress allergic airway inflammation. *Mol. Nutr. Food Res.* **2017**, *61*. [CrossRef] [PubMed]

15. Guo, D.; Wang, J.; Wang, X.; Luo, H.; Zhang, H.; Cao, D.; Chen, L.; Huang, N. Double directional adjusting estrogenic effect of naringin from *Rhizoma drynariae* (Gusuibu). *J. Ethnopharmacol.* **2011**, *138*, 451–457. [CrossRef] [PubMed]

16. Lee, Y.M.; Kim, J.B.; Bae, J.H.; Lee, J.S.; Kim, P.S.; Jang, H.H.; Kim, H.R. Estrogen-like activity of aqueous extract from *Agrimonia pilosa* Ledeb. in MCF-7 cells. *BMC Complement. Altern. Med.* **2012**, *12*, 260. [CrossRef] [PubMed]

17. Limer, J.L.; Speirs, V. Phyto-oestrogens and breast cancer chemoprevention. *Breast Cancer Res.* **2004**, *6*, 119–127. [CrossRef] [PubMed]

18. Mahmoud, A.M.; Yang, W.; Bosland, M.C. Soy isoflavones and prostate cancer: A review of molecular mechanisms. *J. Steroid Biochem. Mol. Biol.* **2014**, *140*, 116–132. [CrossRef] [PubMed]

19. Schmitt, E.; Stopper, H. Estrogenic activity of naturally occurring anthocyanidins. *Nutr. Cancer* **2001**, *41*, 145–149. [CrossRef] [PubMed]

20. Nanashima, N.; Horie, K.; Tomisawa, T.; Chiba, M.; Nakano, M.; Fujita, T.; Maeda, H.; Kitajima, M.; Takamagi, S.; Uchiyama, D.; et al. Phytoestrogenic activity of blackcurrant (*Ribes nigrum*) anthocyanins is mediated through estrogen receptor alpha. *Mol. Nutr. Food Res.* **2015**, *59*, 2419–2431. [CrossRef] [PubMed]

21. Kuiper, G.G.; Lemmen, J.G.; Carlsson, B.; Corton, J.C.; Safe, S.H.; van der Saag, P.T.; van der Burg, B.; Gustafsson, J.A. Interaction of estrogenic chemicals and phytoestrogens with estrogen receptor beta. *Endocrinology* **1998**, *139*, 4252–4263. [CrossRef] [PubMed]

22. Mersereau, J.E.; Levy, N.; Staub, R.E.; Baggett, S.; Zogovic, T.; Chow, S.; Ricke, W.A.; Tagliaferri, M.; Cohen, I.; Bjeldanes, L.F.; et al. Liquiritigenin is a plant-derived highly selective estrogen receptor beta agonist. *Mol. Cell. Endocrinol.* **2008**, *283*, 49–57. [CrossRef] [PubMed]

23. Setchell, K.D.; Clerici, C.; Lephart, E.D.; Cole, S.J.; Heenan, C.; Castellani, D.; Wolfe, B.E.; Nechemias-Zimmer, L.; Brown, N.M.; Lund, T.D.; et al. S-equol, a potent ligand for estrogen receptor beta, is the exclusive enantiomeric form of the soy isoflavone metabolite produced by human intestinal bacterial flora. *Am. J. Clin. Nutr.* **2005**, *81*, 1072–1079. [PubMed]

24. Sareddy, G.R.; Vadlamudi, R.K. Cancer therapy using natural ligands that target estrogen receptor beta. *Chin. J. Nat. Med.* **2015**, *13*, 801–807. [CrossRef]

25. Warner, M.; Huang, B.; Gustafsson, J.A. Estrogen receptor beta as a pharmaceutical target. *Trends Pharmacol. Sci.* **2017**, *38*, 92–99. [CrossRef] [PubMed]

26. Moreira, A.C.; Silva, A.M.; Santos, M.S.; Sardao, V.A. Phytoestrogens as alternative hormone replacement therapy in menopause: What is real, what is unknown. *J. Steroid Biochem. Mol. Biol.* **2014**, *143*, 61–71. [CrossRef] [PubMed]

27. Wuttke, W.; Jarry, H.; Westphalen, S.; Christoffel, V.; Seidlova-Wuttke, D. Phytoestrogens for hormone replacement therapy? *J. Steroid Biochem. Mol. Biol.* **2002**, *83*, 133–147. [CrossRef]

28. Manas, E.S.; Xu, Z.B.; Unwalla, R.J.; Somers, W.S. Understanding the selectivity of genistein for human estrogen receptor-beta using X-ray crystallography and computational methods. *Structure* **2004**, *12*, 2197–2207. [CrossRef] [PubMed]

29. Matsumura, A.; Ghosh, A.; Pope, G.S.; Darbre, P.D. Comparative study of oestrogenic properties of eight phytoestrogens in MCF7 human breast cancer cells. *J. Steroid Biochem. Mol. Biol.* **2005**, *94*, 431–443. [CrossRef] [PubMed]

30. Mocklinghoff, S.; Rose, R.; Carraz, M.; Visser, A.; Ottmann, C.; Brunsveld, L. Synthesis and crystal structure of a phosphorylated estrogen receptor ligand binding domain. *ChemBioChem* **2010**, *11*, 2251–2254. [CrossRef] [PubMed]

31. Guex, N.; Peitsch, M.C.; Schwede, T. Automated comparative protein structure modeling with SWISS-MODEL and Swiss-PdbViewer: A historical perspective. *Electrophoresis* **2009**, *30* (Suppl. 1), S162–S173. [CrossRef] [PubMed]

32. Fang, H.; Tong, W.; Shi, L.M.; Blair, R.; Perkins, R.; Branham, W.; Hass, B.S.; Xie, Q.; Dial, S.L.; Moland, C.L.; et al. Structure-activity relationships for a large diverse set of natural, synthetic, and environmental estrogens. *Chem. Res. Toxicol.* **2001**, *14*, 280–294. [CrossRef] [PubMed]

33. Bourguet, W.; Germain, P.; Gronemeyer, H. Nuclear receptor ligand-binding domains: Three-dimensional structures, molecular interactions and pharmacological implications. *Trends Pharmacol. Sci.* **2000**, *21*, 381–388. [CrossRef]

34. Bruning, J.B.; Parent, A.A.; Gil, G.; Zhao, M.; Nowak, J.; Pace, M.C.; Smith, C.L.; Afonine, P.V.; Adams, P.D.; Katzenellenbogen, J.A.; et al. Coupling of receptor conformation and ligand orientation determine graded activity. *Nat. Chem. Biol.* **2010**, *6*, 837–843. [CrossRef] [PubMed]

35. Brzozowski, A.M.; Pike, A.C.; Dauter, Z.; Hubbard, R.E.; Bonn, T.; Engstrom, O.; Ohman, L.; Greene, G.L.; Gustafsson, J.A.; Carlquist, M. Molecular basis of agonism and antagonism in the oestrogen receptor. *Nature* **1997**, *389*, 753–758. [CrossRef] [PubMed]

36. Sacks, F.M. Dietary phytoestrogens to prevent cardiovascular disease: Early promise unfulfilled. *Circulation* **2005**, *111*, 385–387. [CrossRef] [PubMed]

37. Crisafulli, A.; Altavilla, D.; Marini, H.; Bitto, A.; Cucinotta, D.; Frisina, N.; Corrado, F.; D'Anna, R.; Squadrito, G.; Adamo, E.B.; et al. Effects of the phytoestrogen genistein on cardiovascular risk factors in postmenopausal women. *Menopause* **2005**, *12*, 186–192. [CrossRef] [PubMed]

38. Lissin, L.W.; Cooke, J.P. Phytoestrogens and cardiovascular health. *J. Am. Coll. Cardiol.* **2000**, *35*, 1403–1410. [CrossRef]

39. Sirotkin, A.V.; Harrath, A.H. Phytoestrogens and their effects. *Eur. J. Pharmacol.* **2014**, *741*, 230–236. [CrossRef] [PubMed]

40. Carruba, G. Estrogens and mechanisms of prostate cancer progression. *Ann. N. Y. Acad. Sci.* **2006**, *1089*, 201–217. [CrossRef] [PubMed]
41. Sciarra, F.; Toscano, V. Role of estrogens in human benign prostatic hyperplasia. *Arch. Androl.* **2000**, *44*, 213–220. [PubMed]

Sample Availability: Samples of the compounds are available from the authors.

molecules

MDPI

Article

Antinociceptive and Antibacterial Properties of Anthocyanins and Flavonols from Fruits of Black and Non-Black Mulberries

Hu Chen, Wansha Yu, Guo Chen, Shuai Meng, Zhonghuai Xiang and Ningjia He *

State Key Laboratory of Silkworm Genome Biology, Southwest University, Beibei, Chongqing 400715, China; c9249@email.swu.edu.cn (H.C.); summer92@email.swu.edu.cn (W.Y.); chenguo1992@email.swu.edu.cn (G.C.); xndxms@163.com (S.M.); xbxzh@swu.edu.cn (Z.X.)
* Correspondence: hejia@swu.edu.cn; Tel.: +86-23-6825-0797; Fax: +86-23-6825-1128

Received: 21 November 2017; Accepted: 19 December 2017; Published: 21 December 2017

Abstract: Anthocyanins and flavones are important pigments responsible for the coloration of fruits. Mulberry fruit is rich in anthocyanins and flavonols, which have multiple uses in traditional Chinese medicine. The antinociceptive and antibacterial activities of total flavonoids (TF) from black mulberry (MnTF, TF of *Morus nigra*) and non-black mulberry (MmTF, TF of *Morus mongolica*; and MazTF, TF of *Morus alba* 'Zhenzhubai') fruits were studied. MnTF was rich in anthocyanins (11.3 mg/g) and flavonols (0.7 mg/g) identified by ultra-performance liquid chromatography–tunable ultraviolet/mass single-quadruple detection (UPLC–TUV/QDa). Comparatively, MmTF and MazTF had low flavonol contents and MazTF had no anthocyanins. MnTF showed significantly higher antinociceptive and antibacterial activities toward *Escherichia coli*, *Pseudomonas aeruginosa* and *Staphylococcus aureus* than MmTF and MazTF. MnTF inhibited the expression of interleukin 6 (IL-6), inducible nitric oxide synthase (iNOS), phospho-p65 (p-p65) and phospho-IκBα (p-IκBα), and increased interleukin 10 (IL-10). Additionally, mice tests showed that cyanidin-3-*O*-glucoside (C3G), rutin (Ru) and isoquercetin (IQ) were the main active ingredients in the antinociceptive process. Stronger antinociceptive effect of MnTF was correlated with its high content of anthocyanins and flavonols and its inhibitory effects on proinflammatory cytokines, iNOS and nuclear factor-κB (NF-κB) pathway-related proteins.

Keywords: anthocyanins; antibacterial; antinociceptive; flavonols; *Morus*

1. Introduction

Mulberry is a deciduous tree or shrub of the genus *Morus* in the family *Moraceae* [1]. It has been cultivated and used in traditional medicine by humans for more than 5000 years [2–4]. Mulberry originated in China and is grown throughout Korea, Japan, Mongolia, Southwest Asia, Central Asia, Russia, Europe and South America [5,6]. Eight species of *Morus* were identified by phylogenetic analysis of internal transcribed spacer sequences [7]. The species can also be divided into black mulberry, white mulberry and red mulberry [1].

Previous research has shown that mulberry fruits are rich in anthocyanins, which are responsible for the fruit color [5,8,9]. Anthocyanins have important roles in plants and animals, such as protecting plants from damage caused by UV light, attracting pollinators and serving as antioxidants [10–12]. They also have pharmacological activities, including anti-inflammatory, antitumor, and blood lipid-regulating activities [3,13–15]. Mulberry fruits also contain many active substances, such as flavonols, polyphenols, alkaloids and polysaccharides [1,3,8,16]. Modern pharmacological studies have shown that mulberry fruits may provide health benefits through antioxidant, hypoglycemic, antiobesity, anti-inflammatory, analgesic and immunomodulatory effects [16–20].

Inflammatory pain is a very common and important basic pathological process. Pain is a key diagnostic criterion in many acute and chronic medical conditions [21,22]. The inflammatory pathways (such as arachidonic acid metabolic, NF-κB and nitric oxide (NO) pathways) and inflammatory biomarkers (such as IL-6, IL-10 and iNOS) are associated with pain [3,22]. The most common diseases associated with trauma and infection are inflammatory diseases. Mechanical damage, bacterial infections (e.g., *E. coli*, *P. aeruginosa* or *S. aureus*), viral infections and some drugs can cause pain. Steroidal anti-inflammatory drugs, such as dexamethasone (Dex), and non-steroidal antinociceptive drugs, such as aspirin (Asp), are used to combat inflammation and pain. However, they are also associated with significant side effects, such as weight loss and gastrointestinal disorders [3,23,24]. Asp is one of the world's top three classic drugs and is widely used in analgesic and anti-inflammatory applications. However, millions of people suffer from its side effects every year [25,26]. The most widely used antibiotics, such as penicillin and cefoperazone, have limited recognition of adverse consequences [27,28]. Anthocyanins and flavonols, as natural products, have not yet displayed side effects, which is essential for drug development.

Our previous study had identified nine putative genes involved in anthocyanin and flavonoid biosynthesis in mulberry plants, and anthocyanin content correlated with the expression levels of these genes during the fruit ripening process [29]. Previous studies of ours have found that total flavonoids (TF) of black mulberry possess anti-inflammatory and antioxidant activities that might be correlated with its high anthocyanin content [3,30]. However, the differences in compositions are not clear, nor is it clear whether the TFs (MnTF [total flavonoids of *Morus nigra*], MmTF [total flavonoids of *Morus mongolica*] and MazTF [total flavonoids of *Morus alba* 'Zhenzhubai']) have antinociceptive and antibacterial activities, and which compound is the main active ingredient. Therefore, the aim of this study was to compare the compositions and antinociceptive and antibacterial activities of TFs from black and non-black mulberry fruits and to explore the main active ingredients of these effects.

2. Results

2.1. Determination of Anthocyanin and Flavonol Contents

Three anthocyanins and five flavonols were detected in the TFs by UPLC with tunable ultraviolet (TUV) and quadrupole dalton (QDa) detectors. As shown in Figure 1 and Supplementary Material Figure S1, this method completely resolved all eight compounds within 7 min. As shown in Table 1, all of the calibration curves had good linearity ($r^2 > 0.99$). MnTF and MmTF contained all eight compounds, while MazTF did not contain any anthocyanins. MnTF had more anthocyanins and flavonols than MmTF and MazTF. The C3G, cyanidin-3-*O*-rutinoside (C3R) and pelargonidin-3-*O*-glucoside (Pg3G) contents of MnTF were 8.2, 2.9 and 0.3 mg/g, respectively. All five flavonols were scarce (<0.5 mg/g), especially morin hydrate (Mh) and kaempferol (Ka), which were present in very low amounts or could not be detected. The flavonol contents of MazTF were lower than those of MnTF and MmTF, except for Qu, which was more abundant in MazTF (0.0036 mg/g) than in MmTF (0.0029 mg/g). In general, black mulberries were rich in anthocyanins and flavonols, while non-black mulberries had low amounts of flavonols and few anthocyanins.

2.2. Toxicity Assessment of TFs

Changes in weight and cytotoxicity were used to evaluate the toxicity of drugs in vivo and in vitro. The weight of mice decreased gradually after administration of Dex. Side effects were shown in mice when administered at a dose of 1.5 mg/kg Dex. No significant differences in the weight of mice were detected between the control group and groups treated by TFs (MnTF, MmTF and MazTF) at a dose of 5 g crude extract/kg (Figure 2a). The levels of cytotoxicity were assayed by RAW 264.7 cells in vitro. As shown in Figure 2b, none of the drugs were cytotoxic to cells at the administered dose.

Table 1. Information of chromatography and MS of *M. nigra*, *M. mongolica* and *M. alba* 'Zhenzhubai'.

Compounds	RT [a] (min)	Regression Equation [b]	r^2	Content (mg/g) [c]			Selected Ions by QDa (m/z)
				MnTF	MmTF	MazTF	
C3G	2.186	$y = (12.499x + 1.239) \times 10^3$	0.9999	8.2168 ± 0.0238	0.2220 ± 0.0024	ND	449.18
C3R	2.627	$y = (8.765x + 1.550) \times 10^3$	0.9999	2.8578 ± 0.0146	0.0610 ± 0.0013	ND	595.33
P3G	2.983	$y = (5.230x + 0.770) \times 10^3$	0.9999	0.2539 ± 0.0047	0.0057 ± 0.0003	ND	433.24
Ru	4.556	$y = (6.065x + 2.362) \times 10^3$	0.9999	0.4498 ± 0.0075	0.2723 ± 0.0013	0.0816 ± 0.0015	302.93
IQ	4.624	$y = (2.560x + 0.080) \times 10^3$	0.9999	0.1639 ± 0.0006	0.2459 ± 0.0059	0.0631 ± 0.0033	303.06
Mh	5.405	$y = (6.880x + 0.226) \times 10^3$	0.9999	0.0002 ± 0.0001	<0.0001	<0.0001	303.04
Qu	5.786	$y = (4.870x - 0.074) \times 10^3$	0.9993	0.0716 ± 0.0045	0.0029 ± 0.0002	0.0036 ± 0.0004	303.11
Ka	6.497	$y = (2.710x + 0.487) \times 10^3$	0.9986	<0.0001	<0.0001	ND	287.03

[a] RT, retention time; [b] y, peak area; x, concentration injected (µg/mL); [c] mg/g, weight of the dry powder; ND, not detected.

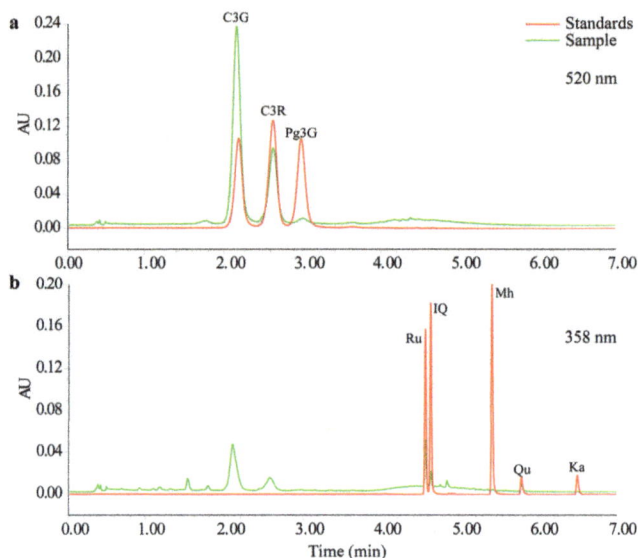

Figure 1. Chromatograms of anthocyanins (**a**) and flavonols (**b**) obtained by UPLC–TUV/QDa.

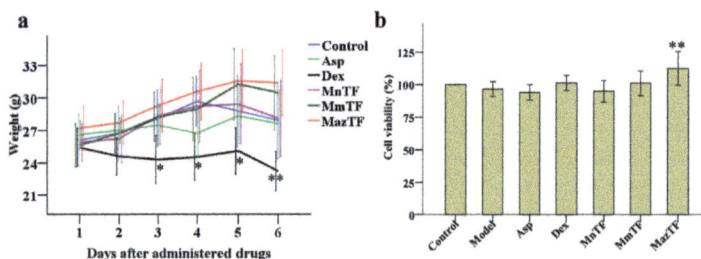

Figure 2. Effects of TFs on weight of mice (**a**) and cytotoxicity of RAW 264.7 cells (**b**). Groups of mice were pretreated (p.o.) with reverse-osmosis water (control and model groups, 20 mL/kg), Asp (aspirin, 150 mg/kg), Dex (dexamethasone, 3 mg/kg), or TFs (5 g crude extract/kg). Data are means ± SD ($n = 10$). RAW 264.7 cells were treated with DMEM (control group), 1 µg/mL lipopolysaccharide (LPS, model group), 1 µg/mL LPS + 0.1 mg/mL Asp, 1 µg/mL LPS + 0.1 mg/mL Dex, or 1 µg/mL LPS + 50 mg crude TF extract/mL. Cell viability is expressed as a percentage of that in the control group, which was set at 100%. Data are means ± SD ($n = 3$).Values with asterisks are significantly different (* $p < 0.05$; ** $p < 0.01$) from those in the control group in (**a**) or the model group in (**b**).

2.3. Antinociceptive Activities of TFs

The response pattern in the formalin-induced pain test consists of two phases, a neurogenic pain phase (0–5 min) and an inflammatory pain phase (15–30 min). As shown in Figure 3a and Supplementary Material Table S1, Asp, an antinociceptive drug, significantly reduced the duration of both phases, while Dex significantly reduced the duration of the secondary phase. The secondary phase in the MnTF and MazTF groups (60 ± 20 s and 48 ± 52 s, respectively) was significantly shorter ($p < 0.05$) than that in the control group (122 ± 49 s). MmTF did not show antinociceptive activity, as it did not significantly reduce the licking (licking, biting or flinching) time of either phase.

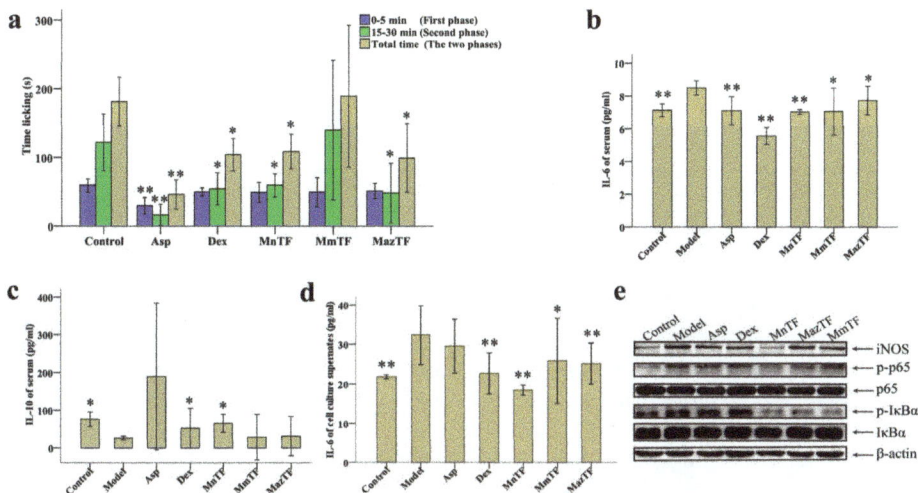

Figure 3. Effects of TFs on pain (**a**); levels of IL-6 (**b**) and IL-10 (**c**) in mice; and levels of IL-6 (**d**) and the expression of pain-related proteins (**e**) in RAW 264.7 cells. Values with asterisks are significantly different (* $p < 0.05$; ** $p < 0.01$) from those in the control group in (**a**); or the model group in (**b–d**).

2.4. Effects of TFs on Cytokines and Pain-Related Proteins

To study the mechanism of antinociceptive effects of TFs, we measured levels of an inflammatory cytokine (IL-6) and an anti-inflammatory cytokine (IL-10). As shown in Figure 3b, the IL-6 level in serum was significantly reduced by MnTF (7.0 pg/mL), MmTF (7.1 pg/mL) and MazTF (7.7 pg/mL) compared with the model group (8.5 pg/mL). There was a similar trend in cell culture supernates (Figure 3d). Only MnTF significantly increased the serum level of IL-10, from 26.7 pg/mL to 66.0 pg/mL ($p < 0.05$), after injury (Figure 3c). In summary, TF of black mulberries significantly reduced the level of an early and mid-term development of inflammatory cytokine (IL-6) and increased the level of an anti-inflammatory cytokine (IL-10), while TF of non-black mulberries had significant effects on IL-6 production.

Western blotting was used to investigate the effects of TFs on the expression of inflammation-related proteins (iNOS, p65, IκBα, p-p65 and p-IκBα) in RAW 264.7 cells. As shown in Figure 3e and Supporting Information Figure S2, the expression levels of p65 and IκBα were not significantly different among groups, while expression of the phosphorylated products (p-p65 and p-IκBα) decreased significantly in the TF-treated groups, especially the MnTF group. The expression level of iNOS was increased in all groups except the MnTF group. Thus, TFs of black mulberries had stronger effects on p-p65, p-IκBα and iNOS expression than TFs of non-black mulberries. A schematic representation of the inhibitory effect of MnTF on the NF-κB and NO pathways is shown in Supplementary Material Figure S3.

2.5. Antinociceptive Activities of C3G, Ru and IQ

The mice tests of three main flavonoids of TFs were performed to learn the main active ingredients in the antinociceptive process. As shown in Figure 4 and Supplementary Material Table S1, neither C3G, Ru nor IQ individually reduced the duration of both phases, while the mix (C3G, Ru and IQ) significantly reduced the duration of the secondary phase (inflammatory pain phase). Therefore, anthocyanins and flavonols work together to yield more effective antinociceptive activity.

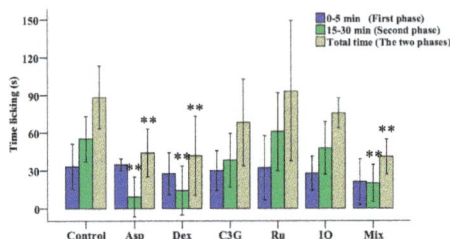

Figure 4. Antinociceptive activities of C3G, Ru and IQ in mice. Values with asterisks are significantly different (** $p < 0.01$) from those in the control group.

2.6. Antibacterial Activities of TFs

E. coli, P. aeruginosa and *S. aureus* are three species of inflammatory pain-causing bacteria. As shown in Figure 5a, MnTF strongly inhibited all three bacteria. While bacteria barely grew on Luria–Bertani (LB) plates after being treated with MnTF at 1.8 mg/mL, MmTF-treated and MazTF-treated bacteria covered the plates. Additionally, Figure 5b shows that the antibacterial activity of MnTF was stronger than that of MmTF ($p < 0.01$) or MazTF ($p < 0.01$). In general, TFs of black mulberries had stronger antibacterial activity than TFs of non-black mulberries.

In the minimum bactericidal concentration (MBC) test, MnTF showed strong, dose-dependent antibacterial activities against all three bacteria (Figure 5c). The MBCs of MnTF for *E. coli, P. aeruginosa* and *S. aureus* were 2, 2 and 1.8 mg/mL, respectively (Figure 5c and Figure S4).

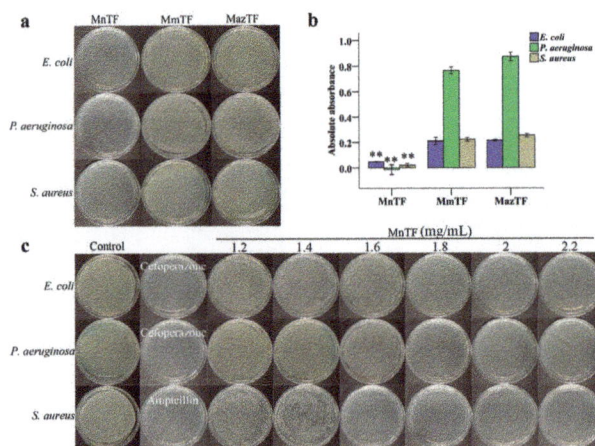

Figure 5. Antibacterial activities of TFs shown on plates (**a**) and by absorbance (**b**); and minimum bactericidal concentration (MBC) of MnTF against *E. coli, P. aeruginosa* and *S. aureus* (**c**). The concentration of TFs was 1.8 mg/mL (**a**). Data are means ± SD ($n = 3$). Values with asterisks (**) are significantly different ($p < 0.01$) from those in the MmTF and MazTF groups.

3. Discussion

Mulberry fruits have many bioactive components, such as anthocyanins, flavonols and polysaccharides [3,9,18]. In this study, the chromatography method we used enabled detection of three anthocyanins in mulberry fruits (Figure 1a). The key factor in the UPLC method was the proportion of acetonitrile (ACN) in mobile phase B. The contents of C3G (8.2 mg/g) and C3R (2.9 mg/g) in MnTF were consistent with those in our previous study (8.3 mg/g and 2.9 mg/g, respectively) [3]. However, one other anthocyanin (Pg3G) and four additional important flavonols were identified by UPLC–TUV/QDa in MnTF in this study. Thus, most of the flavonoids identified in mulberry fruits were flavonols. Although five anthocyanins and many flavonols were previously identified in mulberry extracts by high-performance liquid chromatography (HPLC) or UPLC, it was difficult to achieve good resolution of the anthocyanin peaks [5,31]. C3G and C3R are the main anthocyanins responsible for the color of mulberry fruits. The anthocyanin and flavonol contents of black mulberries were about 40-times and 1.3-times higher, respectively, than those of *M. mongolica* fruits (Table 1).

Weight variation in mice and cytotoxicity are widely used to evaluate the drug toxicity. Our study showed that Dex had significant side effects in mice (Figure 2a), which is consistent with other studies [23,24,32]. In contrast, TFs were not toxic to mice and cells as can be seen by the unchangeable weight and high levels of cell viability similar to the control group (Figure 2a,b). Furthermore, mulberry is a traditional fruit that has long been consumed by humans. Therefore, mulberry TFs are considered to be safe when used for functional development as health products.

Formalin-induced pain-like behavior in mice was a classic model of inflammation and pain [33,34]. Inflammatory pain is divided into a neurogenic phase (initial phase, 0–5 min) and an inflammatory phase (secondary phase, 15–30 min) in the formalin-induced pain model [3,34]. In this study, MnTF and MazTF showed stable antinociceptive effect (Figure 3a,b,d). Moreover, MazTF had low flavonol contents and had no anthocyanins by UPLC–TUV/QDa. This means that flavonols play an important role in the analgesic effect. Some previous studies have also proved this view [35–37]. Meanwhile, treatment of mice with MnTF at a dose of 5 g crude extract/kg reduced the duration of the secondary phase (Figure 3a and Table S1). A previous study showed that MnTF at a dose of 2.5 g crude extract/kg could also reduce the duration of the neurogenic phase [3].

Pain and inflammation are related to the arachidonic acid metabolic (AAM) pathway, the NF-κB pathway and the NO pathway [38–40]. Dex (a steroidal anti-inflammatory drug) and Asp (a non-steroidal antinociceptive drug) combat pain and inflammation by inhibiting leukotrienes and prostaglandins of the AAM pathway, but they are associated with significant side effects. Previous studies [38,41,42] and this study showed that MnTF regulated the inflammatory process mainly by affecting the NF-κB and NO pathways. The expression of IL-6 is regulated by the NF-κB [43]. NF-κB is normally present in cells in a p50/p65 heterodimeric inhibitory state, which can be activated by phosphorylation, thereby promoting the production of IL-6, TNF-α, IL-1β and so on [44–46]. In the early stage of inflammatory infection, IL-6 is induced by TNF-α and keeps a rising trend for a long time in the later period [34,47]. IL-6 may be a feedback inhibition factor of TNF-α and may negatively regulate the production and release of endogenous TNF-α [48–50]. As shown in Figure 3e and Supplementary Material Figure S2, MnTF inhibited the expression of iNOS, p-IκBα and p-p65, reduced the levels of pain-related cytokines (TNF-α, IL-1β, IL-6, IFN-γ and NO), and increased the level of an anti-inflammatory cytokine (IL-10). Asp and Dex regulate the inflammatory and antinociceptive process by affecting the AAM pathway. The inhibition of AAM blocked the activity of cyclooxygenase 1 (COX-1), which resulted in the intestinal lesions and the weight changes in mice [40,51]. MnTF did not have this side effect (Figure 2a,b).

The mixed reagents of C3G, Ru and IQ showed significant antinociceptive activity (Figure 4 and Table S1). Although MazTF had no anthocyanins, containing Ru, IQ and Qu (Table 1), it showed excellent non-toxic and antinociceptive activities (Figures 2 and 3a). Therefore, the fundamental reason why MnTF showed excellent anti-inflammatory activities was that MnTF contains the three main active ingredients, and the flavonols might be the key active substances to play an analgesic effect with anthocyanins, synergistically.

In this study, we investigated the antibacterial properties of anthocyanins and flavonols from mulberry fruits against *E. coli*, *P. aeruginosa* and *S. aureus*. The antibacterial activity of TFs from black mulberries was stronger than that of TFs from non-black mulberries (Figure 5a,b). In the MBC assay, we observed that MnTF showed stronger inhibitory activity against *S. aureus* than against *E. coli* (Figure 5c and Figure S4). Similarly, Wang, Li and Jiang (2010) reported that mulberry polysaccharides had antibacterial activities against *Bacillus subtilis*, *E. coli* and *S. aureus*, and that the antibacterial activity against *E. coli* was especially strong [52]. Morin of mulberry fruits moderately inhibited *Streptococcus mutans*. Moreover, stem bark of *M. alba* var. *alba*, *M. alba* var. *rosa* and *Morus rubra* were potent antimicrobial agents against bacteria that cause infections in humans (*S. aureus*, *Enterococcus faecalis*, *Staphylococcus epidermis*, *E. coli* and *Salmonella Typhimurium bacteria*) [53]. Thus, various parts of mulberry, such as fruits, leaves and stem bark, have antibacterial activities against a variety of bacteria.

Bacteria cause the oxidative stress reaction and produce reactive oxygen species in the body [54,55]. Polyhydroxy compounds inhibit oxidative stress and induce the body's release of the related inflammatory factors [56]. Black mulberry is rich in anthocyanins and flavonols, which might be responsible for its antioxidant activity. More in-depth experiments need to be done in the future.

4. Materials and Methods

4.1. Mulberry Fruits and Animals

Fruits of *M. nigra* were collected at the Hetian Sericultural Research Institute (37°08′50.85″ N, 79°54′26.99″ E; Xinjiang, China) in July 2016. Fruits of *M. mongolica* and *M. alba* 'Zhenzhubai' (a mulberry cultivar) were collected from the mulberry breeding center at Southwest University (29°49′36.72″ N, 106°25′29.19″ E; Chongqing, China) in May 2016. The voucher specimens of *M. nigra* (Mn-20160720), *M. mongolica* (Mm-20160515) and *M. alba* 'Zhenzhubai' (Maz-20160515) were kept at our laboratory. The fruits were homogenized, oven-dried to a constant mass, and ground into a powder. The powder was sieved with a 60-mesh sieve and stored at −40 °C for further analysis. Male Kunming mice (about 18–22 g) were purchased from Chongqing Medical University, Chongqing. Experiments were carried out according to the guidelines of the International Association for the Study of Pain on the use of animals in pain research. This research was approved by the Animal Care Committee of Southwest University (License number: SCXK (JUN) 2012-0011).

4.2. Chemicals and Reagents

C3G, C3R, Pg3G, and LPS were purchased from Sigma-Aldrich (St. Louis, MO, USA). Ru, IQ, Mh, Qu, and Ka were obtained from ChromaBio (Chengdu, China). Phosphoric acid, ACN, and methanol for UPLC were purchased from Thermo Fisher Scientific (Waltham, MA, USA). Asp and Dex for animal experiments were purchased from Original (Shenyang, Liaoning, China) and Xianju Pharma (Hangzhou, Zhejiang, China), respectively. Ultrapure water was prepared using a Milli-Q system (Millipore, Billerica, MA, USA).

The strains of *E. coli* (CMCC44102), *P. aeruginosa* (CMCC10104), and *S. aureus* (CMCC26003) were provided by the National Center for Medical Culture (CMCC) (Beijing, China) and kept at our laboratory. Enzyme-linked immunosorbent assay (ELISA) kits for IL-6 and IL-10 were purchased from CUSABIO (Wuhan, China). Fetal bovine serum, antibiotics (streptomycin and penicillin), trypsin, and DMEM were purchased from Gibco (Grand Island, NY, USA). MTT and BeyoECL Plus and Enhanced BCA Protein Assay kits were purchased from Beyotime (Shanghai, China). Western blotting reagents were purchased from Cell Signaling Technology (Boston, MA, USA).

4.3. Extraction of TFs

TFs were extracted and measured as previously described [30] except that the final extraction volume was increased from 200 mL to 400 mL. MnTF, MmTF, and MazTF represent flavonoid extracts of *M. nigra*, *M. mongolica*, and *M. alba* 'Zhenzhubai', respectively.

4.4. UPLC–TUV/QDa Conditions and Determination of TFs

Chromatographic separation was carried out on a Waters Acquity UPLC I-Class system coupled with TUV and a single-quadrupole mass detector (QDa) with an electrospray ionization source and an Acquity UPLC BEH C_{18} column (1.0 × 100 mm, 1.7 μm, Waters, Milford, MA, USA). The solvent system consisted of a binary mobile phase: solution A was Milli-Q water with 0.2% (*v/v*) H_3PO_4, and solution B was 40% (*v/v*) ACN with 0.2% (*v/v*) H_3PO_4. The linear elution gradient profile was as follows: 0–3 min, 20–27% B, curve 6; 3–6.5 min, 27–84% B, curve 5; 6.5–7 min, 84–20% B, curve 1. The flow rate was 0.17 mL/min, the column temperature was kept at 40 °C, and the injection volume was 1 μL. The detection wavelengths were 520 nm and 358 nm for anthocyanins and flavonoids, respectively. Concentration detection range: C3G (1.56–100 μg/mL), C3R (0.78–50 μg/mL), P3G, Ru, IQ, Mh, Qu, and Ka (1.56–100 μg/mL).

QDa detector was achieved in SIR and positive electrospray ionization mode (ESI+) mode. The capillary voltage was 0.8 kV and the probe temperature was 600 °C. The sampling frequency was 5 Hz. The 287 *m/z* (Ka), 303 *m/z* (Ru, IQ, Mh, and Qu), 433 *m/z* (P3G), 449 *m/z* (C3G), and 595 *m/z* (C3R) ions were monitored with a cone voltage of 21 kV.

4.5. Evaluation of Toxicity and Antinociceptive Activity in Mice

Male Kunming mice were divided into groups (*n* = 10 for each group) and treated with reverse-osmosis water (20 mL/kg; control), Asp (150 mg/kg), Dex (3 mg/kg), TFs (5 g crude extract/kg), C3G, Ru, or IQ (40 mg/kg), and Mix (Mixed solution of 40 mg/kg C3G, 2.33 mg/kg Ru, and 0.87 mg/kg IQ). The mice were fed adaptively for three days and after that drugs were administered by gavage once per day for seven days. The weights of mice were measured and recorded before gavage and on each of the last six days of drug administration.

Antinociceptive effect was studied in a formalin-induced mouse pain-like behavior model according to previously described methods [3,33,34]. Ten microliters of 2.5% (*v/v*) formalin solution was injected into the left hind paw of mice in all groups. Then, the total number of mice that exhibited pain-like behaviors (licking, biting, and flinching) was recorded in the neurogenic phase (initial phase, 0–5 min after formalin injection) and the inflammatory phase (secondary phase, 15–30 min after formalin injection).

4.6. Immunological Procedures

4.6.1. Blood Collection for Cytokines Analysis

Male Kunming mice were divided into seven groups and treated: the six groups described in Section 4.5, and a model group was added and treated with reverse-osmosis water (20 mL/kg). After the last treatment of drugs, the dorsal side of the right ear of all mice, except those in the control group, was treated with 0.2 mL of xylene and 0.2 mL of 0.4% (*v/v*) acetic acid (intraperitoneal injection). Three hours later, an eyeball was extirpated and blood was collected. Blood samples were clotted overnight at 4 °C and then centrifuged at 1000× *g* for 15 min. The serum was immediately assayed or stored at −40 °C [3].

4.6.2. Cell Culture, Cytotoxicity and Western Blot Analysis

RAW 264.7 cells were cultured in 24-well plates and treated with DMEM (control group), 1 μg/mL LPS (model group), 1 μg/mL LPS + 0.1 mg/mL Asp, 1 μg/mL LPS + 0.1 mg/mL Dex, or 1 μg/mL

LPS + 50 mg crude TFs extract/mL. After addition of the drugs, cells were incubated for 24 h and then IL-6 was assessed by ELISA, and cytotoxicity was assessed by the MTT method [38,41].

Expression of key regulatory proteins, including iNOS, p65, IκBα, and the phosphorylated products p-p65 and p-IκBα, was analyzed by Western blotting [42,57]. RAW 264.7 cell lysates (25 μg of protein) were subjected to electrophoresis on 12% sodium dodecyl sulfate–polyacrylamide gels. The resolved proteins were transferred to polyvinylidene fluoride membranes (240 mA for 90 min). The membranes were incubated with Tris-buffered saline (TBS) containing 5% (w/v) nonfat milk at 4 °C overnight, washed with TBS–Tween 20 (TBST) for 5 min, incubated with specific antibodies against iNOS, p65, IκBα, p-p65, p-IκBα, or β-actin for 1 h, and washed three times with TBST for 10 min. Then, the membranes were incubated with horseradish peroxidase-conjugated secondary antibody for 1 h and washed three times with TBST for 10 min. Blots were visualized using the BeyoECL Plus kit.

4.7. Antibacterial Assays

4.7.1. Comparison of Antibacterial Activities

Three bacteria (*E. coli*, *P. aeruginosa*, and *S. aureus*) were used to evaluate antibacterial activity of TFs. Briefly, 90 mg/mL TFs was diluted 50 fold by LB medium (the effective concentrations of C3G, Ru and Qu were 6.4 μg/mL, 0.8 μg/mL and 0.4 μg/mL, respectively). 0.2 mL of the diluted TFs was mixed with 0.02 mL of bacterial suspension (OD_{600} = 0.6) and incubated at 37 °C for 24 h. The antibacterial activity was evaluated by measuring the OD_{600} of the bacterial suspension. Then, 0.1 mL of the culture was spread on LB plates. The inoculated plates were incubated at 37 °C for 24 h and then the numbers of colonies were recorded.

4.7.2. MBC of MnTF

MBC is the minimum drug concentration required to kill 99.9% of the test microorganisms or inhibit the growth of colonies to not more than five. For *E. coli*, *P. aeruginosa*, and *S. aureus*, the positive control drugs were cefoperazone (100 μg/mL), cefoperazone (200 μg/mL), and ampicillin (100 μg/mL), respectively. Six isocratic solutions of MnTF (60 mg/mL to 110 mg/mL, MnTF was diluted 50 fold by LB medium) were prepared. Then, 0.2 mL of drug or MnTF was mixed with 0.02 mL of bacterial suspension (OD_{600} = 0.6) and incubated at 37 °C for 24 h. The OD_{600} of the bacterial suspension was measured and then 0.1 mL of the culture was spread on LB plates. The inoculated plates were incubated at 37 °C for 24 h and then the numbers of colonies were recorded [58].

4.8. Statistical Analyses

Results were expressed as means ± standard deviation (SD). Data were analyzed using SPSS Statistics version 17.0. One-way analysis of variance was used for intergroup comparisons; p values < 0.05 were considered statistically significant.

5. Conclusions

Three anthocyanins and five flavonols were identified in mulberry fruits. TF of black mulberry exhibited stronger antinociceptive and antibacterial effects than that of non-black mulberries. One of the conclusions made from the present study demonstrated that anthocyanins (C3G) and flavonols (Ru and IQ) were responsible for, or at least would be correlated with, the antinociceptive effect of black mulberry. Evidence of inhibitory effects on proinflammatory cytokines, iNOS and NF-κB pathway-related proteins underlying antinociceptive and antibacterial effects of mulberry TF will expand our knowledge of anthocyanins and flavonols, and could be incorporated into alternatives for analgesic and antibacterial drugs.

Supplementary Materials: The following are available online, Figure S1: Spectra of anthocyanins and flavonols by UPLC-TUV/QDa; Figure S2. Western blotting of TFs on the expression of inflammation-related proteins. The grayscale of β-actin is set to 1, and the other groups use it as a reference; Figure S3: Schematic representation of the inhibitory effect of MnTF on the cytokines, NF-κB, and NO pathways in RAW 264.7 cells; Figure S4: MBC of MnTF against *E. coli* (a), *P. aeruginosa* (b), and *S. aureus* (c) shown by absorbance. Data are means ± SD ($n = 3$). Values with asterisks (**) are significantly different ($p < 0.01$) from values in the control group; Table S1: Data of antinociceptive activities in mice.

Acknowledgments: This project was funded by the research grants from the National Hi-Tech Research and Development Program of China [grant numbers 2013AA100605-3]; Natural Science Foundation of China [grant numbers 31572323]; China Postdoctoral Science Foundation Funded Projects [grant numbers 2013M540694, 2014T70845 and 2016M592622].

Author Contributions: H.C. and N.H. conceived and designed the experiments. H.C., W.Y., G.C., and S.M. performed the experiments. H.C., N.H., and Z.X. analyzed the data. H.C. and N.H. wrote the paper.

Conflicts of Interest: The authors declare no conflict of interest.

Abbreviations

ACN	acetonitrile
Asp	aspirin
C3G	cyanidin-3-*O*-glucoside
C3R	cyanidin-3-*O*-rutinoside
Dex	dexamethasone
ELISA	enzyme-linked immunosorbent assay
HPLC	high-performance liquid chromatography
iNOS	inducible nitric oxide synthase
IQ	isoquercetin
IFN-γ	interfron γ
IL-6	interleukin 6
IL-10	interleukin 10
Ka	kaempferol
LPS	lipopolysaccharide
MazTF	total flavonoid of *Morus alba* 'Zhenzhubai'
MBC	minimum bactericidal concentration
MnTF	total flavonoid of *Morus nigra*
MmTF	total flavonoid of *Morus mongolica*
MeOH	methanol
Mh	morin hydrate
NF-κB	nuclear factor-κB
NO	nitric oxide
p-IκBα	phospho-IκBα
p-p65	phospho-p65
Pg3G	pelargonidin-3-*O*-glucoside
Qu	quercetin
Ru	quercetin-3-*O*-rutinlside
SD	standard deviation
TF	total flavonoids
TFs	total flavonoids of *M. nigra*, *M. mongolica*, and *M. alba* 'Zhenzhubai'
TNF-α	tumor necrosis factor α
UPLC	ultra-performance liquid chromatographic
UPLC-TUV/QDa	ultra-performance liquid chromatography–tunable ultraviolet/mass single-quadrupole detector

References

1. Gundogdu, M.; Muradoglu, F.; Sensoy, R.I.G.; Yilmaz, H. Determination of fruit chemical properties of *Morus nigra* L.; *Morus alba* L. and *Morus rubra* L. by HPLC. *Sci. Hortic.* **2011**, *132*, 37–41. [CrossRef]
2. He, N.; Zhang, C.; Qi, X.; Zhao, S.; Tao, Y.; Yang, G.; Lee, T.H.; Wang, X.; Cai, Q.; Li, D.; et al. Draft genome sequence of the mulberry tree *Morus notabilis*. *Nat. Commun.* **2013**, *4*, 2445. [CrossRef] [PubMed]

3. Chen, H.; Pu, J.; Liu, D.; Yu, W.; Shao, Y.; Yang, G.; Xiang, Z.; He, N. Anti-inflammatory and antinociceptive properties of flavonoids from the fruits of black mulberry (*Morus nigra* L.). *PLoS ONE* **2016**, *11*, e0153080. [CrossRef] [PubMed]

4. Lim, H.J.; Jin, H.G.; Woo, E.R.; Lee, S.K.; Kim, H.P. The root barks of *Morus alba* and the flavonoid constituents inhibit airway inflammation. *J. Ethnopharmacol.* **2013**, *149*, 169–175. [CrossRef] [PubMed]

5. Ozgen, M.; Serce, S.; Kaya, C. Phytochemical and antioxidant properties of anthocyanin-rich *Morus nigra* and *Morus rubra* fruits. *Sci. Hortic.* **2009**, *119*, 275–279. [CrossRef]

6. Queiroz, G.T.; Santos, T.R.; Macedo, R.; Peters, V.M.; Leite, M.N.; de Cassia da Silveira e Sa, R.; de Oliveira Guerra, M. Efficacy of *Morus nigra* L. on reproduction in female Wistar rats. *Food Chem. Toxicol.* **2012**, *50*, 816–822. [CrossRef] [PubMed]

7. Zeng, Q.; Chen, H.; Zhang, C.; Han, M.; Li, T.; Qi, X.; Xiang, Z.; He, N. Definition of eight mulberry species in the genus *Morus* by internal transcribed spacer-based phylogeny. *PLoS ONE* **2015**, *10*, e0135411. [CrossRef] [PubMed]

8. Chen, Y.; Zhang, W.; Zhao, T.; Li, F.; Zhang, M.; Li, J.; Zou, Y.; Wang, W.; Cobbina, S.J.; Wu, X.; et al. Adsorption properties of macroporous adsorbent resins for separation of anthocyanins from mulberry. *Food Chem.* **2016**, *194*, 712–722. [CrossRef] [PubMed]

9. Ercisli, S.; Orhan, E. Chemical composition of white (*Morus alba*), red (*Morus rubra*) and black (*Morus nigra*) mulberry fruits. *Food Chem.* **2007**, *103*, 1380–1384. [CrossRef]

10. Kim, S.B.; Chang, B.Y.; Hwang, B.Y.; Kim, S.Y.; Lee, M.K. Pyrrole alkaloids from the fruits of *Morus alba*. *Bioorg. Med. Chem. Lett.* **2014**, *24*, 5656–5659. [CrossRef] [PubMed]

11. Levyadun, S.; Gould, K.S. Role of anthocyanins in plant defence. *Anthocyanins* **2008**, 22–28. [CrossRef]

12. Ravindra, P.V.; Narayan, M.S. Antioxidant activity of the anthocyanin from carrot (*Daucus carota*) callus culture. *Int. J. Food Sci. Nutr.* **2003**, *54*, 349–355. [CrossRef] [PubMed]

13. Kolodziejczyk, J.; Saluk-Juszczak, J.; Posmyk, M.; Janas, K.; Wachowicz, B. Red cabbage anthocyanins may protect blood plasma proteins and lipids. *Open Life Sci.* **2011**, *6*, 565–574. [CrossRef]

14. Chen, T.; Hu, S.; Zhang, H.; Guan, Q.; Yang, Y.; Wang, X. Anti-inflammatory effects of *Dioscorea alata* L. anthocyanins in a TNBS-induced colitis model. *Food Funct.* **2017**, *8*, 659–669. [CrossRef] [PubMed]

15. Cremonini, E.; Mastaloudis, A.; Hester, S.N.; Verstraeten, S.V.; Anderson, M.; Wood, S.M.; Waterhouse, A.L.; Fraga, C.G.; Oteiza, P.I. Anthocyanins inhibit tumor necrosis alpha-induced loss of Caco-2 cell barrier integrity. *Food Funct.* **2017**, *8*, 2915–2923. [CrossRef] [PubMed]

16. Choi, J.W.; Synytsya, A.; Capek, P.; Bleha, R.; Pohl, R.; Park, Y.I. Structural analysis and anti-obesity effect of a pectic polysaccharide isolated from Korean mulberry fruit Oddi (*Morus alba* L.). *Carbohydr. Polym.* **2016**, *146*, 187–196. [CrossRef] [PubMed]

17. Yimam, M.; Lee, Y.C.; Moore, B.; Jiao, P.; Hong, M.; Nam, J.B.; Kim, M.R.; Hyun, E.J.; Chu, M.; Brownell, L.; et al. Analgesic and anti-inflammatory effects of UP1304, a botanical composite containing standardized extracts of *Curcuma longa* and *Morus alba*. *J. Integr. Med.* **2016**, *14*, 60–68. [CrossRef]

18. Chen, C.; You, L.J.; Abbasi, A.M.; Fu, X.; Liu, R.H. Optimization for ultrasound extraction of polysaccharides from mulberry fruits with antioxidant and hyperglycemic activity in vitro. *Carbohydr. Polym.* **2015**, *130*, 122–132. [CrossRef] [PubMed]

19. Liu, C.J.; Lin, J.Y. Anti-inflammatory effects of phenolic extracts from strawberry and mulberry fruits on cytokine secretion profiles using mouse primary splenocytes and peritoneal macrophages. *Int. Immunopharmacol.* **2013**, *16*, 165–170. [CrossRef] [PubMed]

20. Lee, J.S.; Synytsya, A.; Kim, H.B.; Choi, D.J.; Lee, S.; Lee, J.; Kim, W.J.; Jang, S.; Park, Y.I. Purification, characterization and immunomodulating activity of a pectic polysaccharide isolated from Korean mulberry fruit Oddi (*Morus alba* L.). *Int. Immunopharmacol.* **2013**, *17*, 858–866. [CrossRef] [PubMed]

21. Walker, J.; Ley, J.P.; Schwerzler, J.; Lieder, B.; Beltran, L.; Ziemba, P.M.; Hatt, H.; Hans, J.; Widder, S.; Krammer, G.E.; et al. Nonivamide, a capsaicin analogue, exhibits anti-inflammatory properties in peripheral blood mononuclear cells and U-937 macrophages. *Mol. Nutr. Food Res.* **2017**, *61*. [CrossRef] [PubMed]

22. DeVon, H.A.; Piano, M.R.; Rosenfeld, A.G.; Hoppensteadt, D.A. The association of pain with protein inflammatory biomarkers: A review of the literature. *Nurs. Res.* **2014**, *63*, 51–62. [CrossRef] [PubMed]

23. Lopes-de-Araujo, J.; Neves, A.R.; Gouveia, V.M.; Moura, C.C.; Nunes, C.; Reis, S. Oxaprozin-loaded lipid nanoparticles towards overcoming NSAIDs side-effects. *Pharm. Res.* **2016**, *33*, 301–314. [CrossRef] [PubMed]

24. Roth, S.H. Coming to terms with nonsteroidal anti-inflammatory drug gastropathy. *Drugs* **2012**, *72*, 873–879. [CrossRef] [PubMed]

25. Jacob, J.N.; Badyal, D.K. Biological studies of turmeric oil, part 3: Anti-inflammatory and analgesic properties of turmeric oil and fish oil in comparison with aspirin. *Nat. Prod. Commun.* **2014**, *9*, 225–228. [PubMed]

26. McQuay, H.J.; Moore, R.A. Dose-response in direct comparisons of different doses of aspirin, ibuprofen and paracetamol (acetaminophen) in analgesic studies. *Br. J. Clin. Pharmacol.* **2007**, *63*, 271–278. [CrossRef] [PubMed]

27. White, A.C., Jr.; Kang, G. Antibiotics, microbiota and health: Are there dangers hiding in plain sight? *Curr. Opin. Infect. Dis.* **2015**, *28*, 455–456. [CrossRef] [PubMed]

28. Dellit, T.H.; Owens, R.C.; McGowan, J.E., Jr.; Gerding, D.N.; Weinstein, R.A.; Burke, J.P.; Huskins, W.C.; Paterson, D.L.; Fishman, N.O.; Carpenter, C.F.; et al. Infectious Diseases Society of America and the Society for Healthcare Epidemiology of America guidelines for developing an institutional program to enhance antimicrobial stewardship. *Clin. Infect. Dis.* **2007**, *44*, 159–177. [CrossRef] [PubMed]

29. Qi, X.; Shuai, Q.; Chen, H.; Fan, L.; Zeng, Q.; He, N. Cloning and expression analyses of the anthocyanin biosynthetic genes in mulberry plants. *Mol. Genet. Genom.* **2014**, *289*, 783–793. [CrossRef] [PubMed]

30. Chen, H.; Pu, J.S.; Xiang, Z.H.; He, N.J. Extraction and antioxidant activity of total flavonoids from black mulberry fruit. *Food Sci.* **2014**, *35*, 7–12.

31. Mena, P.; Sánchez-Salcedo, E.M.; Tassotti, M.; Martínez, J.J.; Hernández, F.; Rio, D.D. Phytochemical evaluation of eight white (*Morus alba* L.) and black (*Morus nigra* L.) mulberry clones grown in Spain based on UHPLC-ESI-MS^n metabolomic profiles. *Food Res. Int.* **2016**, *89*, 1116–1122. [CrossRef]

32. Rainsford, K.D. Profile and mechanisms of gastrointestinal and other side effects of nonsteroidal anti-inflammatory drugs (NSAIDs). *Am. J. Med.* **1999**, *107*, 27S–35S. [CrossRef]

33. Hunskaar, S.; Hole, K. The formalin test in mice: Dissociation between inflammatory and non-inflammatory pain. *Pain* **1987**, *30*, 103–114. [CrossRef]

34. Alvarez Perez Gil, A.L.; Barbosa Navarro, L.; Patipo Vera, M.; Petricevich, V.L. Anti-inflammatory and antinociceptive activities of the ethanolic extract of *Bougainvillea xbuttiana*. *J. Ethnopharmacol.* **2012**, *144*, 712–719. [CrossRef] [PubMed]

35. Wang, Y.; Chen, P.; Tang, C.; Wang, Y.; Li, Y.; Zhang, H. Antinociceptive and anti-inflammatory activities of extract and two isolated flavonoids of *Carthamus tinctorius* L. *J. Ethnopharmacol.* **2014**, *151*, 944–950. [CrossRef] [PubMed]

36. Azevedo, M.I.; Pereira, A.F.; Nogueira, R.B.; Rolim, F.E.; Brito, G.A.; Wong, D.V.; Lima-Junior, R.C.; de Albuquerque Ribeiro, R.; Vale, M.L. The antioxidant effects of the flavonoids rutin and quercetin inhibit oxaliplatin-induced chronic painful peripheral neuropathy. *Mol. Pain* **2013**, *9*, 53. [CrossRef] [PubMed]

37. Lapa Fda, R.; Gadotti, V.M.; Missau, F.C.; Pizzolatti, M.G.; Marques, M.C.; Dafre, A.L.; Farina, M.; Rodrigues, A.L.; Santos, A.R. Antinociceptive properties of the hydroalcoholic extract and the flavonoid rutin obtained from *Polygala paniculata* L. in mice. *Basic Clin. Pharmacol.* **2009**, *104*, 306–315. [CrossRef] [PubMed]

38. Wang, L.; Xu, M.L.; Liu, J.; Wang, Y.; Hu, J.H.; Wang, M.H. Sonchus asper extract inhibits LPS-induced oxidative stress and pro-inflammatory cytokine production in RAW264.7 macrophages. *Nutr. Res. Pract.* **2015**, *9*, 579–585. [CrossRef] [PubMed]

39. Yoon, J.H.; Baek, S.J. Molecular targets of dietary polyphenols with anti-inflammatory properties. *Yonsei Med. J.* **2005**, *46*, 585–596. [CrossRef] [PubMed]

40. Meng, H.; McClendon, C.L.; Dai, Z.; Li, K.; Zhang, X.; He, S.; Shang, E.; Liu, Y.; Lai, L. Discovery of novel 15-lipoxygenase activators to shift the human arachidonic acid metabolic network toward inflammation resolution. *J. Med. Chem.* **2016**, *59*, 4202–4209. [CrossRef] [PubMed]

41. Yang, D.J.; Lin, J.T.; Chen, Y.C.; Liu, S.C.; Lu, F.J.; Chang, T.J.; Wang, M.; Lin, H.W.; Chang, Y.Y. Suppressive effect of carotenoid extract of *Dunaliella salina* alga on production of LPS-stimulated pro-inflammatory mediators in RAW264.7 cells via NF-κB and JNK inactivation. *J. Funct. Foods* **2013**, *5*, 607–615. [CrossRef]

42. Chan, K.C.; Huang, H.P.; Ho, H.H.; Huang, C.N.; Lin, M.C.; Wang, C.J. Mulberry polyphenols induce cell cycle arrest of vascular smooth muscle cells by inducing NO production and activating AMPK and p53. *J. Funct. Foods* **2015**, *15*, 604–613. [CrossRef]

43. Matsui, H.; Ihara, Y.; Fujio, Y.; Kunisada, K.; Akira, S.; Kishimoto, T.; Yamauchitakihara, K. Induction of interleukin (IL)-6 by hypoxia is mediated by nuclear factor (NF)-κB and NF-IL6 in cardiac myocytes. *Cardiovasc. Res.* **1999**, *270*, 11463–11471. [CrossRef]

44. Vallabhapurapu, S.; Karin, M. Regulation and function of NF-κB transcription factors in the immune system. *Annu. Rev. Immunol.* **2009**, *27*, 693–733. [CrossRef] [PubMed]

45. Karin, M.; Cao, Y.; Greten, F.R.; Li, Z.W. NF-κB in cancer: From innocent bystander to major culprit. *Nat. Rev. Cancer* **2002**, *2*, 301–310. [CrossRef] [PubMed]

46. Menghini, L.; Ferrante, C.; Leporini, L.; Recinella, L.; Chiavaroli, A.; Leone, S.; Pintore, G.; Vacca, M.; Orlando, G.; Brunetti, L. An hydroalcoholic chamomile extract modulates inflammatory and immune response in HT29 cells and isolated rat colonic inflammation. *Phytother. Res.* **2016**, *30*, 1513–1518. [CrossRef] [PubMed]

47. Bian, A.L.; Hu, H.Y.; Rong, Y.D.; Wang, J.; Wang, J.X.; Zhou, X.Z. A study on relationship between elderly sarcopenia and inflammatory factors IL-6 and TNF-α. *Eur. J. Med. Res.* **2017**, *22*, 25. [CrossRef] [PubMed]

48. Menghini, L.; Leporini, L.; Pintore, G.; Ferrante, C.; Recinella, L.; Orlando, G.; Vacca, M.; Brunetti, L. A natural formulation (imoviral) increases macrophage resistance to LPS-induced oxidative and inflammatory stress in vitro. *J. Biol. Regul. Homeost. Agents* **2014**, *28*, 775–782. [PubMed]

49. Kyrkanides, S.; Olschowka, J.A.; Williams, J.P.; Hansen, J.T.; O'Banion, M.K. TNFα and IL-1β mediate intercellular adhesion molecule-1 induction via microglia–astrocyte interaction in CNS radiation injury. *J. Neuroimmunol.* **1999**, *95*, 95–106. [CrossRef]

50. Starkie, R.; Ostrowski, S.R.; Jauffred, S.; Febbraio, M.; Pedersen, B.K. Exercise and IL-6 infusion inhibit endotoxin-induced TNF-α production in humans. *FASEB J.* **2003**, *17*, 884–886. [CrossRef] [PubMed]

51. Khanapure, S.P.; Garvey, D.S.; Janero, D.R.; Letts, L.G. Eicosanoids in inflammation: Biosynthesis, pharmacology, and therapeutic frontiers. *Curr. Top. Med. Chem.* **2007**, *7*, 311–340. [CrossRef] [PubMed]

52. Wang, F.; Jianrong, L.I.; Jiang, Y. Polysaccharides from mulberry leaf in relation to their antioxidant activity and antibacterial ability. *J. Food Process Eng.* **2010**, *33*, 39–50. [CrossRef]

53. Thabti, I.; Elfalleh, W.; Tlili, N.; Ziadi, M.; Campos, M.G.; Ferchichi, A. Phenols, flavonoids, and antioxidant and antibacterial activity of leaves and stem bark of *Morus* species. *Int. J. Food Prop.* **2014**, *17*, 842–854. [CrossRef]

54. Vazquezguillamet, M.C.; Vazquez, R.; Micek, S.T.; Kollef, M.H. Predicting resistance to Piperacillin-Tazobactam, Cefepime and Meropenem in Septic Patients with bloodstream infection due to Gram-negative bacteria. *Clin. Infect. Dis.* **2017**, *65*, 1607–1614. [CrossRef] [PubMed]

55. Doernberg, S.B.; Lodise, T.P.; Thaden, J.T.; Munita, J.M.; Cosgrove, S.E.; Arias, C.A.; Boucher, H.W.; Corey, G.R.; Lowy, F.D.; Murray, B.; et al. Gram-positive bacterial infections: Research priorities, accomplishments, and future directions of the antibacterial resistance leadership group. *Clin. Infect. Dis.* **2017**, *64* (Suppl. 1), S24–S29. [CrossRef] [PubMed]

56. Huang, W.H.; Lee, A.R.; Yang, C.H. Antioxidative and anti-Inflammatory activities of polyhydroxy flavonoids of GEORGI. *Biosci. Biotechnol. Biochem.* **2014**, *70*, 2371–2380. [CrossRef] [PubMed]

57. Pham, T.H.; Kim, M.S.; Le, M.Q.; Song, Y.S.; Bak, Y.; Ryu, H.W.; Oh, S.R.; Yoon, D.Y. Fargesin exerts anti-inflammatory effects in THP-1 monocytes by suppressing PKC-dependent AP-1 and NF-kB signaling. *Phytomed. Int. J. Phytother. Phytopharmacol.* **2017**, *24*, 96–103. [CrossRef] [PubMed]

58. Omara, S.T. MIC and MBC of Honey and Gold Nanoparticles against methicillin-resistant (MRSA) and vancomycin-resistant (VRSA) coagulase-positive *S. aureus* isolated from contagious Bovine clinical Mastitis. *J. Genetic Eng. Biotechnol.* **2017**, *15*, 219–230. [CrossRef]

Sample Availability: Samples of MnTF, MmTF, and MazTF are available from the authors.

molecules

MDPI

Communication

Cardiovascular Mechanisms of Action of Anthocyanins May Be Associated with the Impact of Microbial Metabolites on Heme Oxygenase-1 in Vascular Smooth Muscle Cells

Emily F. Warner [1], Ildefonso Rodriguez-Ramiro [1], Maria A. O'Connell [2] and Colin D. Kay [1,*]

[1]   Department of Nutrition and Preventative Medicine, Norwich Medical School, Bob Champion Research
    and Education Building, University of East Anglia, Norwich NR4 7UQ, UK;
    emily.warner@uea.ac.uk (E.F.W.); I.Rodriguez-Ramiro@uea.ac.uk (I.R.-R.)
[2]   School of Pharmacy, University of East Anglia, Norwich NR4 7TJ, UK; m.oconnell@uea.ac.uk
*   Correspondence: cdkay@ncsu.edu; Tel.: +1-704-250-5452

Received: 22 March 2018; Accepted: 11 April 2018; Published: 13 April 2018

Abstract: Anthocyanins are reported to have cardio-protective effects, although their mechanisms of action remain elusive. We aimed to explore the effects of microbial metabolites common to anthocyanins and other flavonoids on vascular smooth muscle heme oxygenase-1 (HO-1) expression. Thirteen phenolic metabolites identified by previous anthocyanin human feeding studies, as well as 28 unique mixtures of metabolites and their known precursor structures were explored for their activity on HO-1 protein expression in rat aortic smooth muscle cells (RASMCs). No phenolic metabolites were active when treated in isolation; however, five mixtures of phenolic metabolites significantly increased HO-1 protein expression (127.4–116.6%, $p \leq 0.03$). The present study demonstrates that phenolic metabolites of anthocyanins differentially affect HO-1 activity, often having additive, synergistic or nullifying effects.

Keywords: anthocyanin; smooth muscle cells; metabolism; antioxidant; atherosclerosis

---

1. Introduction

It is now accepted that microbial metabolites (phenolic and aromatic ring-fission catabolites) of dietary flavonoids are more bioavailable than their precursor structures [1]. We have previously reported that anthocyanin metabolites have additive activity in inflammatory systems [2,3], and others have also demonstrated this in vascular smooth muscle cells [4]. As the mechanisms of action of flavonoids have remained unresolved for decades, the need to elucidate the activity of phenolic metabolites has become the focus of recent flavonoid research.

Human feeding studies have shown beneficial effects of flavonoid-rich foods on vascular function, including blood flow and flow-mediated vasodilation (FMD) [5]. Vascular smooth muscle cells contain various sources of reactive oxygen species (ROS), such as NAPDH oxidase (NOX), which under conditions of stress lead to proliferation, migration, and cytokine production, events which are central to the progression of atherosclerosis [6]. We have previously demonstrated that a number of anthocyanin metabolites decreased superoxide ions in cultured endothelial cells but had no effect on the enzyme responsible for their generation, NOX [7]. However, anthocyanin metabolites increased the expression of oxidant-response protein, heme oxygenase-1 (HO-1), which may induce antioxidant activity through the production of biliverdin (converted into cellular antioxidant bilirubin), a by-product of heme degradation [8].

In the present study, we therefore explored the action of microbial-derived metabolites of anthocyanins on HO-1 production as a means to rationalise our previous observations. Many of these metabolites are also common to other precursor flavonoids [1]. The aim of the present study was to determine the activity of 13 common phenolic metabolites, identified from human feeding studies [1], relative to six precursor flavonoids (Figure 1), as well as 28 unique mixtures (Table S1), on HO-1 expression in rat aortic smooth muscle cells (RASMCs).

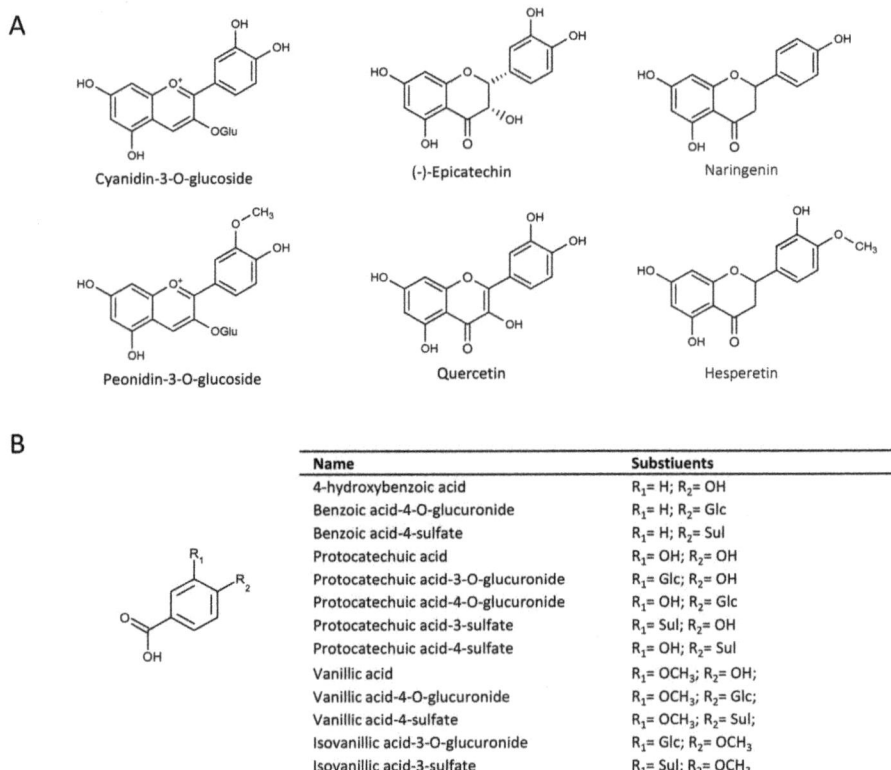

A

Cyanidin-3-O-glucoside

(-)-Epicatechin

Naringenin

Peonidin-3-O-glucoside

Quercetin

Hesperetin

B

Name	Substiuents
4-hydroxybenzoic acid	R_1= H; R_2= OH
Benzoic acid-4-O-glucuronide	R_1= H; R_2= Glc
Benzoic acid-4-sulfate	R_1= H; R_2= Sul
Protocatechuic acid	R_1= OH; R_2= OH
Protocatechuic acid-3-O-glucuronide	R_1= Glc; R_2= OH
Protocatechuic acid-4-O-glucuronide	R_1= OH; R_2= Glc
Protocatechuic acid-3-sulfate	R_1= Sul; R_2= OH
Protocatechuic acid-4-sulfate	R_1= OH; R_2= Sul
Vanillic acid	R_1= OCH$_3$; R_2= OH;
Vanillic acid-4-O-glucuronide	R_1= OCH$_3$; R_2= Glc;
Vanillic acid-4-sulfate	R_1= OCH$_3$; R_2= Sul;
Isovanillic acid-3-O-glucuronide	R_1= Glc; R_2= OCH$_3$
Isovanillic acid-3-sulfate	R_1= Sul; R_2= OCH$_3$

Figure 1. Structures of (**A**) flavonoids and (**B**) metabolites included in treatments. OH, hydroxyl; Glc, oxygen-linked-glucuronide; OGlu, oxygen-linked glycoside; Sul, sulfate; OCH$_3$, oxygen-linked methyl group.

2. Results

2.1. Effect of Flavonoids and Their Metabolites on HO-1 Protein Expression

Thirteen phenolic metabolites and six precursor flavonoids were screened for their effect at 10 μM on RASMC HO-1 protein expression after 24 h incubation (Figure 2). HO-1 expression increased in response to two precursor flavonoids, quercetin (200.87% ± 28.82%, p = 0.009) and peonidin-3-glucoside (164.05% ± 32.88%, $p \leq$ 0.001) while no significant activity (p > 0.05) was observed for single metabolite treatments. Protocatechuic acid (PCA) appeared to have a modest, but non-significant, effect (121.87% ± 15.47%, p = 0.18).

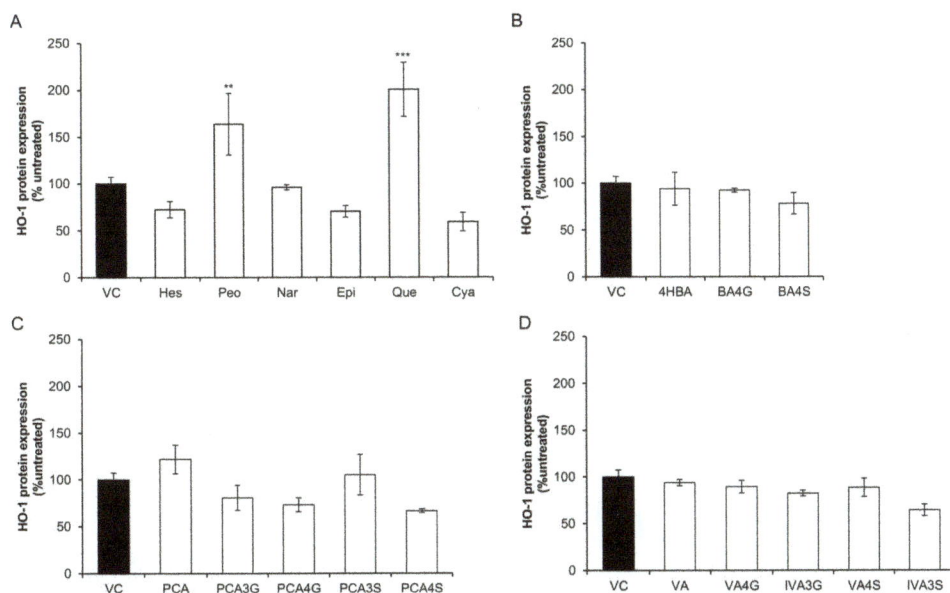

Figure 2. Effect of 10 µM flavonoids and phenolic acid metabolites on HO-1 protein expression in RASMCs. (**A**) precursor flavonoids; (**B**) benzoic acid metabolites; (**C**) protocatechuic acid metabolites; and (**D**) vanillic acid metabolites. Data are presented as a percentage of an untreated control (media only). Filled columns represent vehicle control (0.02% DMSO) and clear columns represent treatments (10 µM). All columns are representative of the mean ± SD, *n* = 3 independent samples, ** *p* ≤ 0.01, *** *p* ≤ 0.001 (ANOVA with post hoc Dunnett relative to vehicle control (0.02% DMSO)). 4HBA, 4-hydroxybenzoic acid; BA4G, benzoic acid-4-O-glucuronide; BA4S, benzoic acid-4-sulfate; C3G, cyanidin-3-O-glucoside; EPI, (−)-epicatechin; HES, hesperetin; IVA, isovanillic acid; IVA3G, isovanillic acid-3-O-glucuronide; IVA3S, isovanillic acid-3-sulfate; NAR, naringenin; P3G, peonidin-3-O-glucoside; PCA, protocatechuic acid; PCA3G, protocatechuic acid-3-O-glucuronide; PCA4G, protocatechuic acid-4-O-glucuronide; PCA3S, protocatechuic acid-3-sulfate; PCA4S, protocatechuic acid-4-sulfate; QUE, quercetin; VA, vanillic acid; VA4G, vanillic acid-4-O-glucuronide; VA4S, vanillic acid-4-sulfate.

2.2. Effect of Mixtures of Flavonoids and Their Metabolites on HO-1 Expression

Twenty-one mixtures of conjugated and unconjugated phenolic metabolites and seven mixtures of precursor flavonoids, designed based upon their structural similarity, were screened at cumulative concentrations of 10 µM for their effect on RASMC HO-1 protein expression after 24 h treatment (Figure 3). HO-1 expression was increased following treatment with one mixture of precursor flavonoids, consisting of equimolar concentrations of hesperetin and peonidin-3-O-glucoside (181.15% ± 13.46%, *p* ≤ 0.001; Figure 3A). Five mixtures of conjugated and unconjugated flavonoid metabolites, including: protocatechuic acid + vanillic acid (127.06% ± 5.83%, *p* = 0.005; Figure 3B); 4-hydroxybenzoic acid + benzoic acid-4-sulfate (129.20% ± 4.00%, *p* = 0.02; Figure 3C); protocatechuic acid + protocatechuic acid-3-O-glucuronide (127.43% ± 6.55%, *p* = 0.001; Figure 3D); protocatechuic acid + protocatechuic acid-3-O-glucuronide + protocatechuic acid-4-O-glucuronide (116.58% ± 4.77%, *p* = 0.03; Figure 3D); vanillic acid + isovanillic acid-3-O-glucuronide (128.02% ± 15.26%, *p* = 0.009; Figure 3E), were also active.

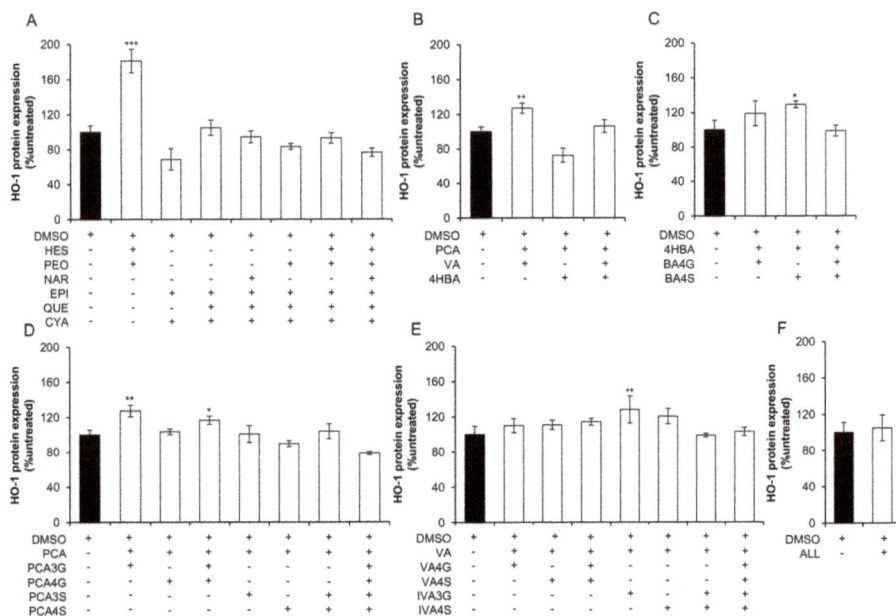

Figure 3. Effect of 10 μM mixtures of flavonoids and phenolic acid metabolites on HO-1 protein expression in RASMCs. (**A**) precursor flavonoids; (**B**) phenolic acids; (**C**) benzoic acid metabolites; (**D**) protocatechuic acid metabolites; (**E**) vanillic acid metabolites; and (**F**) all metabolites. Data are presented as a percentage of an untreated control (media only). Filled columns represent vehicle control (0.02% DMSO) and clear columns represent treatments (10 μM). All columns are representative of the mean ± SD, n = 3 independent samples. * $p \leq 0.05$, ** $p \leq 0.01$, *** $p \leq 0.001$ (ANOVA with post hoc Dunnett relative to vehicle control (0.02% DMSO)). 'ALL' is a mixture composed of 13 conjugated and unconjugated phenolic acids at equimolar concentrations to a cumulative concentration of 10 μM. 4HBA, 4-hydroxybenzoic acid; BA4G, benzoic acid-4-*O*-glucuronide; BA4S, benzoic acid-4-sulfate; C3G, cyanidin-3-*O*-glucoside; DMSO, dimethyl sulfoxide; EPI, (−)-epicatechin; HES, hesperetin; IVA, isovanillic acid; IVA3G, isovanillic acid-3-*O*-glucuronide; IVA3S, isovanillic acid-3-sulfate; NAR, naringenin; P3G, peonidin-3-*O*-glucoside; PCA, protocatechuic acid; PCA3G, protocatechuic acid-3-*O*-glucuronide; PCA4G, protocatechuic acid-4-*O*-glucuronide; PCA3S, protocatechuic acid-3-sulfate; PCA4S, protocatechuic acid-4-sulfate; QUE, quercetin; VA, vanillic acid; VA4G, vanillic acid-4-*O*-glucuronide; VA4S, vanillic acid-4-sulfate.

3. Discussion

Bacterial catabolism of flavonoids reduces the bioavailability of the precursor flavonoid, while producing a number of bioavailable phenolic metabolites [1]. Notwithstanding previous works suggesting that precursor flavonoids have additive or synergistic effects in combination [2], the present study of the combined effects of flavonoid metabolites is contemporary, owing to the recent availability of synthetic standards. The present study is the first to observe the effects of conjugated phenolic metabolites in combination in vascular smooth muscle cells and suggest that activity of anthocyanins and other flavonoids may be the result of a cumulative effect of multiple metabolites upregulating antioxidant-response protein, HO-1. Here, we found that five mixtures consisting of conjugated and unconjugated phenolic metabolites, one mixture consisting of hesperetin and peonidin-3-*O*-glucoside (Figure 3) and two single flavonoid treatments (quercetin and peonidin-3-*O*-glucoside) (Figure 2) actively upregulated HO-1 protein in RASMCs. These data suggest that conjugated metabolites of flavonoids may not actively increase HO-1 protein in isolation, but act additively or synergistically in combination.

HO-1 protein was increased in response to five mixtures of phenolic metabolites, which is of particular interest as the concentration used (10 μM) is within the range of cumulative metabolite concentrations reported in vivo (0.80–13.18 μM) [9,10]. Keane et al. reported no effect in response to single metabolite treatments protocatechuic acid (PCA) and vanillic acid (VA) on vascular smooth muscle cell (VSMC) migration, whereas mixtures of PCA and VA increased VSMC migration, suggestive of an additive effect [4]. Interestingly, the present study also observed that PCA and VA in isolation did not significantly increase HO-1 expression, but a combination consisting of 5 μM of each metabolite (to a cumulative concentration of 10 μM) significantly upregulated HO-1 protein. This supports the hypothesis that these abundant metabolites act additively on HO-1 expression and provides a rationale for the lack of effect observed in our previous study where no activity was seen in endothelial cells in response to PCA or VA in isolation on protein expression [7].

Quercetin significantly induced HO-1 protein expression in the present study, which is in accordance with previous studies, though it should be noted that quercetin circulates as its unconjugated, aglycone structure at negligible concentrations post-consumption [11]. Given that phenolic metabolites exist at much higher concentrations for longer periods of time in the circulation [1], greater focus should be given to their bioactivity in future studies. Similarly, the anthocyanin peonidin-3-*O*-glucoside (P3G) is unstable and rapidly degrades to phenolic acid derivatives at physiological pH and therefore has low plasma bioavailability [10]. The apparent reduction of activity between P3G and its B-ring derivative, VA, suggests that the activity of some anthocyanins may be lost in vivo due to chemical degradation or bacterial catabolism, implying metabolism differentially impacts the activity of anthocyanins.

The present study has provided a novel insight into the effects of anthocyanin metabolites on HO-1 in RASMCs, though further work is required to elucidate the underlying mechanisms of these treatments. Validation of these effects is required at the mRNA level and including Nrf2 localisation and, ultimately, conformation in animal and human studies. In addition, the use of rat-derived cells may be seen as a limitation, as the phenotypes and expression levels of cellular proteins may not be conserved between species, and, even though costs of using human vascular smooth muscle cells are prohibitive, future work should validate these finding in human cells such as human coronary artery smooth muscle cells (HCASMCs). In addition, the individual 10 μM treatments utilised were beyond the physiological concentrations for single precursor flavonoids [10], but necessary as a comparison to the combination treatments, which were well within physiologically achievable concentrations. In a previously published study, we observed effects on HO-1 expression in response to 10 μM VA in human endothelial cells [7]. Therefore, prior to the present study, a handful of compounds were tested in RASMC at 1, 10, and 100 μM for their effect on HO-1, and 10 μM was identified as most effective (data not shown) and therefore utilised for the present screen. It is important to note that treatments may be more active at concentrations below 10 μM, as in a previous study by our group treatment concentrations as low as 0.19 μM were active [3], and therefore HO-1 expression in response to varied concentrations should be explored in future studies.

4. Materials and Methods

4.1. Treatments

The conjugated metabolites: benzoic acid-4-*O*-glucuronide, benzoic acid-4-sulfate, isovanillic acid-3-*O*-glucuronide (4-methoxybenzoic acid-3-*O*-glucuronide), isovanillic acid-3-sulfate (4-methoxybenzoic acid-3-sulfate), protocatechuic acid-3-*O*-glucuronide (4-hydroxybenzoic acid-3-*O*-glucuronide), protocatechuic acid-4-*O*-glucuronide (3-hydroxybenzoic acid-4-*O*-glucuronide), protocatechuic acid-3-sulfate (4-hydroxybenzoic acid-3-sulfate), protocatechuic acid-4-sulfate (3-hydroxybenzoic acid-4-sulfate), vanillic acid-4-*O*-glucuronide (3-methoxybenzoic acid-4-*O*-glucuronide), and vanillic acid-4-sulfate

Molecules **2018**, *23*, 898

(3-methoxybenzoic acid-4-sulfate), were previously synthesised at the University of St. Andrews (UK) [12]. Flavonoids (cyanidin-3-*O*-glucoside, (−)-epicatechin, hesperetin, naringenin, and quercetin) and unconjugated phenolic acids: 4-hydroxybenzoic acid, protocatechuic acid (3,4-dihydroxybenzoic acid), and vanillic acid (4-hydroxy-3-methoxybenzoic acid), were obtained from Sigma Aldrich (Dorset, UK), with the exception of peonidin-3-*O*-glucoside (Extrasynthase, Genay, France). Stock solutions of all compounds were prepared in 100% DMSO at 200 mM, with the exception of cyanidin-3-*O*-glucoside and peonidin-3-*O*-glucoside, which were prepared at 40 mM, and sulfate-conjugated phenolic acids at 25 mM in 50% DMSO (50% PBS) to maintain stability whilst reducing final DMSO concentrations in working solutions. All stock solutions were stored at 80 °C. Working solutions of individual treatments were added to supplemented media at 10 μM concentrations immediately prior to treatment. Treatments containing mixtures of compounds consisted of equimolar concentrations of the constituent treatment compounds (Table S1) to a cumulative concentration of 10 μM. For example, a combination comprising four constituents required 2.5 μM of each, equating to a total concentration of 10 μM. No treatments were cytotoxic as established utilising the WST-1 cytotoxicity assay (Roche, West Sussex, UK) (data not shown).

4.2. Cell Culture

Cryopreserved, second passage, pooled Clonetics rat aortic smooth muscle cells (RASMCs) from adult Sprague Dawley (Lonza Biologics, Manchester, UK), were maintained in Dulbecco's modified Eagle's medium: F12 (DMEM) containing 0.1% gentamycin/amphotericin and 20% FBS (Lonza Biologics, Manchester, UK). Cells were used between passages 3 and 6.

4.3. HO-1 Protein Expression

RASMC were seeded at 300,000 cells/well in fibronectin-coated 6-well plates. Supplemented media was replaced by serum-free media 24 h prior to experiment commencement. Cells were treated with media only (untreated control), 10 μM treatment, 0.02% DMSO (vehicle control), and incubated for 24 h at 37 °C, 5% CO_2, in a humidified atmosphere. Cells were washed 3x with PBS and cells lysed with Extraction Reagent Buffer (Enzo Lifesciences, City, UK) and stored at −80 °C until required, undergoing one freeze-thaw cycle. HO-1 protein in cell lysates was determined by use of Rat Hmox-1 ELISA Kits (Enzo Lifesciences, Exeter, UK) according to the manufacturer's instructions. HO-1 protein concentrations were normalised to the total cell protein content using the Pierce Protein BCA protein assay (Thermo Fisher Scientific, Loughborough, UK).

4.4. Data Analysis

HO-1 protein values were presented as a percentage of an untreated control (media only) and reported as the mean ± standard deviation of 3 independent samples. Treatment effects were determined relative to the vehicle control (0.02% DMSO) and established by one-way analysis of variance (ANOVA) with post hoc Dunnett. Analyses were conducted using SPSS for Windows (version 22.0; IBM, New York, NY, USA). Data were considered significant where $p \leq 0.05$.

5. Conclusions

In conclusion, the present study has demonstrated that the bioactivity of common phenolic metabolites is increased when in combination, indicating additive or synergistic effects.

Supplementary Materials: The following are available online at http://www.mdpi.com/1420-3049/23/4/898/s1: Table S1: Combination treatment constituents and relative concentrations.

Acknowledgments: We thank Qingzhi Zhang, Saki Raheem, and David O'Hagan of the University of St Andrews (UK) who undertook the chemical synthesis of phenolic conjugate standards and the late Nigel Botting (also of the University of St. Andrews, UK) who helped establish this research collaboration, including the design of the synthesis project objectives. This study was supported by funding from the UK Biotechnology and Biological

Sciences Research Council Diet and Health Research Industry Club (BBSRC-DRINC; BB/I006028/1). I.R.R. was funded by a BBSRC-DRINC Post-Doctoral Fellowship grant (BB/I006028/1).

Author Contributions: C.D.K. and M.O.C. conceived and managed the project. E.F.W., I.R.-R., C.K., and M.O.C. designed the culture experiments and methodology. I.R.-R. conducted the cell culture work and E.F.W. collected and graphed the data. C.K., M.O.C., I.R.-R., and E.F.W. contributed to the analysis and interpretation of the data. E.F.W. and C.K. drafted the initial manuscript and all authors reviewed and approved the final content.

Conflicts of Interest: The authors declare no conflict of interest.

References

1. Kay, C.D.; Pereira-Caro, G.; Ludwig, I.A.; Clifford, M.N.; Crozier, A. Anthocyanins and Flavanones Are More Bioavailable than Previously Perceived: A Review of Recent Evidence. *Annu. Rev. Food Sci. Technol.* **2017**, *8*, 155–180. [CrossRef] [PubMed]

2. Di Gesso, J.L.; Kerr, J.S.; Zhang, Q.; Raheem, S.; Yalamanchili, S.K.; O'Hagan, D.; Kay, C.D.; O'Connell, M.A. Flavonoid metabolites reduce tumor necrosis factor-alpha secretion to a greater extent than their precursor compounds in human THP-1 monocytes. *Mol. Nutr. Food Res.* **2015**, *59*, 1143–1154. [CrossRef] [PubMed]

3. Warner, E.F.; Smith, M.J.; Zhang, Q.; Raheem, K.S.; O'Hagan, D.; O'Connell, M.A.; Kay, C.D. Signatures of anthocyanin metabolites identified in humans inhibit biomarkers of vascular inflammation in human endothelial cells. *Mol. Nutr. Food Res.* **2017**, *61*. [CrossRef] [PubMed]

4. Keane, K.M.; Bell, P.G.; Lodge, J.K.; Constantinou, C.L.; Jenkinson, S.E.; Bass, R.; Howatson, G. Phytochemical uptake following human consumption of Montmorency tart cherry (L. *Prunus cerasus*) and influence of phenolic acids on vascular smooth muscle cells in vitro. *Eur. J. Nutr.* **2016**, *55*, 1695–1705. [CrossRef] [PubMed]

5. Hooper, L.; Kroon, P.A.; Rimm, E.B.; Cohn, J.S.; Harvey, I.; Le Cornu, K.A.; Ryder, J.J.; Hall, W.L.; Cassidy, A. Flavonoids, flavonoid-rich foods, and cardiovascular risk: A meta-analysis of randomized controlled trials. *Am. J. Clin. Nutr.* **2008**, *88*, 38–50. [CrossRef] [PubMed]

6. Lusis, A.J. Atherosclerosis. *Nature* **2000**, *407*, 233–241. [CrossRef] [PubMed]

7. Edwards, M.; Czank, C.; Woodward, G.M.; Cassidy, A.; Kay, C.D. Phenolic metabolites of anthocyanins modulate mechanisms of endothelial function. *J. Agric. Food Chem.* **2015**, *63*, 2423–2431. [CrossRef] [PubMed]

8. Gozzelino, R.; Jeney, V.; Soares, M.P. Mechanisms of cell protection by heme oxygenase-1. *Annu. Rev. Pharmacol. Toxicol.* **2010**, *50*, 323–354. [CrossRef] [PubMed]

9. Czank, C.; Cassidy, A.; Zhang, Q.; Morrison, D.J.; Preston, T.; Kroon, P.A.; Botting, N.P.; Kay, C.D. Human metabolism and elimination of the anthocyanin, cyanidin-3-glucoside: A (13)C-tracer study. *Am. J. Clin. Nutr.* **2013**, *97*, 995–1003. [CrossRef] [PubMed]

10. De Ferrars, R.M.; Czank, C.; Zhang, Q.; Botting, N.P.; Kroon, P.A.; Cassidy, A.; Kay, C.D. The pharmacokinetics of anthocyanins and their metabolites in humans. *Br. J. Pharmacol.* **2014**, *171*, 3268–3282. [CrossRef] [PubMed]

11. Nabavi, S.F.; Russo, G.L.; Daglia, M.; Nabavi, S.M. Role of quercetin as an alternative for obesity treatment: You are what you eat! *Food Chem.* **2015**, *179*, 305–310. [CrossRef] [PubMed]

12. Zhang, Q.; Raheem, K.S.; Botting, N.P.; Slawin, A.M.Z.; Kay, C.D.; O'Hagan, D. Flavonoid metabolism: The synthesis of phenolic glucuronides and sulfates as candidate metabolites for bioactivity studies of dietary flavonoids. *Tetrahedron* **2012**, *68*, 4194–4201. [CrossRef]

Sample Availability: Samples are not available from the authors.

MDPI

St. Alban-Anlage 66

4052 Basel

Switzerland

Tel. +41 61 683 77 34

Fax +41 61 302 89 18

www.mdpi.com

Molecules Editorial Office

E-mail: molecules@mdpi.com

www.mdpi.com/journal/molecules